AN ELEMENTARY TREATISE
ON THE
DYNAMICS OF A PARTICLE
AND OF RIGID BODIES

BY THE SAME AUTHOR

ELEMENTS OF HYDROSTATICS

ELEMENTS OF TRIGONOMETRY

ELEMENTARY TREATISE ON DYNAMICS OF A PARTICLE AND OF RIGID BODIES SOLUTIONS

ELEMENTARY TREATISE ON STATICS

MECHANICS AND HYDROSTATICS

PLANE TRIGONOMETRY

Part I. An Elementary Course

Part II. Analytical Trigonometry

Solutions, Part I

Solutions to Elements of Statics and Dynamics

ELEMENTS OF STATICS (*separately*)

ELEMENTS OF DYNAMICS (*separately*)

AN ELEMENTARY TREATISE ON THE DYNAMICS OF A PARTICLE AND OF RIGID BODIES

BY

S. L. LONEY

CAMBRIDGE
AT THE UNIVERSITY PRESS
1960

CAMBRIDGE
UNIVERSITY PRESS

University Printing House, Cambridge CB2 8BS, United Kingdom

Cambridge University Press is part of the University of Cambridge.

It furthers the University's mission by disseminating knowledge in the pursuit of education, learning and research at the highest international levels of excellence.

www.cambridge.org
Information on this title: www.cambridge.org/9781316633335

Reprinted 1960
First paperback edition 2016

A catalogue record for this publication is available from the British Library

ISBN 978-1-316-63333-5 Paperback

CONTENTS

DYNAMICS OF A PARTICLE

DYNAMICS OF A RIGID BODY

PREFACE

In the following work I have tried to write an elementary class-book on those parts of Dynamics of a Particle and Rigid Dynamics which are usually read by Students attending a course of lectures in Applied Mathematics for a Science or Engineering Degree, and by Junior Students for Mathematical Honours. Within the limits with which it professes to deal, I hope it will be found to be fairly complete.

I assume that the Student has previously read some such course as is included in my Elementary Dynamics. I also assume that he possesses a fair working knowledge of Differential and Integral Calculus; the Differential Equations, with which he will meet, are solved in the Text, and in an Appendix he will find a summary of the methods of solution of such equations.

In Rigid Dynamics I have chiefly confined myself to two-dimensional motion, and I have omitted all reference to moving axes.

I have included in the book a large number of Examples, mostly collected from University and College Examination Papers; I have verified every question, and hope that there will not be found a large number of serious errors.

Solutions of the Examples have now been published.

S. L. LONEY

December, 1926

CHAPTER I

FUNDAMENTAL DEFINITIONS AND PRINCIPLES

1. The velocity of a point is the rate of its displacement, so that, if P be its position at time t and Q that at time $t+\Delta t$, the limiting value of the quantity $\frac{PQ}{\Delta t}$, as Δt is made very small, is its velocity.

Since a displacement has both magnitude and direction, the velocity possesses both also; the latter can therefore be represented in magnitude and direction by a straight line, and is hence called a vector quantity.

2. A point may have two velocities in different directions at the same instant; they may be compounded into one velocity by the following theorem known as the Parallelogram of Velocities;

If a moving point possess simultaneously velocities which are represented in magnitude and direction by the two sides of a parallelogram drawn from a point, they are equivalent to a velocity which is represented in magnitude and direction by the diagonal of the parallelogram passing through the point.

Thus two component velocities AB, AC are equivalent to the resultant velocity AD, where AD is the diagonal of the parallelogram of which AB, AC are adjacent sides.

If BAC be a right angle and $BAD=\theta$, then $AB=AD\cos\theta$, $AC=AD\sin\theta$, and a velocity v along AD is equivalent to the two component velocities $v\cos\theta$ along AB and $v\sin\theta$ along AC.

Triangle of Velocities. If a point possess two velocities completely represented (*i.e.* represented in magnitude, direction and sense) by two straight lines AB and BC, their resultant is completely represented by AC. For completing the parallelogram $ABCD$, the velocities AB, BC are equivalent to AB, AD whose resultant is AC.

Parallelopiped of Velocities. If a point possess three velocities completely represented by three straight lines OA, OB, OC their resultant is, by successive applications of the parallelogram of velocities, completely represented by OD, the diagonal of the parallelopiped of which OA, OB, OC are conterminous edges.

Similarly OA, OB, and OC are the component velocities of OD.

If OA, OB, and OC are mutually at right angles and u, v, w are the velocities of the moving point along these directions, the resultant velocity is $\sqrt{u^2+v^2+w^2}$ along a line whose direction cosines are proportional to u, v, w and are thus equal to

$$\frac{u}{\sqrt{u^2+v^2+w^2}},\ \frac{v}{\sqrt{u^2+v^2+w^2}}\ \text{and}\ \frac{w}{\sqrt{u^2+v^2+w^2}}.$$

Similarly, if OD be a straight line whose direction cosines referred to three mutually perpendicular lines OA, OB, OC are l, m, n, then a velocity V along OD is equivalent to component velocities lV, mV, nV along OA, OB, and OC respectively.

3. Change of Velocity. Acceleration. If at any instant the velocity of a moving point be represented by OA, and at any subsequent instant by OB, and if the parallelogram $OABC$ be completed whose diagonal is OB, then OC or AB represents the velocity which must be compounded with OA to give OB, *i.e.* it is the change in the velocity of the moving point.

Acceleration is the rate of change of velocity, *i.e.* if OA, OB represent the velocities at times t and $t+\Delta t$, then the limiting value of $\frac{BA}{\Delta t}$ (*i.e.* the limiting value of the ratio of the change in the velocity to the change in the time), as Δt becomes indefinitely small, is the acceleration of the moving point. As in the case of velocities, a moving point may possess simultaneously accelerations in different directions, and they may be compounded into one by a theorem known as the Parallelogram of Accelerations similar to the Parallelogram of Velocities.

As also in the case of velocities an acceleration may be resolved into two component accelerations.

The results of Art. 2 are also true for accelerations as well as velocities.

4. Relative Velocity. When the distance between two moving points is altering, either in direction or in magnitude or in both, each point is said to have a velocity relative to the other.

Suppose the velocities of two moving points A and B to be represented by the two lines AP and BQ (which are not necessarily in the same plane), so that in the unit of time the positions of the points would change from A and B to P and Q. Draw BR equal and parallel to AP. The velocity BQ is, by the Triangle of Velocities, equivalent to the velocities BR, RQ, *i.e.* the velocity of B is equivalent to the velocity of A together with a velocity RQ.

The velocity of B relative to A is thus represented by RQ.

Now the velocity RQ is equivalent to velocities RB and BQ (by the

Triangle of Velocities), *i.e.* to velocities completely represented by BQ and PA.

Hence *the velocity of B relative to A is obtained by compounding the absolute velocity of B with a velocity equal and opposite to that of A.*

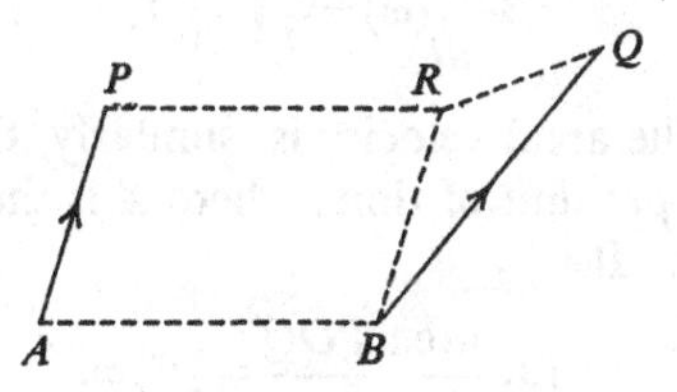

Conversely, since the velocity BQ is equivalent to the velocities BR and RQ, *i.e.* to the velocity of A together with the velocity of B relative to A, therefore *the velocity of any point B is obtained by compounding together its velocity relative to any other point A and the velocity of A.*

The same results are true for accelerations, since they also are vector quantities and therefore follow the parallelogram law.

5. Angular velocity of a point whose motion is in one plane.

If a point P be in motion in a plane, and if O be a fixed point and Ox a fixed line in the plane, the rate of increase of the angle xOP per unit of time is called the angular velocity of P about O.

Hence, if at time t the angle xOP be θ, the angular velocity about O is $\frac{d\theta}{dt}$.

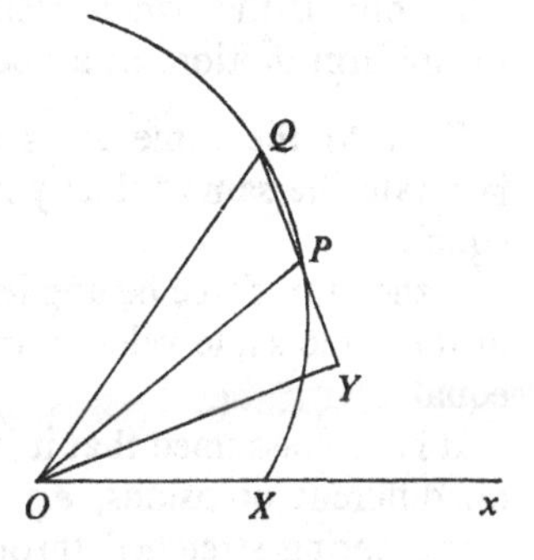

If Q be the position of the point P at time $t+\Delta t$, where Δt is small, and v the velocity of the point at time t, then

$$v=\text{Lt.}\ \frac{PQ}{\Delta t}.$$

If

$$\angle POQ=\Delta\theta, \text{ and } OP=r,\ OQ=r+\Delta r,$$

then

$$r(r+\Delta r)\sin\Delta\theta=2\Delta POQ=PQ.OY,$$

where OY is the perpendicular on PQ.

Hence, dividing by Δt, and proceeding to the limit when Δt is very small, we have

$$r^2\frac{d\theta}{dt}=v.p,$$

where p is the perpendicular from O upon the tangent at P to the path of the moving point.

Hence, if ω be the angular velocity, we have $r^2\omega = v.p$.

The angular acceleration is the rate at which the angular velocity increases per unit of time, and

$$=\frac{d}{dt}(\omega)=\frac{d}{dt}\left(\frac{v.p}{r^2}\right).$$

Areal velocity. The areal velocity is, similarly, the rate at which the area XOP increases per unit of time, where X is the point in which the path of P meets Ox. It

$$=\text{Lt.}\ \frac{\text{area } POQ}{\Delta t}=\tfrac{1}{2}r^2.\omega.$$

6. Mass and Force. Matter has been defined to be " that which can be perceived by the senses " or " that which can be acted upon by, or can exert, force." It is like time and space a primary conception, and hence it is practically impossible to give it a precise definition. A body is a portion of matter bounded by surfaces.

A particle is a portion of matter which is indefinitely small in all its dimensions. It is the physical correlative of a geometrical point. A body which is incapable of any rotation, or which moves without any rotation, may for the purposes of Dynamics, be often treated as a particle.

The mass of a body is the quantity of matter it contains.

A force is that which changes, or tends to change, the state of rest, or uniform motion, of a body.

7. If to the same mass we apply forces in succession, and they generate the same velocity in the same time, the forces are said to be equal.

If the same force be applied to two different masses, and if it produce in them the same velocity in the same time, the masses are said to be equal.

It is here assumed that it is possible to create forces of equal intensity on different occasions, *e.g.* that the force necessary to keep a given spiral spring stretched through the same distance is always the same when other conditions are unaltered.

Hence by applying the same force in succession we can obtain a number of masses each equal to a standard unit of mass.

8. Practically, different units of mass are used under different conditions and in different countries.

The British unit of mass is called the Imperial Pound, and consists of a lump of platinum deposited in the Exchequer Office.

The French, or Scientific, unit of mass is called a gramme, and is the one-thousandth part of a certain quantity of platinum, called a Kilogramme, which is deposited in the Archives.

One gramme = about 15·432 grains
= 0·0022046 lb.
One Pound = 453·59 grammes.

9. The units of length employed are, in general, either a foot or a centimetre.

A centimetre is the one-hundredth part of a metre which

= 39·37 inches
= 3·2809 ft. approximately.

The unit of time is a second. 86,400 seconds are equal to a mean solar day, which is the mean or average time taken by the Earth to revolve once on its axis with regard to the Sun.

The system of units in which a centimetre, gramme, and second are respectively the units of length, mass, and time is called the C.G.S. system of units.

10. The density of a body, when uniform, is the mass of a unit volume of the body, so that, if m is the mass of a volume V of a body whose density is ρ, then $m = V\rho$. When the density is variable, its value at any point of the body is equal to the limiting value of the ratio of the mass of a very small portion of the body surrounding the point to the volume of that portion, so that

$$\rho = \text{Lt.}\ \frac{m}{V}, \text{ when } V \text{ is taken to be indefinitely small.}$$

The weight of a body at any place is the force with which the Earth attracts the body. The body is assumed to be of such finite size, compared with the Earth, that the weights of its component parts may be assumed to be in parallel directions.

If m be the mass and v the velocity of a particle, its Momentum is mv and its Kinetic Energy is $\frac{1}{2}mv^2$. The former is a vector quantity depending on the direction of the velocity. The latter does not depend on this direction and such a quantity is called a Scalar quantity.

11. Newton's Laws of Motion.

Law I. Every body continues in its state of rest, or of uniform motion in a straight line, except in so far as it be compelled by impressed force to change that state.

Law II. The rate of change of momentum is proportional to the impressed force, and takes place in the direction in which the force acts.

Law III. To every action there is an equal and opposite reaction.

These laws were first formally enunciated by Newton in his *Principia* which was published in the year 1686.

12. If P be the force which in a particle of mass m produces an acceleration f, then Law II states that

$$P=\lambda\frac{d}{dt}(mv), \text{ where } \lambda \text{ is some constant,}$$
$$=\lambda mf.$$

If the unit of force be so chosen that it shall in unit mass produce unit acceleration, this becomes $P=mf$.

If the mass be not constant we must have, instead,

$$P=\frac{d}{dt}(mv).$$

The unit of force, for the Foot-Pound-Second system, is called a Poundal, and that for the C.G.S. system is called a Dyne.

13. The acceleration of a freely falling body at the Earth's surface is denoted by g, which has slightly different values at different points. In feet-second units the value of g varies from 32·09 to 32·25 and in the C.G.S. system from 978·10 to 983·11. For the latitude of London these values are 32·2 and 981 very approximately, and in numerical calculations these are the values generally assumed.

If W be the weight of a mass of one pound, the previous article gives that

$$W=1.g \text{ poundals,}$$

so that the weight of a lb. = 32·2 poundals approximately.

So the weight of a gramme = 981 dynes nearly.

A poundal and a dyne are absolute units, since their values are the same everywhere.

14. Since, by the Second Law, the change of motion produced by a force is in the direction in which the force acts, we find the combined effect of a set of forces on the motion of a particle by finding the effect of each force just as if the other forces did not exist, and then compounding these effects. This is the principle known as that of the *Physical Independence of Forces*.

From this principle, combined with the Parallelogram of Accelerations, we can easily deduce the Parallelogram of Forces.

15. Impulse of a force. Suppose that at time t the value of a force, whose direction is constant, is P. Then the impulse of the force in time τ is defined to be $\int_0^\tau P.dt$.

From Art. 12 it follows that the impulse

$$=\int_0^\tau m\frac{dv}{dt}dt=\Big[mv\Big]_0^\tau$$

= the momentum generated in the direction of the force in time τ.

Sometimes, as in the case of blows and impacts, we have to deal with forces which are very great and act for a very short time, and we cannot measure the magnitude of the forces. We measure the effect of such forces by the momentum each produces, or by its impulse during the time of its action.

16. Work. The work done by a force is equal to the product of the force and the distance through which the point of application is moved in the direction of the force, or, by what is the same thing, the product of the element of distance described by the point of application and the resolved part of the force in the direction of this element. It therefore $=\int Pds$, where ds is the element of the path of the point of application of the force during the description of which the force in the direction of ds was P.

If X, Y, Z be the components of the force parallel to the axes when its point of application is (x, y, z), so that $X=P\frac{dx}{ds}$, etc. then

$$\int(Xdx+Ydy+Zdz)=\int(P\frac{dx}{ds}dx+\ldots+\ldots)$$

$$=\int P\left\{\left(\frac{dx}{ds}\right)^2+\left(\frac{dy}{ds}\right)^2+\left(\frac{dz}{ds}\right)^2\right\}ds=\int Pds$$

= the work done by the force P.

The theoretical units of work are a Foot-Poundal and an Erg. The former is the work done by a poundal during a displacement of one foot in the direction of its action; the latter is that done by a Dyne during a similar displacement of one cm.

One Foot-Poundal = 421390 Ergs nearly. One Foot-Pound is the work done in raising one pound vertically through one foot.

17. Power. The rate of work, or Power, of an agent is the work that would be done by it in a unit of time.

The unit of power used by Engineers is a Horse-Power. An agent is said to be working with one Horse-Power, or H.P., when it would raise 33,000 pounds through one foot per minute.

18. The Potential Energy of a body due to a given system of forces is the work the system can do on the body as it passes from its present configuration to some standard configuration, usually called the zero position.

For example, since the attraction of the Earth (considered as a uniform sphere of radius a and density ρ) is known to be $\gamma . \frac{4\pi a^3 \rho}{3} . \frac{1}{x^2}$ at a distance x from the centre, the potential energy of a unit particle at a distance y from the centre of the Earth, $(y > a)$,

$$= \int_y^a \left(-\frac{4\pi\gamma\rho a^3}{3x^2} \right) dx = \frac{4}{3}\pi\gamma\rho a^3 \left\{ \frac{1}{a} - \frac{1}{y} \right\},$$

the surface of the earth being taken as the zero position.

19. From the definitions of the following physical quantities in terms of the units of mass, length, and time, it is clear that their dimensions are as stated.

Quantity	Mass	Dimensions in length	Time
Volume Density	1	−3	
Velocity		1	−1
Acceleration		1	−2
Force	1	1	−2
Momentum	1	1	−1
Impulse	1	1	−1
Kinetic Energy	1	2	−2
Power or Rate of Work	1	2	−3
Angular Velocity			−1

CHAPTER II

MOTION IN A STRAIGHT LINE

20. Let the distance of a moving point P from a fixed point O be x at any time t. Let its distance similarly at time $t+\Delta t$ be $x+\Delta x$, so that $PQ=\Delta x$.

The velocity of P at time t

$$=\text{Limit, when } \Delta t=0, \text{ of } \frac{PQ}{\Delta t}$$

$$=\text{Limit, when } \Delta t=0, \text{ of } \frac{\Delta x}{\Delta t}=\frac{dx}{dt}.$$

Hence the velocity $v=\dfrac{dx}{dt}$.

Let the velocity of the moving point at time $t+\Delta t$ be

$$v+\Delta v.$$

Then the acceleration of P at time t

$$=\text{limit, when } \Delta t=0, \text{ of } \frac{\Delta v}{\Delta t}$$

$$=\frac{dv}{dt}$$

$$=\frac{d^2x}{dt^2}.$$

21. Motion in a straight line with constant acceleration f.

Let x be the distance of the moving point at time t from a fixed point in the straight line.

Then $$\frac{d^2x}{dt^2}=f \quad \text{..............................(1).}$$

Hence, on integration, $$v=\frac{dx}{dt}=ft+A \quad \text{.........................(2),}$$

where A is an arbitrary constant.

Integrating again, we have

$$x=\tfrac{1}{2}ft^2+At+B \quad\text{.........................(3)},$$

where B is an arbitrary constant.

Again, on multiplying (1) by $2\frac{dx}{dt}$, and integrating with respect to t, we have

$$v^2=\left(\frac{dx}{dt}\right)^2=2fx+C \quad\text{........................(4)},$$

where C is an arbitrary constant.

These three equations contain the solution of all questions on motion in a straight line with constant acceleration. The arbitrary constants A, B, C are determined from the initial conditions.

Suppose, for example, that the particle started at a distance a from a fixed point O on the straight line with velocity u in a direction away from O, and suppose that the time t is reckoned from the instant of projection.

We then have that when $t=0$, then $v=u$ and $x=a$. Hence the equations (2), (3), and (4) give

$$u=A,\ a=B, \text{ and } u^2=C+2fa.$$

Hence we have $\qquad v=u+ft,$

$$x-a=ut+\tfrac{1}{2}ft^2,$$

and $\qquad v^2=u^2+2f(x-a),$

the three standard equations of Elementary Dynamics.

22. *A particle moves in a straight line OA starting from rest at A and moving with an acceleration which is always directed towards O and varies as the distance from O; to find the motion.*

Let x be the distance OP of the particle from O at any time t; and let the acceleration at this distance be μx.

The equation of motion is then

$$\frac{d^2x}{dt^2}=-\mu x \quad\text{............................(1)}.$$

[We have a negative sign on the right-hand side because $\frac{d^2x}{dt^2}$ is the acceleration in the direction of x increasing, *i.e.* in the direction OP; whilst μx is the acceleration towards O, *i.e.* in the direction PO.]

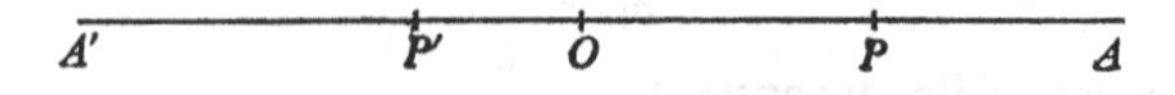

Multiplying by $2\frac{dx}{dt}$ and integrating, we have

$$\left(\frac{dx}{dt}\right)^2 = -\mu x^2 + C.$$

If OA be a, then $\frac{dx}{dt}=0$ when $x=a$, so that $0=-\mu a^2+C$,

and $$\therefore \left(\frac{dx}{dt}\right)^2 = \mu(a^2-x^2).$$

$$\therefore \frac{dx}{dt} = -\sqrt{\mu}\sqrt{a^2-x^2} \text{ .... (2).}$$

[The negative sign is put on the right-hand side because the velocity is clearly negative so long as OP is positive and P is moving towards O.]

Hence, on integration,

$$t\sqrt{\mu} = -\int \frac{dx}{\sqrt{a^2-x^2}} = \cos^{-1}\frac{x}{a} + C_1,$$

where $$0 = \cos^{-1}\frac{a}{a} + C_1, \quad \textit{i.e. } C_1 = 0,$$

if the time be measured from the instant when the particle was at A.

$$\therefore x = a\cos\sqrt{\mu t} \text{ .... (3).}$$

When the particle arrives at O, x is zero; and then, by (2), the velocity $= -a\sqrt{\mu}$. The particle thus passes through O and immediately the acceleration alters its direction and tends to diminish the velocity; also the velocity is destroyed on the left-hand side of O as rapidly as it was produced on the right-hand side; hence the particle comes to rest at a point A' such that OA and OA' are equal. It then retraces its path, passes through O, and again is instantaneously at rest at A. The whole motion of the particle is thus an oscillation from A to A' and back, continually repeated over and over again.

The time from A to O is obtained by putting x equal to zero in (3). This gives $\cos(\sqrt{\mu t})=0$, *i.e.* $t=\frac{\pi}{2\sqrt{\mu}}$.

The time from A to A' and back again, *i.e.* the time of a complete oscillation, is four times this, and therefore $=\frac{2\pi}{\sqrt{\mu}}$.

This result is independent of the distance a, *i.e. is independent of the distance from the centre at which the particle started.* It depends solely on the quantity μ which is equal to the acceleration at unit distance from the centre.

23. Motion of the kind investigated in the previous article is called **Simple Harmonic Motion.**

The time, $\frac{2\pi}{\sqrt{\mu}}$, for a complete oscillation is called the **Periodic Time** of the motion, and the distance, OA or OA', to which the particle vibrates on either side of the centre of the motion is called the **Amplitude** of its motion.

The **Frequency** is the number of complete oscillations that the particle makes in a second, and hence $=\frac{1}{\text{Periodic time}}=\frac{\sqrt{\mu}}{2\pi}$.

24. The equation of motion when the particle is on the left-hand side of O is

$$\frac{d^2x}{dt^2}=\text{acceleration in the direction } P'A$$
$$=\mu . P'O=\mu(-x)=-\mu x.$$

Hence the same equation that holds on the right hand of O holds on the left hand also.

As in Art. 22 it is easily seen that the most general solution of this equation is

$$x=a \cos [\sqrt{\mu t}+\epsilon] \quad \text{.......................(1)},$$

which contains two arbitrary constants a and ϵ.

This gives $$\frac{dx}{dt}=-a\sqrt{\mu} \sin (\sqrt{\mu t}+\epsilon) \quad \text{.....................(2)}.$$

(1) and (2) both repeat when t is increased by $\frac{2\pi}{\sqrt{\mu}}$, since the sine and cosine of an angle always have the same value when the angle is increased by 2π.

Using the standard expression (1) for the displacement in a simple harmonic motion, the quantity ϵ is called the **Epoch,** the angle $\sqrt{\mu t}+\epsilon$ is called the **Argument,** whilst the **Phase** of the motion is the time that has elapsed since the particle was at its maximum distance in the positive direction. Clearly x is a maximum at time t_0 where $\sqrt{\mu t_0}+\epsilon=0$.

Hence the phase at time t

$$=t-t_0=t+\frac{\epsilon}{\sqrt{\mu}}=\frac{\sqrt{\mu t}+\epsilon}{\sqrt{\mu}}.$$

Motion of the kind considered in this article, in which the time of falling to a given point is the same whatever be the distance through which the particle falls, is called **Tautochronous.**

25. In Art. 22 if the particle, instead of being at rest initially, be projected from A with velocity V in the positive direction, we have

$$V^2=-\mu a^2+C.$$

Hence $$\left(\frac{dx}{dt}\right)^2=V^2+\mu(a^2-x^2)$$

$$=\mu(b^2-x^2), \text{ where } b^2=a^2+\frac{V^2}{\mu} \qquad \text{.........(1)},$$

$$\therefore \frac{dx}{dt}=\sqrt{\mu}\sqrt{b^2-x^2},$$

and $$t\sqrt{\mu}=-\cos^{-1}\frac{x}{b}+C_1, \text{ where } 0=-\cos^{-1}\frac{a}{b}+C_1.$$

$$\therefore t\sqrt{\mu}=\cos^{-1}\frac{a}{b}-\cos^{-1}\frac{x}{b} \qquad \text{.........(2).}$$

From (1), the velocity vanishes when

$$x=b=\sqrt{a^2+\frac{V^2}{\mu}},$$

and then, from (2),

$$t\sqrt{\mu}=\cos^{-1}\frac{a}{b}, \text{ i.e. } t=\frac{1}{\sqrt{\mu}}\cos^{-1}\frac{a}{\sqrt{a^2+\frac{V^2}{\mu}}}.$$

The particle then retraces its path, and the motion is the same as in Art. 22 with b substituted for a.

26. Compounding of two simple harmonic motions of the same period and in the same straight line.

The most general displacements of this kind are given by $a\cos(nt+\epsilon)$ and $b\cos(nt+\epsilon')$, so that

$$x=a\cos(nt+\epsilon)+b\cos(nt+\epsilon')$$
$$=\cos nt\,(a\cos\epsilon+b\cos\epsilon')-\sin nt\,(a\sin\epsilon+b\sin\epsilon').$$

Let and $$\left.\begin{aligned} a\cos\epsilon+b\cos\epsilon'&=A\cos E\\ a\sin\epsilon+b\sin\epsilon'&=A\sin E\end{aligned}\right\} \qquad \text{.........(1)},$$

so that

$$A=\sqrt{a^2+b^2+2ab\cos(\epsilon-\epsilon')}, \text{ and } \tan E=\frac{a\sin\epsilon+b\sin\epsilon'}{a\cos\epsilon+b\cos\epsilon'}.$$

Then $$x=A\cos(nt+E),$$

so that the composition of the two given motions gives a similar motion of the same period whose amplitude and epoch are known.

If we draw OA $(=a)$ at an angle ϵ to a fixed line, and OB $(=b)$ at an angle ϵ' and complete the parallelogram $OACB$ then by equations (1) we see that OC represents A and that it is inclined at an angle E to the fixed line. The line representing the resultant of the two given motions is therefore the geometrical resultant of the lines representing the two component motions.

So with more than two such motions of the same period.

27. We cannot compound two simple harmonic motions of different periods.

The case when the periods are nearly but not quite equal, is of some considerable importance.

In this case we have

$$x=a\cos(nt+\epsilon)+b\cos(n't+\epsilon'),$$

where $n'-n$ is small, $=\lambda$ say.

Then $$x=a\cos(nt+\epsilon)+b\cos[nt+\epsilon_1'],$$

where $$\epsilon_1'=\lambda t+\epsilon'.$$

By the last article

$$x=A\cos(nt+E)\quad\text{.......................(1)},$$

where $$A^2=a^2+b^2+2ab\cos(\epsilon-\epsilon_1')$$
$$=a^2+b^2+2ab\cos[\epsilon-\epsilon'-(n'-n)t]\quad\text{...........(2)},$$

and $$\tan E=\frac{a\sin\epsilon+b\sin\epsilon_1'}{a\cos\epsilon+b\cos\epsilon_1'}$$
$$=\frac{a\sin\epsilon+b\sin[\epsilon'+(n'-n)t]}{a\cos\epsilon+b\cos[\epsilon'+(n'-n)t]}\quad\text{.................(3)}.$$

The quantities A and E are now not constant, but they vary *slowly* with the time, since $n'-n$ is very small.

The greatest value of A is when $\epsilon-\epsilon'-(n'-n)t=$ any even multiple of π and then its value is $a+b$.

The least value of A is when $\epsilon-\epsilon'-(n'-n)t=$ any odd multiple of π and then its value is $a-b$.

At any given time therefore the motion may be taken to be a simple harmonic motion of the same approximate period as either of the given component motions, but with its amplitude A and epoch E gradually changing from definite minimum to definite maximum values, the periodic times of these changes being $\dfrac{2\pi}{n'-n}$.

[The Student who is acquainted with the theory of Sound may compare the phenomenon of Beats.]

28. *Ex.* 1. Shew that the resultant of two simple harmonic vibrations in the same direction and of equal periodic time, the amplitude of one being twice that of the other and its phase a quarter of a period in advance, is a simple harmonic vibration of amplitude $\sqrt{5}$ times that of the first and whose phase is in advance of the first by $\frac{\tan^{-1} 2}{2\pi}$ of a period.

Ex. 2. A particle is oscillating in a straight line about a centre of force O, towards which when at a distance r the force is $m . n^2 r$, and a is the amplitude of the oscillation; when at a distance $\frac{a\sqrt{3}}{2}$ from O, the particle receives a blow in the direction of motion which generates a velocity na. If this velocity be away from O, shew that the new amplitude is $a\sqrt{3}$.

Ex. 3. A particle P, of mass m, moves in a straight line Ox under a force $m\mu$ (distance) directed towards a point A which moves in the straight line Ox with constant acceleration a. Shew that the motion of P is simple harmonic, of period $\frac{2\pi}{\sqrt{\mu}}$, about a moving centre which is always at a distance $\frac{a}{\mu}$ behind A.

Ex. 4. An elastic string without weight, of which the unstretched length is l and the modulus of elasticity is the weight of n ozs., is suspended by one end, and a mass of m ozs. is attached to the other; shew that the time of a vertical oscillation is $2\pi\sqrt{\frac{ml}{ng}}$.

Ex. 5. One end of an elastic string, whose modulus of elasticity is λ and whose unstretched length is a, is fixed to a point on a smooth horizontal table and the other end is tied to a particle of mass m which is lying on the table. The particle is pulled to a distance where the extension of the string is b and then let go; shew that the time of a complete oscillation is $2\left(\pi+\frac{2a}{b}\right)\sqrt{\frac{am}{\lambda}}$.

Ex. 6. An endless cord consists of two portions, of lengths $2l$ and $2l'$ respectively, knotted together, their masses per unit of length being m and m'. It is placed in stable equilibrium over a small smooth peg and then slightly displaced. Shew that the time of a complete oscillation is

$$2\pi\sqrt{\frac{ml+m'l'}{(m-m')g}}.$$

Ex. 7. Assuming that the earth attracts points inside it with a force which varies as the distance from its centre, shew that, if a straight frictionless airless tunnel be made from one point of the earth's surface to any other point, a train would traverse the tunnel in slightly less than three-quarters of an hour.

29. *Motion when the motion is in a straight line and the acceleration is proportional to the distance from a fixed point O in the straight line and is always away from O.*

Here the equation of motion is

$$\frac{d^2x}{dt^2}=\mu x \quad \text{.............................(1).}$$

Suppose the velocity of the particle to be zero at a distance a from O at time zero.

The integral of (1) is

$$\left(\frac{dx}{dt}\right)^2=\mu x^2+A, \text{ where } 0=\mu a^2+A.$$

$$\therefore \frac{dx}{dt}=\sqrt{\mu(x^2-a^2)} \quad \text{........................(2),}$$

the positive sign being taken in the right-hand member since the velocity is positive in this case.

$$\therefore t\sqrt{\mu}=\int\frac{dx}{\sqrt{x^2-a^2}}=\log\left[x+\sqrt{x^2-a^2}\right]+B,$$

where $$0=\log[a]+B.$$

$$\therefore t\sqrt{\mu}=\log\frac{x+\sqrt{x^2-a^2}}{a}.$$

$$\therefore x+\sqrt{x^2-a^2}=ae^{t\sqrt{\mu}}.$$

$$\therefore x-\sqrt{x^2-a^2}=\frac{a^2}{x+\sqrt{x^2-a^2}}=ae^{-t\sqrt{\mu}}.$$

Hence, by addition.

$$x=\frac{a}{2}e^{\sqrt{\mu}\cdot t}+\frac{a}{2}e^{-\sqrt{\mu}t} \quad \text{......................(3).}$$

As t increases, it follows from (3) that x continually increases, and then from (2) that the velocity continually increases also.

Hence the particle would continually move along the positive direction of the axis of x and with continually increasing velocity.

Equation (3) may be written in the form

$$x=a\cosh(\sqrt{\mu}t),$$

and then (2) gives $\qquad v=a\sqrt{\mu}\sinh(\sqrt{\mu}t).$

30. In the previous article suppose that the particle were initially projected *towards* the origin O with velocity V; then we should have $\frac{dx}{dt}$ equal to $-V$ when $x=a$; and equation (2) would be more complicated. We may however take the most general solution of (1) in the form)

$$x=Ce^{\sqrt{\mu}t}+De^{-\sqrt{\mu}t} \qquad \text{......................(4)},$$

where C and D are any constants.

Since, when $t=0$, we have $x=a$ and $\frac{dx}{dt}=-V$, this gives

$$a=C+D, \text{ and } -V=\sqrt{\mu}C-\sqrt{\mu}D.$$

Hence $\qquad C=\frac{1}{2}\left(a-\frac{V}{\sqrt{\mu}}\right)$ and $D=\frac{1}{2}\left(a+\frac{V}{\sqrt{\mu}}\right).$

$\therefore$ (4) gives

$$x=\frac{1}{2}\left(a-\frac{V}{\sqrt{\mu}}\right)e^{\sqrt{\mu}t}+\frac{1}{2}\left(a+\frac{V}{\sqrt{\mu}}\right)e^{-\sqrt{\mu}t} \qquad \text{...........(5)}$$

$$=a\cosh(\sqrt{\mu}t)-\frac{V}{\sqrt{\mu}}\sinh(\sqrt{\mu}t) \qquad \text{.................(6)}.$$

In this case the particle will arrive at the origin O when

$$0=\frac{1}{2}\left(a-\frac{V}{\sqrt{\mu}}\right)e^{\sqrt{\mu}t}+\frac{1}{2}\left(a+\frac{V}{\sqrt{\mu}}\right)e^{-\sqrt{\mu}t},$$

i.e. when $\qquad e^{2\sqrt{\mu}t}=\frac{V+a\sqrt{\mu}}{V-a\sqrt{\mu}},$

i.e. when $\qquad t=\frac{1}{2\sqrt{\mu}}\log\frac{V+a\sqrt{\mu}}{V-a\sqrt{\mu}}.$

In the particular case when $V=a\sqrt{\mu}$ this value of t is infinity.

If therefore the particle were projected at distance a towards the origin with the velocity $a\sqrt{\mu}$, it would not arrive at the origin until after an infinite time.

Also, putting $V=a\sqrt{\mu}$ in (5), we have

$$x=ae^{-\sqrt{\mu}t}, \text{ and } v=\frac{dx}{dt}=-a\sqrt{\mu}e^{-\sqrt{\mu}t}.$$

The particle would therefore always be travelling towards O with a continually decreasing velocity, but would take an infinite time to get there.

31. *A particle moves in a straight line OA with an acceleration which is always directed towards O and varies inversely as the square of its distance from O; if initially the particle were at rest at A, find the motion.*

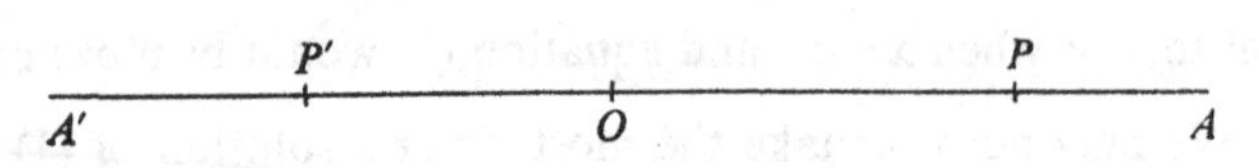

Let OP be x, and let the acceleration of the particle when at P be $\frac{\mu}{x^2}$ in the direction PO. The equation of motion is therefore

$$\frac{d^2x}{dt^2}=\text{acceleration along } OP=-\frac{\mu}{x^2} \quad \text{..............(1).}$$

Multiplying both sides by $2\frac{dx}{dt}$ and integrating, we have

$$\left(\frac{dx}{dt}\right)^2=\frac{2\mu}{x}+C,$$

where $0=\frac{2\mu}{a}+C$, from the initial conditions.

Subtracting, $$\left(\frac{dx}{dt}\right)^2=2\mu\left(\frac{1}{x}-\frac{1}{a}\right).$$

$$\therefore \frac{dx}{dt}=-\sqrt{2\mu}\sqrt{\frac{a-x}{ax}} \quad \text{......................(2),}$$

the negative sign being prefixed because the motion of P is towards O, *i.e.* in the direction of x decreasing.

Hence $$\sqrt{\frac{2\mu}{a}}.t=-\int\sqrt{\frac{x}{a-x}}dx.$$

To integrate the right-hand side, put $x=a\cos^2\theta$, and we have

$$\sqrt{\frac{2\mu}{a}}.t=\int\frac{\cos\theta}{\sin\theta}.2a\cos\theta\sin\theta d\theta$$

$$=a\int(1+\cos 2\theta)\,d\theta=a\,(\theta+\tfrac{1}{2}\sin 2\theta)+C_1$$

$$=a\cos^{-1}\sqrt{\frac{x}{a}}+\sqrt{ax-x^2}+C_1,$$

where $\qquad 0=a\cos^{-1}(1)+0+C, \quad i.e.\ C_1=0.$

$$\therefore t=\sqrt{\frac{a}{2\mu}}\left[\sqrt{ax-x^2}+a\cos^{-1}\sqrt{\frac{x}{a}}\right] \quad \text{...(3).}$$

Equation (2) gives the velocity at any point P of the path, and (3) gives the time from the commencement of the motion.

The velocity on arriving at the origin O is found, by putting $x=0$ in (2), to be infinite.

Also the corresponding time, from (3),

$$=\sqrt{\frac{a}{2\mu}}[a\cos^{-1}.0]=\frac{\pi}{2}\frac{a^{\frac{3}{2}}}{\sqrt{2\mu}}.$$

The equation of motion (1) will not hold after the particle has passed through O; but it is clear that then the acceleration, being opposite to the direction of the velocity, will destroy the velocity, and the latter will be diminished at the same rate as it was produced on the positive side of O. The particle will therefore, by symmetry, come to rest at a point A' such that AO and OA' are equal. It will then return, pass again through O and come to rest at A.

The total time of the oscillation = four times the time from A to O = $2\pi\frac{a^{\frac{3}{2}}}{\sqrt{2\mu}}$.

32. By the consideration of Dimensions only we can shew that the time $\propto\frac{a^{\frac{3}{2}}}{\sqrt{\mu}}$. For the only quantities that can appear in the answer are a and μ. Let then the time be $a^p\mu^q$.

Since $\frac{\mu}{(\text{distance})^2}$ is an acceleration, whose dimensions are $[L]\,[T]^{-2}$, the dimensions of μ are $[L]^3\,[T]^{-2}$; hence the dimensions of $a^p\mu^q$ are

$$[L]^{p+3q}[T]^{-2q}.$$

Since this is a time, we have $p+3q=0$ and $-2q=1$.

$$\therefore q=-\frac{1}{2} \text{ and } p=\frac{3}{2}. \quad \text{Hence the required time} \propto\frac{a^{\frac{3}{2}}}{\sqrt{\mu}}.$$

33. As an illustration of Art. 31 let us consider the motion of a particle let fall towards the earth (assumed at rest) from a point outside it. It is shewn in treatises on Attractions that the attraction on a particle outside the earth (assumed to be a homogeneous sphere), varies inversely as the square of its distance from the centre. The acceleration of a particle outside the earth at distance x may therefore be taken to be $\frac{\mu}{x^2}$.

If a be the radius of the earth this quantity at the earth's surface is equal to g, and hence $\frac{\mu}{a^2}=g$, *i.e.* $\mu=ga^2$.

For a point P outside the earth the equation of motion is therefore

$$\frac{d^2x}{dt^2}=-\frac{ga^2}{x^2} \quad \text{.........................(1)},$$

$$\therefore \left(\frac{dx}{dt}\right)^2=\frac{2ga^2}{x}+C.$$

If the particle started from rest at a distance b from the centre of the earth, this gives

$$\left(\frac{dx}{dt}\right)^2=2ga^2\left(\frac{1}{x}-\frac{1}{b}\right) \quad \text{.......................(2)},$$

and hence the square of the velocity on reaching the surface of the earth $=2ga\left(1-\frac{a}{b}\right)$ ....(3).

Let us now assume that there is a hole going down to the earth's centre just sufficient to admit of the passing of the particle.

On a particle inside the earth the attraction can be shewn to vary directly as the distance from the centre, so that the acceleration at distance x from its centre is $\mu_1 x$, where $\mu_1 a=$ its value at the earth's surface$=g$.

The equation of motion of the particle when inside the earth therefore is

$$\frac{d^2x}{dt^2}=-\frac{g}{a}x,$$

and therefore

$$\left(\frac{dx}{dt}\right)^2=-\frac{g}{a}x^2+C_1.$$

Now when $x=a$, the square of the velocity is given by (3), since there was no instantaneous change of velocity at the earth's surface.

$$\therefore 2ga\left(1-\frac{a}{b}\right)=-\frac{g}{a}.a^2+C_1,$$

$$\therefore \left(\frac{dx}{dt}\right)^2=-\frac{g}{a}x^2+ga\left[3-\frac{2a}{b}\right].$$

On reaching the centre of the earth the square of the velocity is therefore $ga\left(3-\frac{2a}{b}\right)$.

34. *Ex.* 1. A particle falls towards the earth from infinity; shew that its velocity on reaching the earth is the same as it would have acquired in falling with constant acceleration g through a distance equal to the earth's radius.

Ex. 2. Shew that the velocity with which a body falling from infinity reaches the surface of the earth (assumed to be a homogeneous sphere of radius 4000 miles) is about 7 miles per second.

In the case of the sun shew that it is about 380 miles per second, the radius of the sun being 440,000 miles and the distance of the earth from it 92,500,000 miles.

Ex. 3. If the earth's attraction vary inversely as the square of the distance from its centre, and g be its magnitude at the surface, the time of falling from a height h above the surface to the surface is

$$\sqrt{\frac{a+h}{2g}}\left[\frac{a+h}{a}\sin^{-1}\sqrt{\frac{h}{a+h}}+\sqrt{\frac{h}{a}}\right],$$

where a is the radius of the earth and the resistance of the air is neglected.

If h be small compared with a, shew that this result is approximately

$$\sqrt{\frac{2h}{g}}\left[1+\frac{5}{6}\frac{h}{a}\right].$$

35. It is clear that equations (2) and (3) of Art. 31 cannot be true after the particle has passed O; for on giving x negative values these equations give impossible values for v and t.

When the particle is at P', to the left of O, the acceleration is $\frac{\mu}{OP'^2}$, *i.e.* $\frac{\mu}{x^2}$, towards the right. Now $\frac{d^2x}{dt^2}$ means the acceleration towards the positive direction of x. Hence, when P' is on the left of O, the equation of motion is

$$\frac{d^2x}{dt^2}=\frac{\mu}{x^2},$$

giving a different solution from (2) and (3).

The general case can be easily considered. Let the acceleration be

$\mu(\text{distance})^n$ towards O. The equation of motion when the particle is on the right hand of O is clearly

$$\frac{d^2x}{dt^2} = -\mu . x^n.$$

When P' is on the left of O, the equation is

$$\frac{d^2x}{dt^2} = \text{acceleration in direction } OA$$

$$= \mu(P'O)^n = \mu(-x)^n.$$

These two equations are the same if

$$-\mu . x^n = \mu(-x)^n, \text{ i.e. if } (-1)^n = -1,$$

i.e. if n be an odd integer, or if it be of the form $\dfrac{2p+1}{2q+1}$, where p and q are integers; in these cases the same equation holds on both sides of the origin; otherwise it does not.

36. *Ex. A small bead, of mass m, moves on a straight rough wire under the action of a force equal to mμ times the distance of the bead from a fixed point A outside the wire at a perpendicular distance a from it. Find the motion if the bead start from rest at a distance c from the foot, O, of the perpendicular from A upon the wire.*

Let P be the position of the bead at any time t, where

$$OP = x \text{ and } AP = y.$$

Let R be the normal reaction of the wire and μ_1 the coefficient of friction.

Resolving forces perpendicular to the wire, we have

$$R = m\mu y \sin OPA = m\mu a.$$

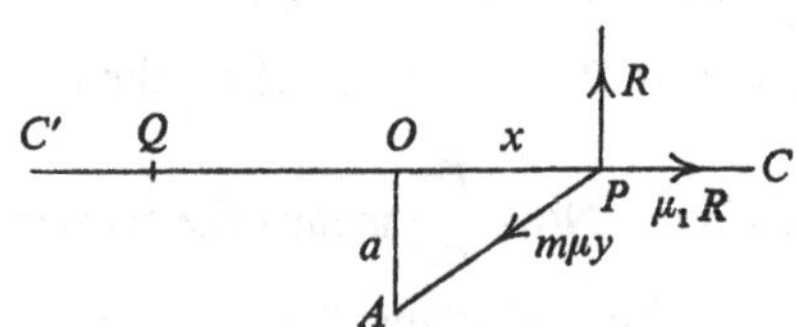

Hence the friction $\qquad \mu_1 R = m\mu\mu_1 a.$

The resolved part of the force $m\mu y$ along the wire

$$= m\mu y \cos OPA = m\mu x.$$

Hence the total acceleration $= \mu\mu_1 a - \mu x.$

The equation of motion is thus

$$\frac{d^2x}{dt^2}=\mu\mu_1 a-\mu x=-\mu(x-\mu_1 a) \quad \text{.................(1)},$$

so long as P is to the right of O.

[If P be to the left of O and moving towards the left, the equation of motion is

$$\frac{d^2x}{dt^2}=\text{acceleration in the direction } OC$$

$$=\mu\mu_1 a+\mu(-x), \text{ as in the last article,}$$

and this is the same as (1) which therefore holds on both sides of O.]

Integrating, we have

$$\left(\frac{dx}{dt}\right)^2=-\mu(x-\mu_1 a)^2+C,$$

where $$0=-\mu(c-\mu_1 a)^2+C.$$

$$\therefore\ v^2=\left(\frac{dx}{dt}\right)^2=\mu[(c-\mu_1 a)^2-(x-\mu_1 a)^2] \quad \text{...........(2)},$$

and therefore, as in Art. 22, $\sqrt{\mu}t=\cos^{-1}\dfrac{x-\mu_1 a}{c-\mu_1 a}+C_1$,

where $$0=\cos^{-1}\frac{c-\mu_1 a}{c-\mu_1 a}+C_1, \textit{ i.e. } C_1=0.$$

$$\therefore\ \sqrt{\mu}t=\cos^{-1}\frac{x-\mu_1 a}{c-\mu_1 a} \quad \text{.....................(3)},$$

(2) and (3) give the velocity and time for any position.

From (2) the velocity vanishes when $x-\mu_1 a=\pm(c-\mu_1 a)$,

i.e. when $x=c=OC$, and when $x=-(c-2a\mu_1)$,

i.e. at the point C', where $OC'=c-2a\mu_1$,

and then from (3) the corresponding time

$$=\frac{1}{\sqrt{\mu}}\cos^{-1}\frac{-c+\mu_1 a}{c-\mu_1 a}=\frac{1}{\sqrt{\mu}}\cos^{-1}(-1)=\frac{\pi}{\sqrt{\mu}}.$$

The motion now reverses and the particle comes to rest at a point C'' on the right of O where $OC''=OC'-2\mu_1 a=OC-4\mu_1 a$.

Finally, when one of the positions of instantaneous rest is at a distance which is equal to or less than $\mu_1 a$ from O, the particle remains at rest. For at this point the force towards the centre is less than the limiting friction and therefore only just sufficient friction will be exerted to keep the particle at rest.

It will be noted that the periodic time $\frac{2\pi}{\sqrt{\mu}}$ is not affected by the friction, but the amplitude of the motion is altered by it.

37. *Ex. A particle, of mass m, rests in equilibrium at a point N, being attracted by two forces equal to $m\mu^n$ (distance)^n and $m\mu'^n$ (distance)^n towards two fixed centres O and O′. If the particle be slightly displaced from N, and if n be positive, shew that it oscillates, and find the time of a small oscillation.*

x P

O N O'

Let $OO'=a$, $ON=d$, and $NO'=d'$, so that

$$\mu^n.d^n=\mu'^n.d'^n \quad \text{.........................(1)},$$

since there is equilibrium at N.

$$\therefore \frac{d}{\mu'}=\frac{d'}{\mu}=\frac{a}{\mu+\mu'} \quad \text{......................(2)}.$$

Let the particle be at a distance x from N towards O'.

The equation of motion is then

$$\frac{d^2x}{dt^2}=-\mu^n.OP^n+\mu'^n.PO'^n=-\mu^n(d+x)^n+\mu'^n(d'-x)^n \quad \text{.........(3)}.$$

If x is positive, the right-hand side is negative; if x is negative, it is positive; the acceleration is towards N in either case.

Expanding by the Binomial Theorem, (3) gives

$$\frac{d^2x}{dt^2}=-\mu^n(d^n+nd^{n-1}x+\ldots)+\mu'^n(d'^n-nd'^{n-1}x+\ldots)$$

$$=-nx[\mu^n d^{n-1}+\mu'^n d'^{n-1}]$$

$$+\text{terms involving higher powers of } x$$

$$=-nxa^{n-1}\frac{(\mu\mu')^{n-1}}{(\mu+\mu')^{n-2}}+\ldots \text{ by (2)}.$$

If x be so small that its squares and higher powers may be neglected, this gives

$$\frac{d^2x}{dt^2}=-n\frac{(\mu\mu'a)^{n-1}}{(\mu+\mu')^{n-2}}x \quad \text{......................(4)}.$$

Hence, as in Art. 22, the time of a small oscillation

$$=2\pi \div \sqrt{n\frac{(\mu\mu' a)^{n-1}}{(\mu+\mu')^{n-2}}} = 2\pi\sqrt{\frac{(\mu+\mu')^{n-2}}{n(\mu\mu' a)^{n-1}}}.$$

If n be negative, the right-hand member of (4) is positive and the motion is not one of oscillation.

EXAMPLES ON CHAPTER II

1. A particle moves towards a centre of attraction starting from rest at a distance a from the centre; if its velocity when at any distance x from the centre vary as $\sqrt{\frac{a^2-x^2}{x^2}}$, find the law of force.

2. A particle moves from rest at a distance a from a centre of force where the repulsion at distance x is μx^{-2}; shew that its velocity at distance x is $\sqrt{\frac{2\mu(x-a)}{ax}}$ and that the time it has taken is

$$\sqrt{\frac{a}{2\mu}}\left[\sqrt{x^2-ax}+a\log_e\left(\sqrt{\frac{x}{a}}+\sqrt{\frac{x}{a}-1}\right)\right].$$

3. Prove that it is impossible for a particle to move from rest so that its velocity varies as the distance described from the commencement of the motion.

If the velocity vary as (distance)^n, shew that n cannot be greater than $\frac{1}{2}$.

4. A point moves in a straight line towards a centre of force

$$\left\{\frac{\mu}{(\text{distance})^3}\right\},$$

starting from rest at a distance a from the centre of force; shew that the time of reaching a point distant b from the centre of force is $\frac{a\sqrt{a^2-b^2}}{\sqrt{\mu}}$, and that its velocity then is $\frac{\sqrt{\mu}}{ab}\sqrt{a^2-b^2}$.

5. A particle falls from rest at a distance a from a centre of force, where the acceleration at distance x is $\mu x^{-\frac{5}{3}}$; when it reaches the centre shew that its velocity is infinite and that the time it has taken is $\frac{2a^{\frac{4}{3}}}{\sqrt{3\mu}}$.

6. A particle moves in a straight line under a force to a point in it varying as (distance)$^{-\frac{4}{3}}$; shew that the velocity in falling from rest at infinity to a distance a is equal to that acquired in falling from rest at a distance a to a distance $\frac{a}{8}$.

7. A particle, whose mass is m, is acted upon by a force $m\mu\left(x+\frac{a^4}{x^3}\right)$ towards the origin; if it start from rest at a distance a, shew that it will arrive at the origin in time $\frac{\pi}{4\sqrt{\mu}}$.

8. A particle moves in a straight line with an acceleration towards a fixed point in the straight line, which is equal to $\frac{\mu}{x^2}-\frac{\lambda}{x^3}$ when the particle is at a distance x from the given point; it starts from rest at a distance a; shew that it oscillates between this distance and the distance $\frac{\lambda a}{2\mu a-\lambda}$, and that its periodic time is $\frac{2\pi\mu a^3}{(2a\mu-\lambda)^{\frac{3}{2}}}$.

9. A particle moves with an acceleration which is always towards, and equal to μ divided by the distance from, a fixed point O. If it start from rest at a distance a from O, shew that it will arrive at O in time

$$a\sqrt{\frac{\pi}{2\mu}}.\quad\left[\text{Assume that }\int_0^\infty e^{-x^2}dx=\frac{\sqrt{\pi}}{2}.\right]$$

10. A particle is attracted by a force to a fixed point varying inversely as the nth power of the distance; if the velocity acquired by it in falling from an infinite distance to a distance a from the centre is equal to the velocity that would be acquired by it in falling from rest at a distance a to a distance $\frac{a}{4}$, shew that $n=\frac{3}{2}$.

11. A particle rests in equilibrium under the attraction of two centres of force which attract directly as the distance, their attractions per unit of mass at unit distance being μ and μ'; the particle is slightly displaced towards one of them; shew that the time of a small oscillation is $\frac{2\pi}{\sqrt{\mu+\mu'}}$.

12. A mass of 100 lbs. hangs freely from the end of a rope. The mass is hauled up vertically from rest by winding up the rope, the pull of which starts at 150 lbs. weight and diminishes uniformly at the rate of 1 lb. wt. for each foot wound up. Neglecting the weight of the rope, shew that the mass has described 50 feet at the end of time $\frac{5\pi\sqrt{2}}{8}$ secs. and that its velocity then is $20\sqrt{2}$ ft. per sec.

13. A particle moves in a straight line with an acceleration equal to $\mu\div$ the nth power of the distance from a fixed point O in the straight line. If it be projected towards O, from a point at a distance a, with the velocity it would have acquired in falling from infinity, shew that it will reach O in time $\frac{2}{n+1}\sqrt{\frac{n-1}{2\mu}}.\,a^{\frac{n+1}{2}}$.

14. In the previous question if the particle started from rest at distance a, shew that it would reach O in time

$$\sqrt{\frac{n-1}{2\mu}}\pi a^{\frac{n+1}{2}}\frac{\Gamma\left(\frac{1}{n-1}+\frac{1}{2}\right)}{\Gamma\left(\frac{1}{n-1}\right)},\text{ or }\sqrt{\frac{\pi}{2\mu(1-n)}}a^{\frac{n+1}{2}}\frac{\Gamma\left(\frac{1}{1-n}\right)}{\Gamma\left(\frac{1}{1-n}+\frac{1}{2}\right)},$$

according as n is $>$ or $<$ unity.

15. A shot, whose mass is 50 lbs., is fired from a gun, 4 inches in diameter and 8 feet in length. The pressure of the powder-gas is inversely proportional to the volume behind the shot and changes from an initial value of 10 tons' weight per square inch to 1 ton wt. per sq. inch as the shot leaves the gun. Shew that the muzzle velocity of the shot is approximately 815 feet per second, having given $\log_e 10=2{\cdot}3026$.

16. If the Moon and Earth were at rest, shew that the least velocity with which a particle could be projected from the Moon, in order to reach the Earth, is about

$1\frac{1}{2}$ miles per second, assuming their radii to be 1100 and 4000 miles respectively, the distance between their centres 240,000 miles, and the mass of the Moon to be $\frac{1}{81}$ that of the Earth.

17. A small bead can slide on a smooth wire AB, being acted upon by a force per unit of mass equal to $\mu \div$ the square of its distance from a point O which is outside AB. Shew that the time of a small oscillation about its position of equilibrium is $\frac{2\pi}{\sqrt{\mu}} b^{\frac{3}{2}}$, where b is the perpendicular distance of O from AB.

18. A solid attracting sphere, of radius a and mass M, has a fine hole bored straight through its centre; a particle starts from rest at a distance b from the centre of the sphere in the direction of the hole produced, and moves under the attraction of the sphere entering the hole and going through the sphere; shew that the time of a complete oscillation is

$$\frac{4}{\sqrt{2\gamma M}}\left[\sqrt{2}a^{\frac{3}{2}} \sin^{-1}\sqrt{\frac{b}{3b-2a}}+b^{\frac{3}{2}} \cos^{-1}\sqrt{\frac{a}{b}}+\sqrt{ab(b-a)}\right],$$

where γ is the constant of gravitation.

19. A circular wire of radius a and density ρ attracts a particle according to the Newtonian law $\gamma\frac{m_1 m_2}{(\text{distance})^2}$; if the particle be placed on the axis of the wire at a distance b from the centre, find its velocity when it is at any distance x.

If it be placed on the axis at a small distance from the centre, shew that the time of a complete oscillation is $a\sqrt{\frac{2\pi}{\gamma\rho}}$.

20. In the preceding question if the wire repels instead of attracting, and the particle be placed in the plane of the wire at a small distance from its centre, shew that the time of an oscillation is $2a\sqrt{\frac{\pi}{\gamma\rho}}$.

21. A particle moves in a straight line with an acceleration directed towards, and equal to μ times the distance from, a point in the straight line, and with a constant acceleration f in a direction opposite to that of its initial motion; shew that its time of oscillation is the same as it is when f does not exist.

22. A particle P moves in a straight line OCP being attracted by a force $m\mu$. PC always directed towards C, whilst C moves along OC with constant acceleration f. If initially C was at rest at the origin O, and P was at a distance c from O and moving with velocity V, prove that the distance of P from O at any time t is

$$\left(\frac{f}{\mu}+c\right) \cos \sqrt{\mu}t+\frac{V}{\sqrt{\mu}} \sin \sqrt{\mu}t-\frac{f}{\mu}+\frac{f}{2}t^2.$$

23. Two bodies, of masses M and M', are attached to the lower end of an elastic string whose upper end is fixed and hang at rest; M' falls off; shew that the distance of M from the upper end of the string at time t is

$$a+b+c \cos\left(\sqrt{\frac{g}{b}}t\right),$$

where a is the unstretched length of the string, and b and c the distances by which it would be stretched when supporting M and M' respectively.

24. A point is performing a simple harmonic motion. An additional acceleration is given to the point which is very small and varies as the cube of the distance from the origin. Shew that the increase in the amplitude of the vibration is proportional to the cube of the original amplitude if the velocity at the origin is the same in the two motions.

25. One end of a light extensible string is fastened to a fixed point and the other end carries a heavy particle; the string is of unstretched length a and its modulus of elasticity is n times the weight of the particle. The particle is pulled down till it is at a depth b below the fixed point and then released. Shew that it will return to this position at the end of time $2\sqrt{\frac{a}{ng}}\left[\frac{\pi}{2}+\operatorname{cosec}^{-1}p+\sqrt{p^2-1}\right]$, where $p=\frac{nb}{a}-(n+1)$, provided that p is not $>\sqrt{1+4n}$.

If $p>\sqrt{1+4n}$, shew how to find the corresponding time.

26. An endless elastic string, whose modulus of elasticity is λ and natural length is $2\pi c$, is placed in the form of a circle on a smooth horizontal plane and is acted upon by a force from the centre equal to μ times the distance per unit mass of the string. Shew that its radius will vary harmonically about a mean length $\frac{2\pi\lambda c}{2\pi\lambda-m\mu c}$, where m is the mass of the string, assuming that $2\pi\lambda>m\mu c$.

Examine the case when $2\pi\lambda=m\mu c$.

27. An elastic string of mass m and modulus of elasticity λ rests unstretched in the form of a circle whose radius is a. It is now acted on by a repulsive force situated in its centre whose magnitude per unit mass of the string is

$$\frac{\mu}{(\text{distance})^2}.$$

Shew that when the circle next comes to rest its radius is a root of the quadratic equation

$$r^2-ar=\frac{m\mu}{\pi\lambda}.$$

28. A smooth block, of mass M, with its upper and lower faces horizontal planes, is free to move in a groove in a parallel plane, and a particle of mass m is attached to a fixed point in the upper face by an elastic string whose natural length is a and modulus E. If the system starts from rest with the particle on the upper face and the string stretched parallel to the groove to $(n+1)$ times its natural length, shew that the block will perform oscillations of amplitude $\frac{(n+1)am}{M+m}$ in the periodic time $2\left(\pi+\frac{2}{n}\right)\sqrt{\frac{aMm}{E(M+m)}}$.

29. A particle is attached to a point in a rough plane inclined at an angle α to the horizon; originally the string was unstretched and lay along a line of greatest slope; shew that the particle will oscillate only if the coefficient of friction is $<\frac{1}{3}\tan\alpha$.

30. A mass of m lbs. moves initially with a velocity of u ft. per sec. A constant power equal to H horse-power is applied so as to increase its velocity; shew that the time that elapses before the acceleration is reduced to $\frac{1}{n}$th of its original value is $\frac{m(n^2-1)u^2}{1100gH}$.

31. Shew that the greatest velocity which can be given to a bullet of mass M fired from a smooth-bore gun is $\sqrt{\frac{2\Pi V}{M}\{m \log m+1-m\}}$, where changes of temperature are neglected, and the pressure Π in front of the bullet is supposed constant, the volume V of the powder in the cartridge being assumed to turn at once, when fired, into gas of pressure $m\Pi$ and of volume V.

32. Two masses, m_1 and m_2, are connected by a spring of such a strength that when m_1 is held fixed m_2 performs n complete vibrations per second. Shew that if m_2 be held fixed, m_1 will make $n\sqrt{\frac{m_2}{m_1}}$, and, if both be free, they will make $n\sqrt{\frac{m_1+m_2}{m_1}}$ vibrations per second, the vibrations in each case being in the line of the spring.

33. A body is attached to one end of an inextensible string, and the other end moves in a vertical line with simple harmonic motion of amplitude a and makes n complete oscillations per second. Shew that the string will not remain tight during the motion unless $n^2<\frac{g}{4\pi^2 a}$.

34. A light spring is kept compressed by the action of a given force; the force is suddenly reversed; prove that the greatest subsequent extension of the spring is three times its initial contraction.

35. Two masses, M and m, connected by a light spring, fall in a vertical line with the spring unstretched until M strikes an inelastic table. Shew that if the height through which M falls is greater than $\frac{M+2m}{2m}l$, the mass M will after an interval be lifted from the table, l being the length by which the spring would be extended by the weight of M.

36. Two uniform spheres, of masses m_1 and m_2 and of radii a_1 and a_2, are placed with their centres at a distance a apart and are left to their mutual attractions; shew that they will have come together at the end of time

$$\sqrt{\frac{2\pi a D R}{3g(m_1+m_2)}}\left[a\cos^{-1}\sqrt{\frac{a_1+a_2}{a}}+\sqrt{(a_1+a_2)(a-a_1-a_2)}\right],$$

where R is the radius, and D the mean density of the Earth.

If $m_1=m_2=4$ lbs., $a_1=a_2=1{\cdot}5$ inches, and a=one foot, shew that the time is about 3 hours, assuming $R=4000$ miles and $D=350$ lbs. per cubic foot.

[When the spheres have their centres at a distance x, the acceleration of m_1 due to the attraction of m_2 is $\gamma\frac{m_2}{x_2}$ and that of m_2 due to m_1 is $\gamma\frac{m_1}{x_2}$. Hence the acceleration of m_2 relative to m_1 is $\gamma\frac{m_1+m_2}{x^2}$ and the equation of relative motion is $\ddot{x}=-\gamma\frac{m_1+m_2}{x^2}$.]

37. Assuming the mass of the Moon to be $\frac{1}{81}$ that of the Earth, that their radii are respectively 1100 and 4000 miles, and the distance between their centres 240,000 miles, shew that, if they were instantaneously reduced to rest and allowed to fall towards one another under their mutual attraction only, they would meet in about $4\frac{1}{2}$ days.

38. A particle is placed at the end of the axis of a thin attracting cylinder of radius a and of infinite length; shew that its kinetic energy when it has described a distance x varies as $\log \dfrac{x+\sqrt{x^2+a^2}}{a}$.

39. AB is a uniform string of mass M and length $2a$; every element of it is repelled with a force, $=\mu$. distance, acting from a point O in the direction of AB produced; shew that the acceleration of the string is the same as that of a particle placed at its middle point, and that the tension at any point P of the string varies as $AP.PB$.

40. Shew that the curve which is such that a particle will slide down each of its tangents to the horizontal axis in a given time is a cycloid whose axis is vertical.

41. Two particles, of masses m and m', are connected by an elastic string whose coefficient of elasticity is λ; they are placed on a smooth table, the distance between them being a, the natural length of the string. The particle m is projected with velocity V along the direction of the string produced; find the motion of each particle, and shew that in the subsequent motion the greatest length of the string is $a+Vp$, and that the string is next at its natural length after time πp, where

$$p^2=\frac{mm'}{m+m'}\frac{a}{\lambda}.$$

42. Two particles, each of mass m, are attached to the ends of an inextensible string which hangs over a smooth pulley; to one of them, A, another particle of mass $2m$ is attached by means of an elastic string of natural length a, and modulus of elasticity $2mg$. If the system be supported with the elastic string just unstretched and be then released, shew that A will descend with acceleration $g\sin^2\left[\sqrt{\frac{g}{2a}}.t\right]$.

43. A weightless elastic string, of natural length l and modulus λ, has two equal particles of mass m at its ends and lies on a smooth horizontal table perpendicular to an edge with one particle just hanging over. Shew that the other particle will pass over at the end of time t given by the equation

$$2l+\frac{mgl}{\lambda}\sin^2\sqrt{\frac{\lambda}{2ml}}t=\tfrac{1}{2}gt^2.$$

CHAPTER III

UNIPLANAR MOTION WHERE THE ACCELERATIONS PARALLEL TO FIXED AXES ARE GIVEN

38. Let the coordinates of a particle referred to axes Ox and Oy be x and y at time t, and let its accelerations parallel to the axes at this instant be X and Y.

The equations of motion are then

$$\frac{d^2x}{dt^2}=X \quad \text{.............................(1),}$$

and

$$\frac{d^2y}{dt^2}=Y \quad \text{.............................(2).}$$

Integrating each of these equations twice, we have two equations containing four arbitrary constants. These latter are determined from the initial conditions, *viz.* the initial values of x, y, $\frac{dx}{dt}$ and $\frac{dy}{dt}$.

From the two resulting equations we then eliminate t, and obtain a relation between x and y which is the equation to the path.

39. *Parabolic motion under gravity, supposed constant, the resistance of the air being neglected.*

Let the axis of y be drawn vertically upward, and the axis of x horizontal. Then the horizontal acceleration is zero, and the vertical acceleration is $-g$.

Hence the equations of motion are

$$\frac{d^2x}{dt^2}=0, \text{ and } \frac{d^2y}{dt^2}=-g \quad \text{......................(1).}$$

Integrating with respect to t, we have

$$\frac{dx}{dt}=A, \text{ and } \frac{dy}{dt}=-gt+C \quad \text{....................(2).}$$

Integrating a second time,

$$x=At+B, \text{ and } y=-g\frac{t^2}{2}+Ct+D \quad \text{..............(3).}$$

If the particle be projected from the origin with a velocity u at an angle α with the horizon, then when $t=0$ we have $x=y=0$, $\frac{dx}{dt}=u \cos\alpha$, and $\frac{dy}{dt}=u\sin\alpha$.

Hence from (2) and (3) we have initially $u\cos\alpha=A$, $u\sin\alpha=C$, $0=B$, and $0=D$.

$\therefore$ (3) gives $x=u\cos\alpha t$, and $y=u\sin\alpha t-\frac{1}{2}gt^2$.

Eliminating t, we have

$$y=x\tan\alpha-\frac{g}{2}\frac{x^2}{u^2\cos^2\alpha},$$

which is the equation to a parabola.

40. *A particle describes a path with an acceleration which is always directed towards a fixed point and varies directly as the distance from it; to find the path.*

Let O be the centre of acceleration and A the point of projection. Take OA as the axis of x and OY parallel to the direction of the initial velocity, V, of projection.

Let P be any point on the path, and let MP be the ordinate of P.

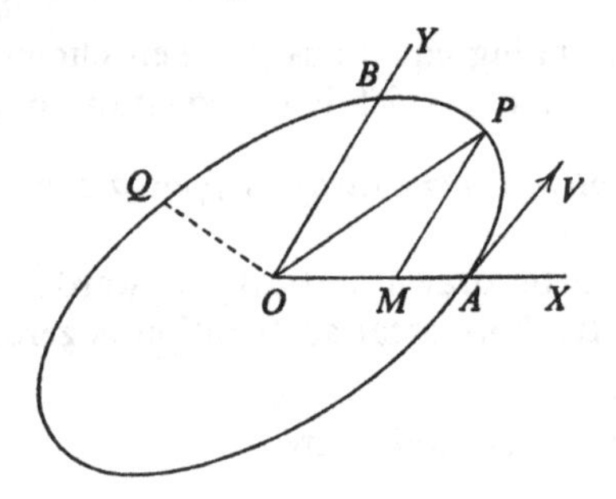

The acceleration, $\mu.PO$, along PO is equivalent, by the triangle of accelerations, to accelerations along PM and MO equal respectively to $\mu.PM$ and $\mu.MO$.

Hence the equations of motion are

$$\frac{d^2x}{dt^2}=-\mu x \quad \text{..........................(1)},$$

and

$$\frac{d^2y}{dt^2}=-\mu y \quad \text{..........................(2)}.$$

The solutions of these equations are, as in Art. 22,

$$x=A\cos[\sqrt{\mu}t+B] \quad \text{.....................(3)},$$

and

$$y=C\cos[\sqrt{\mu}t+D] \quad \text{.....................(4)}.$$

The initial conditions are that when $t=0$, then

$$x=OA=a, \quad \frac{dx}{dt}=0, \quad y=0, \quad \text{and } \frac{dy}{dt}=V.$$

Hence, from (3), $a=A\cos B$ and $0=-A\sin B$.

These give $B=0$ and $A=a$.

Also, from (4), similarly we have

$$0=C\cos D, \text{ and } V=-C\sqrt{\mu}\sin D.$$

$$\therefore D=\frac{\pi}{2}, \text{ and } C=-\frac{V}{\sqrt{\mu}}.$$

$\therefore$ (3) and (4) give

$$x=a\cos(\sqrt{\mu}t) \quad \text{.........................(5)},$$

and

$$y=-\frac{V}{\sqrt{\mu}}\cos\left[\sqrt{\mu}t+\frac{\pi}{2}\right]=\frac{V}{\sqrt{\mu}}\sin(\sqrt{\mu}t) \quad \text{...........(6)}.$$

$$\therefore \frac{x^2}{a^2}+\frac{y^2}{\frac{V^2}{\mu}}=1.$$

The locus of P is therefore an ellipse, referred to OX and OY as a pair of conjugate diameters.

Also, if the ellipse meet OY in B, then $OB=\frac{V}{\sqrt{\mu}}$, *i.e.* $V=\sqrt{\mu}\times$ semi-diameter conjugate to OA. Since any point on the path may be taken as the point of projection, this result will be always true, so that at any point the velocity

$$=\sqrt{\mu}\times\text{semi-conjugate diameter.}$$

[This may be independently derived from (5) and (6). For

$$(\text{Velocity at } P)^2=\dot{x}^2+\dot{y}^2+2\dot{x}\dot{y}\cos\omega$$

$$=a^2\mu\sin^2(\sqrt{\mu}t)+V^2\cos^2(\sqrt{\mu}t)-2aV\sqrt{\mu}\sin(\sqrt{\mu}t)\cos(\sqrt{\mu}t)\cos\omega$$

$$=\mu\left[a^2+\frac{V^2}{\mu}-a^2\cos^2(\sqrt{\mu}t)-\frac{V^2}{\mu}\sin^2\sqrt{\mu}t\right.$$
$$\left.-\frac{2aV}{\sqrt{\mu}}\sin(\sqrt{\mu}t)\cos(\sqrt{\mu}t)\cos\omega\right]$$

$$=\mu\left[a^2+\frac{V^2}{\mu}-x^2-y^2-2xy\cos\omega\right]=\mu\left(a^2+\frac{V^2}{\mu}-OP^2\right)$$

$=\mu\times$ square of semi-diameter conjugate to OP.]

From equations (5) and (6) it is clear that the values of x and y are the same at time $t+\frac{2\pi}{\sqrt{\mu}}$ as they are at time t.

Hence the time of describing the ellipse is $\frac{2\pi}{\sqrt{\mu}}$.

41. If a particle possess two simple harmonic motions, in perpendicular directions and of the same period, it is easily seen that its path is an ellipse.

If we measure the time from the time when the x-vibration has its maximum value, we have

$$x = a \cos nt \quad \text{.........(1)},$$

and

$$y = b \cos (nt+\epsilon) \quad \text{.........(2)},$$

where a, b are constants.

(2) gives $\frac{y}{b} = \cos nt \cos \epsilon - \sin nt \sin \epsilon = \frac{x}{a} \cos \epsilon - \sin \epsilon \sqrt{1-\frac{x^2}{a^2}}$.

$$\therefore \left(\frac{y}{b} - \frac{x}{a} \cos \epsilon\right)^2 = \sin^2 \epsilon \left(1-\frac{x^2}{a^2}\right),$$

i.e.

$$\frac{x^2}{a^2} - \frac{2xy}{ab} \cos \epsilon + \frac{y^2}{b^2} = \sin^2 \epsilon \quad \text{.........(3)}.$$

This always represents an ellipse whose principal axes do not, in general, coincide with the axes of coordinates, but which is always inscribed in the rectangle $x = \pm a$, $y = \pm b$.

The figure drawn is an ellipse where ϵ is equal to about $\frac{\pi}{3}$.

If $\epsilon = 0$, equation (3) gives $\frac{x}{a} - \frac{y}{b} = 0$, *i.e.* the straight line AC.

If $\epsilon = \pi$, it gives $\frac{x}{a} + \frac{y}{b} = 0$, *i.e.* the straight line BD.

In the particular case when $\epsilon = \frac{\pi}{2}$, *i.e.* when the phase of the y-vibration at zero time is one-quarter of the periodic time, equation (3) becomes

$$\frac{x^2}{a^2} + \frac{y^2}{b^2} = 1,$$

i.e. the path is an ellipse whose principal axes are in the direction of the axes of x and y and equal to the amplitudes of the component vibrations in these directions.

If in addition $a=b$, *i.e.* if the amplitudes of the component vibrations are the same, the path is a circle.

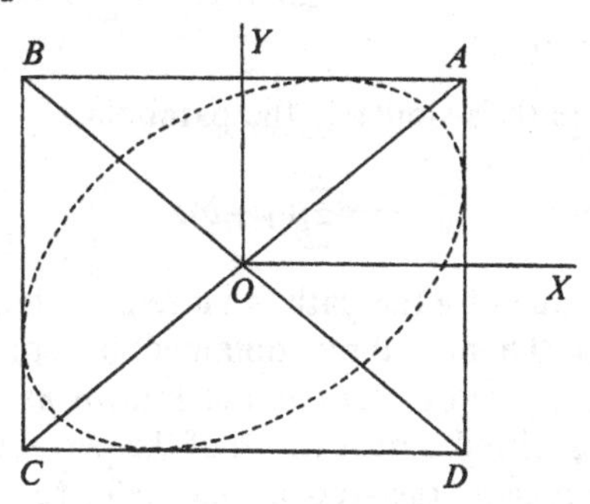

42. If the period of the y-vibration is one-half that of the x-vibration, the equations are

$$x=a \cos nt, \quad \text{and} \quad y=b \cos (2nt+\epsilon).$$

Hence, by eliminating t, we have as the equation to the path

$$\frac{y}{b}=\cos \epsilon . \left[\frac{2x^2}{a^2}-1\right]-\sin \epsilon . \frac{2x}{a}\sqrt{1-\frac{x^2}{a^2}}.$$

On rationalization, this equation becomes one of the fourth degree.

The dotted curve in the figure is the path when $\epsilon=-\frac{\pi}{2}$, *i.e.* when the phase of the y-vibration at zero time is negative and equal to one-quarter of the period of the y-vibration.

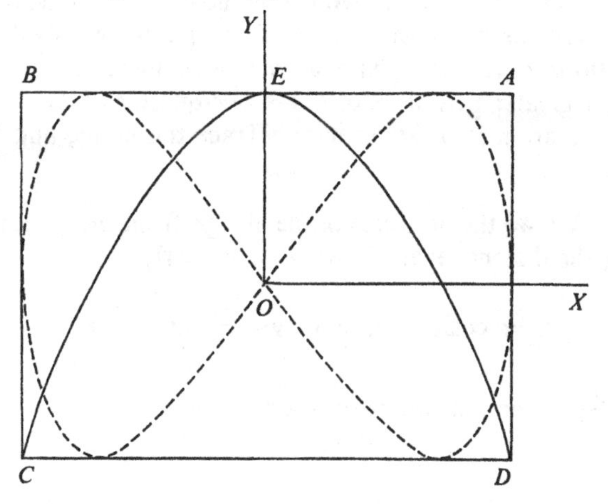

When $\epsilon=\pi$, *i.e.* when the phase of the y-vibration at zero time is one-half of the y-period, the path becomes

$$x^2=-\frac{a^2}{2b}(y-b),$$

i.e. the parabola *CED*.

When $\epsilon=0$, the path is similarly the parabola

$$x^2=\frac{a^2}{2b}(y+b).$$

For any other value of ϵ the path is more complicated.

Curves, such as the preceding, obtained by compounding simple harmonic motions in two directions are known as Lissajous' figures. For other examples with different ratios of the periods, and for different values of the zero phases, the student may refer to any standard book on Physics.

These curves may be drawn automatically by means of a pendulum, or they may be constructed geometrically.

43. Ex. 1. A point moves in a plane so that its projection on the axis of x performs a harmonic vibration of period one second with an amplitude of one foot; also its projection on the perpendicular axis of y performs a harmonic vibration of period two seconds with an amplitude of one foot. It being given that the origin is the centre of the vibrations, and that the point (1, 0) is on the path, find its equation and draw it.

Ex. 2. A point moves in a path produced by the combination of two simple harmonic vibrations in two perpendicular directions, the periods of the components being as 2 : 3; find the paths described (1) if the two vibrations have zero phase at the same instant, and (2) if the vibration of greater period be of phase one-quarter of its period when the other vibration is of zero phase. Trace the paths, and find their equations.

44. If in Art. 40 the acceleration be always from the fixed point and varying as the distance from it, we have similarly

$$x=a\cosh\sqrt{\mu}t,\ \text{and}\ y=\frac{V}{\sqrt{\mu}}\sinh\sqrt{\mu}t.$$

$\therefore\ \dfrac{x^2}{a^2}-\dfrac{y^2}{\dfrac{V^2}{\mu}}=1$, so that the path is a hyperbola.

45. *A particle describes a catenary under a force which acts parallel to its axis; find the law of the force and the velocity at any point of the path.*

Taking the directrix and axis of the catenary as the axes of x and y, we have as the equation to the catenary

$$y=\frac{c}{2}\left(e^{\frac{x}{c}}+e^{-\frac{x}{c}}\right) \quad\text{..........................(1).}$$

Since there is no acceleration parallel to the directrix,

$$\therefore\ \frac{d^2x}{dt^2}=0.$$

$$\therefore\ \frac{dx}{dt}=\text{const.}=u \quad\text{........................(2).}$$

Differentiating equation (1) twice, we have

$$\frac{dy}{dt}=\frac{1}{2}\left(e^{\frac{x}{c}}-e^{-\frac{x}{c}}\right)\cdot\frac{dx}{dt}=\frac{1}{2}\left(e^{\frac{x}{c}}-e^{-\frac{x}{c}}\right).u \quad\text{...........(3),}$$

and

$$\frac{d^2y}{dt^2}=\frac{1}{2c}\left(e^{\frac{x}{c}}+e^{-\frac{x}{c}}\right)\frac{dx}{dt}.u=\frac{u^2}{c^2}y.$$

Also

$$(\text{velocity})^2=\left(\frac{dx}{dt}\right)^2+\left(\frac{dy}{dt}\right)^2=u^2+\frac{u^2}{4}\left(e^{\frac{x}{c}}-e^{-\frac{x}{c}}\right)^2$$

$$=\frac{u^2}{4}\left(e^{\frac{x}{c}}+e^{-\frac{x}{c}}\right)^2=\frac{u^2}{c^2}y^2,$$

so that the velocity $=\frac{u}{c}y$.

Hence the velocity and acceleration at any point both vary as the distance from the directrix.

46. *A particle moves in one plane with an acceleration which is always towards and perpendicular to a fixed straight line in the plane and varies inversely as the cube of the distance from it; given the circumstances of projection, find the path.*

Take the fixed straight line as the axis of x.

Then the equations of motion are

$$\frac{d^2x}{dt^2}=0 \quad\text{.................................(1),}$$

and

$$\frac{d^2y}{dt^2}=-\frac{\mu}{y^3} \quad\text{..............................(2).}$$

(1) gives

$$x=At+B \quad\text{.........................(3).}$$

Multiplying (2) by $\frac{dy}{dt}$ and integrating, we have

$$\left(\frac{dy}{dt}\right)^2=\frac{\mu}{y^2}+C.$$

$$\therefore\ t=\int\frac{ydy}{\sqrt{\mu+Cy^2}}=\frac{1}{C}\sqrt{\mu+Cy^2}+D \quad \text{..............(4).}$$

Let the particle be projected from a point on the axis of y distant b from the origin with component velocities u and v parallel to the axes. Then when $t=0$ we have

$$x=0,\quad y=b,\quad \frac{dx}{dt}=u,\quad \text{and } \frac{dy}{dt}=v.$$

$$\therefore\ A=u,\quad B=0,\quad C=v^2-\frac{\mu}{b^2},\quad \text{and } D=-\frac{b^3v}{b^2v^2-\mu}.$$

$\therefore$ (3) and (4) give

$$x=ut, \text{ and } \left(t+\frac{b^3v}{b^2v^2-\mu}\right)^2=\frac{\mu b^4}{(b^2v^2-\mu)^2}+\frac{y^2b^2}{b^2v^2-\mu}.$$

Eliminating t, we have as the equation to the path

$$\left(\frac{x}{u}-\frac{b^3v}{\mu-b^2v^2}\right)^2+\frac{y^2b^2}{\mu-b^2v^2}=\frac{\mu b^4}{(\mu-b^2v^2)^2}.$$

This is an ellipse or a hyperbola according as $\mu\gtrless b^2v^2$.
If $\mu=b^2v^2$, then $C=0$ and equation (4) becomes

$$t=\int\frac{ydy}{\sqrt{\mu}}=\frac{y^2}{2\sqrt{\mu}}+D=\frac{y^2-b^2}{2\sqrt{\mu}}.$$

Hence the path in this case is $y^2-b^2=2\sqrt{\mu}\frac{x}{u}$, *i.e.* a parabola.

The path is thus an ellipse, parabola, or hyperbola according as $v\lesseqqgtr\sqrt{\frac{\mu}{b^2}}$, *i.e.* according as the initial velocity perpendicular to the given line is less, equal to, or greater than the velocity that would be acquired in falling from infinity to the given point with the given acceleration.

For the square of the latter $=-\int_\infty^b 2\frac{\mu}{y^3}dy=\left[\frac{\mu}{y^2}\right]_\infty^b=\frac{\mu}{b^2}.$

COR. If the particle describe an ellipse and meets the axis of x it will not then complete the rest of the ellipse since the velocity parallel to the axis of x is always constant and in the same direction; it will proceed to describe a portion of another equal ellipse.

47. *If the velocities and accelerations at any instant of particles m_1, m_2, m_3 ... parallel to any straight line fixed in space be v_1, v_2 ... and f_1, f_2 ... to find the velocity and acceleration of their centre of mass.*

If x_1, x_2, ... be the distances of the particles at any instant measured along this fixed line from a fixed point, we have

$$\bar{x}=\frac{m_1x_1+m_2x_2+\dots}{m_1+m_2+\dots}.$$

Differentiating with respect to t, we have

$$\bar{v}=\frac{d\bar{x}}{dt}=\frac{m_1v_1+m_2v_2+\dots}{m_1+m_2+\dots} \quad \text{......................(1)},$$

and

$$\bar{f}=\frac{d^2\bar{x}}{dt^2}=\frac{m_1f_1+m_2f_2+\dots}{m_1+m_2+\dots} \quad \text{......................(2)},$$

where $\bar{v}$ and $\bar{f}$ are the velocity and acceleration required.

Consider any two particles, m_1 and m_2, of the system and the mutual actions between them. These are, by Newton's Third Law, equal and opposite, and therefore their impulses resolved in any direction are equal and opposite. The changes in the momenta of the particles are thus, by Art. 15, equal and opposite, *i.e.* the sum of their momenta in any direction is thus unaltered by their mutual actions. Similarly for any other pair of particles of the system.

Hence the sum of the momenta of the system parallel to any line, and hence by (1) the momentum of the centre of mass, is unaltered by the mutual actions of the system.

If P_1, P_2, ... be the external forces acting on the particles m_1, m_2 ... parallel to the fixed line, we have

$$m_1f_1+m_2f_2+\dots=(P_1+P_2+\dots)+(\text{the sum of the components of the internal actions on the particles})$$

$$=P_1+P_2+\dots,$$

since the internal actions are in equilibrium taken by themselves.

Hence equation (2) gives

$$(m_1+m_2+\dots)\bar{f}=P_1+P_2+\dots,$$

i.e. the motion of the centre of mass in any given direction is the same as if the whole of the particles of the system were collected at it, and all the external forces of the system applied at it parallel to the given direction.

Hence also *If the sum of the external forces acting on any given system of particles parallel to a given direction vanishes, the motion of the centre of gravity in that direction remains unaltered, and the total momentum of the system in that direction remains constant throughout the motion.*

This theorem is known as the Principle of the Conservation of Linear Momentum.

As an example, if a heavy chain be falling freely the motion of its centre of mass is the same as that of a freely falling particle.

EXAMPLES ON CHAPTER III

1. A particle describes an ellipse with an acceleration directed towards the centre; shew that its angular velocity about a focus is inversely proportional to its distance from that focus.

2. A particle is describing an ellipse under a force to the centre; if v, v_1, and v_2 are the velocities at the ends of the latus-rectum and major and minor axes respectively, prove that $v^2v_2^2=v_1^2(2v_2^2-v_1^2)$.

3. The velocities of a point parallel to the axes of x and y are $u+\omega y$ and $v+\omega' x$ respectively, where u, v, ω, and ω' are constants; she that its path is a conic section.

4. A particle moves in a plane under a constant force, the direction of which revolves with a uniform angular velocity; find equations to give the coordinates of the particle at any time t.

5. A small ball is projected into the air; shew that it appears to an observer standing at the point of projection to fall past a given vertical plane with constant velocity.

6. A man starts from a point O and walks, or runs, with a constant velocity u along a straight road, taken as the axis of x. His dog starts at a distance a from O, his starting point being on the axis of y which is perpendicular to Ox, and runs with constant velocity $\frac{u}{\lambda}$ in a direction which is always towards his master. Shew that the equation to his path is

$$2\left[x-\frac{a\lambda}{1-\lambda^2}\right]=y\left[\frac{1}{1+\lambda}\left(\frac{y}{a}\right)^{\lambda}-\frac{1}{1-\lambda}\left(\frac{a}{y}\right)^{\lambda}\right].$$

If $\lambda=1$, shew that the path is the curve $2\left(x+\frac{a}{4}\right)=\frac{y^2}{2a}-a\log\frac{y}{a}$.

[The tangent at any point P of the path of the dog meets Ox at the point where the man then is, so that $\frac{dy}{dx}=-\frac{y}{ut-x}$. Also $\frac{ds}{dt}=\frac{u}{\lambda}$.

$$\therefore\ -y\frac{dx}{dy}=ut-x=\lambda s-x.$$

$$\therefore\ -\frac{d}{dy}\left[y\frac{dx}{dy}\right]=\lambda\frac{ds}{dy}-\frac{dx}{dy},\text{ giving } -y\frac{d^2x}{dy^2}=-\lambda\sqrt{1+\left(\frac{dx}{dy}\right)^2},\text{ etc.}]$$

7. A particle is fastened to one end, B, of a light thread and rests on a horizontal plane; the other end, A, of the thread is made to move on the plane with a given constant velocity in a given straight line; shew that the path of the particle in space is a trochoid.

[Shew that AB turns round A with a constant angular velocity.]

8. Two boats each move with a velocity v relative to the water and both cross a river of breadth a running with uniform velocity V. They start together, one boat crossing by the shortest path and the other in the shortest time. Shew that the difference between the times of arrival is either

$$\frac{a}{v}\left\{\frac{V}{(V^2-v^2)^{\frac{1}{2}}}-1\right\} \text{ or } \frac{a}{v}\left\{\frac{v}{(v^2-V^2)^{\frac{1}{2}}}-1\right\},$$

according as V or v is the greater.

[The angle that v makes with V being θ, the length of the path is $a.\dfrac{\sqrt{V^2+v^2+2Vv\cos\theta}}{v\sin\theta}$ and the corresponding time is $\dfrac{a}{v\sin\theta}$. The condition for a minimum path gives

$$(v\cos\theta+V)(V\cos\theta+v)=0.]$$

9. A particle moves in one plane with an acceleration which is always perpendicular to a given line and is equal to $\mu\div$(distance from the line)^2. Find its path for different velocities of projection.

If it be projected from a point distant $2a$ from the given line with a velocity $\sqrt{\dfrac{\mu}{a}}$ parallel to the given line, shew that its path is a cycloid.

10. If a particle travel with horizontal velocity u and rise to such a height that the variation in gravity must be taken account of as far as small quantities of the first order, shew that the path is given by the equation

$$(h-x)^2=\frac{2u^2}{g}(k-y)\left(1+\frac{5k+y}{6a}\right),$$

where $2a$ is the radius of the earth; the axes of x and y being horizontal and vertical, and h, k being the coordinates of the vertex of the path.

11. A particle moves in a plane with an acceleration which is parallel to the axis of y and varies as the distance from the axis of x; shew that the equation to its path is of the form $y=Aa^x+Ba^{-x}$, when the acceleration is a repulsion.

If the acceleration is attractive, then the equation is of the form

$$y=A\cos[ax+B].$$

12. A particle moves under the action of a repulsive force perpendicular to a fixed plane and proportional to the distance from it. Find its path, and shew that, if its initial velocity be parallel to the plane and equal to that which it would have acquired in moving from rest on the plane to the point of projection, the path is a catenary.

13. A particle describes a rectangular hyperbola, the acceleration being directed from the centre; shew that the angle θ described about the centre in time t after leaving the vertex is given by the equation

$$\tan\theta=\tanh(\sqrt{\mu}t),$$

where μ is acceleration at distance unity.

14. A particle moves freely in a semicircle under a force perpendicular to the bounding diameter; shew that the force varies inversely as the cube of the ordinate to the diameter.

15. Shew that a rectangular hyperbola can be described by a particle under a force parallel to an asymptote which varies as the cube of its distance from the other asymptote.

16. A particle is moving under the influence of an attractive force $m\dfrac{\mu}{y^3}$ towards the axis of x. Shew that, if it be projected from the point $(0, k)$ with component velocities U and V parallel to the axes of x and y, it will not strike the axis of x unless $\mu > V^2k^2$, and that in this case the distance of the point of impact from the origin is $\dfrac{Uk^2}{\mu^{\frac{1}{2}} - Vk}$.

17. A plane has two smooth grooves at right angles cut in it, and two equal particles attracting one another according to the law of the inverse square are constrained to move one in each groove. Shew that the centre of mass of the two particles moves as if attracted to a centre of force placed at the intersection of the grooves and attracting as the inverse square of the distance.

CHAPTER IV

UNIPLANAR MOTION REFERRED TO POLAR COORDINATES CENTRAL FORCES

48. In the present chapter we shall consider cases of motion which are most readily solved by the use of polar coordinates. We must first obtain the velocities and accelerations of a moving point along and perpendicular to the radius vector drawn from a fixed pole.

49. *Velocities and accelerations of a particle along and perpendicular to the radius vector to it from a fixed origin O.*

Let P be the position of the particle at time t, and Q its position at time $t+\Delta t$.

Let $XOP=\theta$, $XOQ=\theta+\Delta\theta$, $OP=r$, $OQ=r+\Delta r$, where OX is a fixed line.

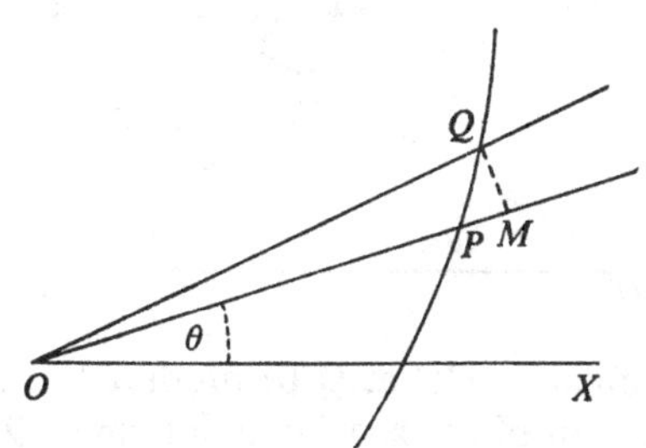

Draw QM perpendicular to OP.

Let u, v be the velocities of the moving point along and perpendicular to OP. Then

$$u=\underset{\Delta t=0}{\text{Lt}}\left[\frac{\text{Distance of particle measured along the line } OP \text{ at time } t+\Delta t - \text{the similar distance at time } t}{\Delta t}\right]$$

$$=\underset{\Delta t=0}{\text{Lt}}\ \frac{OM-OP}{\Delta t}=\underset{\Delta t=0}{\text{Lt}}\frac{(r+\Delta r)\cos\Delta\theta-r}{\Delta t}$$

$$=\underset{\Delta t=0}{\text{Lt}}\frac{(r+\Delta r).1-r}{\Delta t},$$ small quantities above the first order being neglected,

$$=\frac{dr}{dt} \qquad\qquad\qquad (1).$$

Also

$$v = \underset{\Delta t=0}{\text{Lt}} \left[\frac{\text{Distance of particle measured perpendicular to the line } OP \text{ at time } t+\Delta t - \text{the similar distance at time } t}{\Delta t} \right]$$

$$= \underset{\Delta t=0}{\text{Lt}} \frac{QM-0}{\Delta t} = \underset{\Delta t=0}{\text{Lt}} \frac{(r+\Delta r)\sin\Delta\theta}{\Delta t}$$

$$= \underset{\Delta t=0}{\text{Lt}} \frac{(r+\Delta r).\Delta\theta}{\Delta t}, \text{ on neglecting small quantities of the second order,}$$

$$= r\frac{d\theta}{dt}, \text{ in the limit} \qquad \text{.......(2).}$$

The velocities along and perpendicular to OP being u and v, the velocities along and perpendicular to OQ are $u+\Delta u$ and $v+\Delta v$.

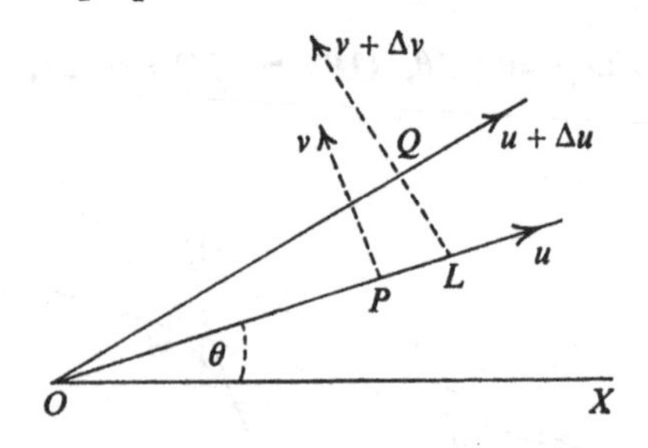

Let the perpendicular to OQ at Q be produced to meet OP at L. Then the acceleration of the moving point along OP

$$= \underset{\Delta t=0}{\text{Lt}} \left[\frac{\text{its velocity along } OP \text{ at time } t+\Delta t - \text{its similar velocity at time } t}{\Delta t} \right]$$

$$= \underset{\Delta t=0}{\text{Lt}} \left[\frac{(u+\Delta u)\cos\Delta\theta - (v+\Delta v)\sin\Delta\theta - u}{\Delta t} \right]$$

$$= \underset{\Delta t=0}{\text{Lt}} \left[\frac{(u+\Delta u).1 - (v+\Delta v).\Delta\theta - u}{\Delta t} \right],$$

on neglecting squares and higher powers of $\Delta\theta$,

$$= \underset{\Delta t=0}{\text{Lt}} \frac{\Delta u - v\Delta\theta}{\Delta t} = \frac{du}{dt} - v\frac{d\theta}{dt}, \text{ in the limit,}$$

$$= \frac{d^2r}{dt^2} - r\left(\frac{d\theta}{dt}\right)^2 \qquad \text{.......(3),}$$

by (1) and (2).

Also the acceleration of the moving point perpendicular to OP in the direction of θ increasing

$$=\underset{\Delta t=0}{\text{Lt}}\left[\frac{\text{its velocity perpendicular to } OP \text{ at time } t+\Delta t-\text{its similar velocity at time } t}{\Delta t}\right]$$

$$=\underset{\Delta t=0}{\text{Lt}}\left[\frac{(u+\Delta u)\sin\Delta\theta+(v+\Delta v)\cos\Delta\theta-v}{\Delta t}\right]$$

$$=\underset{\Delta t=0}{\text{Lt}}\left[\frac{(u+\Delta u).\Delta\theta+(v+\Delta v).1-v}{\Delta t}\right],$$

on neglecting squares and higher powers of $\Delta\theta$,

$$=u\frac{d\theta}{dt}+\frac{dv}{dt}, \text{ in the limit}, =\frac{dr}{dt}\frac{d\theta}{dt}+\frac{d}{dt}\left(r\frac{d\theta}{dt}\right), \text{ by (1) and (2)},$$

$$=2\frac{dr}{dt}\frac{d\theta}{dt}+r\frac{d^2\theta}{dt^2}=\frac{1}{r}\frac{d}{dt}\left[r^2\frac{d\theta}{dt}\right] \quad \ldots\ldots\ldots\ldots(4).$$

COR. If $r=a$, a constant quantity, so that the particle is describing a circle of centre O and radius a, the quantity (3)$=-a\dot{\theta}^2$ and (4)$=a\ddot{\theta}$, so that the accelerations of P along the tangent PQ and the radius PO are $a\ddot{\theta}$ and $a\dot{\theta}^2$.

50. The results of the previous article may also be obtained by resolving the velocities and accelerations along the axes of x and y in the directions of the radius vector and perpendicular to it.

For since $x=r\cos\theta$ and $y=r\sin\theta$,

$$\left.\begin{aligned} \therefore\quad \frac{dx}{dt}&=\frac{dr}{dt}\cos\theta-r\sin\theta\frac{d\theta}{dt}\\ \text{and}\quad \frac{dy}{dt}&=\frac{dr}{dt}\sin\theta+r\cos\theta\frac{d\theta}{dt}\end{aligned}\right\} \quad \ldots\ldots\ldots(1).$$

Also

$$\left.\begin{aligned} \frac{d^2x}{dt^2}&=\frac{d^2r}{dt^2}\cos\theta-2\frac{dr}{dt}\frac{d\theta}{dt}\sin\theta-r\cos\theta\left(\frac{d\theta}{dt}\right)^2-r\sin\theta\frac{d^2\theta}{dt^2}\\ \text{and}\quad =\frac{d^2y}{dt^2}&=\frac{d^2r}{dt^2}\sin\theta+2\frac{dr}{dt}\frac{d\theta}{dt}\cos\theta-r\sin\theta\left(\frac{d\theta}{dt}\right)^2+r\cos\theta\frac{d^2\theta}{dt^2}\end{aligned}\right\} \ldots(2).$$

The component velocity along OP

$$=\frac{dx}{dt}\cos\theta+\frac{dy}{dt}\sin\theta=\frac{dr}{dt}, \text{ by (1)},$$

and perpendicular to OP in the direction of θ increasing it

$$=\frac{dy}{dt}\cos\theta-\frac{dx}{dt}\sin\theta=r\frac{d\theta}{dt}, \text{ by (1)}.$$

The component acceleration along OP

$$=\frac{d^2x}{dt^2}\cos\theta+\frac{d^2y}{dt^2}\sin\theta=\frac{d^2r}{dt^2}-r\left(\frac{d\theta}{dt}\right)^2, \text{ by (2),}$$

and perpendicular to OP it

$$=\frac{d^2y}{dt^2}\cos\theta-\frac{d^2x}{dt^2}\sin\theta=2\frac{dr}{dt}\frac{d\theta}{dt}+r\frac{d^2\theta}{dt^2}, \text{ by (2),}$$

$$=\frac{1}{r}\frac{d}{dt}\left[r^2\frac{d\theta}{dt}\right].$$

51. By the use of Arts. 4 and 49 we can obtain the accelerations of a moving point referred to rectangular axes Ox and Oy, which are not fixed in space, but which revolve in any manner about the origin O in their own plane.

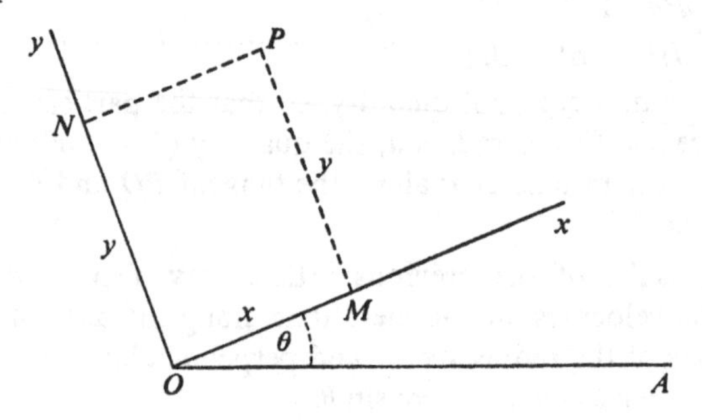

Let OA be a line fixed in space, and, at time t, let θ be the inclination of Ox to OA. Let P be the moving point; draw PM and PN perpendicular to Ox and Oy.

By Art. 49 the velocities of the point M are $\frac{dx}{dt}$ along OM and $x\frac{d\theta}{dt}$ along MP, and the velocities of N are $\frac{dy}{dt}$ along ON and $y\frac{d\theta}{dt}$ along PN produced.

$$\left[\text{For } \frac{d}{dt}(\angle AON)=\frac{d}{dt}(\angle AOM)=\frac{d\theta}{dt}.\right]$$

Hence the velocity of P parallel to Ox

= the velocity of N parallel to Ox + the velocity of P relative to N.

= vel. of N parallel to Ox + the vel. of M along OM

$$=-y\frac{d\theta}{dt}+\frac{dx}{dt} \qquad\qquad (1).$$

So the velocity of P parallel to Oy

$=$vel. of M parallel to $Oy+$the vel. of P relative to M

$=$vel. of M parallel to $Oy+$the vel. of N along ON

$$=x\frac{d\theta}{dt}+\frac{dy}{dt} \qquad\qquad (2).$$

Again, the accelerations of M are, by Art. 49, $\frac{d^2x}{dt^2}-x\left(\frac{d\theta}{dt}\right)^2$ along OM, and $\frac{1}{x}\frac{d}{dt}\left(x^2\frac{d\theta}{dt}\right)$ along MP, and the accelerations of N are $\frac{d^2y}{dt^2}-y\left(\frac{d\theta}{dt}\right)^2$ along ON, and $\frac{1}{y}\frac{d}{dt}\left(y^2\frac{d\theta}{dt}\right)$ along PN produced.

Hence the acceleration of P parallel to Ox

$=$acceleration of N parallel to $Ox+$acceleration of P relative to N

$=$acceleration of N parallel to $Ox+$acceleration of M along OM

$$=-\frac{1}{y}\frac{d}{dt}\left(y^2\frac{d\theta}{dt}\right)+\frac{d^2x}{dt^2}-x\left(\frac{d\theta}{dt}\right)^2 \qquad\qquad (3).$$

Also the acceleration of P parallel to Oy

$=$acceleration of M parallel to $Oy+$acceleration of P relative to M

$=$acceleration of M parallel to $Oy+$acceleration of N along ON

$$=\frac{1}{x}\frac{d}{dt}\left(x^2\frac{d\theta}{dt}\right)+\frac{d^2y}{dt^2}-y\left(\frac{d\theta}{dt}\right)^2 \qquad\qquad (4).$$

Cor. In the particular case when the axes are revolving with a constant angular velocity ω, so that $\frac{d\theta}{dt}=\omega$, these component velocities become

$$\frac{dx}{dt}-y\omega \text{ along } Ox,$$

and

$$\frac{dy}{dt}+x\omega \text{ along } Oy;$$

also the component accelerations are

$$\frac{d^2x}{dt^2}-x\omega^2-2\omega\frac{dy}{dt} \text{ along } Ox,$$

and

$$\frac{d^2y}{dt^2}-y\omega^2+2\omega\frac{dx}{dt} \text{ along } Oy.$$

52. Ex. 1. *Shew that the path of a point P which possesses two constant velocities u and v, the first of which is in a fixed direction and the second of which is perpendicular to the radius OP drawn from a fixed point O, is a conic whose focus is O and whose eccentricity is* $\frac{u}{v}$.

With the first figure of Art. 49, let u be the constant velocity along OX and v the constant velocity perpendicular to OP.

Then we have

$$\frac{dr}{dt}=u\cos\theta,\text{ and }\frac{rd\theta}{dt}=v-u\sin\theta.\qquad\therefore\ \frac{1}{r}\frac{dr}{d\theta}=\frac{u\cos\theta}{v-u\sin\theta}.$$

$$\therefore\ \log r=-\log(v-u\sin\theta)+\text{const.},$$

i.e.

$$r(v-u\sin\theta)=\text{const.}=lv,$$

if the path cut the axis of x at a distance l. Therefore the path is

$$r=\frac{l}{1-\frac{u}{v}\sin\theta},$$

i.e. a conic section whose eccentricity is $\frac{u}{v}$.

Ex. 2. *A smooth straight thin tube revolves with uniform angular velocity ω in a vertical plane about one extremity which is fixed; if at zero time the tube be horizontal, and a particle inside it be at a distance a from the fixed end, and be moving with velocity V along the tube, shew that its distance at time t is*

$$a\cosh(\omega t)+\left(\frac{V}{\omega}-\frac{g}{2\omega^2}\right)\sinh(\omega t)+\frac{g}{2\omega^2}\sin\omega t.$$

At any time t let the tube have revolved round its fixed end O through an angle ωt from the horizontal line OX in an upward direction; let P, where $OP=r$, be the position of the particle then.

By Art. 49,

$$\frac{d^2r}{dt^2}-r\omega^2=\text{acceleration of } P \text{ in the direction } OP$$

$$=-g\sin\omega t,\text{ since the tube is smooth.}$$

The solution of this equation is

$$r=Ae^{\omega t}+Be^{-\omega t}+\frac{1}{D^2-\omega^2}(-g\sin\omega t)$$

$$=L\cosh(\omega t)+M\sinh(\omega t)+\frac{g}{2\omega^2}\sin\omega t,$$

where A and B, and so L and M, are arbitrary constants.

The initial conditions are that $r=a$ and $\dot{r}=V$ when $t=0$.

$$\therefore\ a=L, \text{ and } V=M\omega+\frac{g}{2\omega}.$$

$$\therefore\ r=a\cosh\omega t+\left[\frac{V}{\omega}-\frac{g}{2\omega^2}\right]\sinh(\omega t)+\frac{g}{2\omega^2}\sin\omega t.$$

If R be the normal reaction of the tube, then

$$\frac{R}{m}-g\cos\omega t=\text{the acceleration perpendicular to } OP$$

$$=\frac{1}{r}\frac{d}{dt}(r^2\omega), \text{ by Art. 49, } =2\dot{r}\omega$$

$$=2a\omega^2\sinh(\omega t)+(2V\omega-g)\cosh(\omega t)+g\cos\omega t.$$

EXAMPLES

1. A vessel steams at a constant speed v along a straight line whilst another vessel, steaming at a constant speed V, keeps the first always exactly abeam. Shew that the path of either vessel relatively to the other is a conic section of eccentricity $\frac{v}{V}$.

2. A boat, which is rowed with constant velocity u, starts from a point A on the bank of a river which flows with a constant velocity nu; it points always towards a point B on the other bank exactly opposite to A; find the equation to the path of the boat.

If n be unity, shew that the path is a parabola whose focus is B.

3. An insect crawls at a constant rate u along the spoke of a cartwheel, of radius a, the cart moving with velocity v. Find the acceleration along and perpendicular to the spoke.

4. The velocities of a particle along and perpendicular to the radius from a fixed origin are λr and $\mu\theta$; find the path and shew that the accelerations, along and perpendicular to the radius vector, are

$$\lambda^2 r-\frac{\mu^2\theta^2}{r} \text{ and } \mu\theta\left[\lambda+\frac{\mu}{r}\right].$$

5. A point starts from the origin in the direction of the initial line with velocity $\frac{f}{\omega}$ and moves with constant angular velocity ω about the origin and with constant negative radial acceleration $-f$. Shew that the rate of growth of the radial velocity is never positive, but tends to the limit zero, and prove that the equation of the path is $\omega^2 r=f(1-e^{-\theta})$.

6. A point P describes a curve with constant velocity and its angular velocity about a given fixed point O varies inversely as the distance from O; shew that the curve is an equiangular spiral whose pole is O, and that the acceleration of the point is along the normal at P and varies inversely as OP.

7. A point P describes an equiangular spiral with constant angular velocity about the pole O; shew that its acceleration varies as OP and is in a direction making with the tangent at P the same constant angle that OP makes.

8. A point moves in a given straight line on a plane with constant velocity V, and the plane moves with constant angular velocity ω about an axis perpendicular to itself through a given point O of the plane. If the distance of O from the given straight line be a, shew that the path of the point in space is given by the equation

$$\frac{V\theta}{\omega}=\sqrt{r^2-a^2}+\frac{V}{\omega}\cos^{-1}\frac{a}{r},$$

referred to O as pole.

[If θ be measured from the line to which the given line is perpendicular at zero time, then $r^2=a^2+V^2.t^2$ and $\theta=\omega t+\cos^{-1}\frac{a}{r}$.]

9. A straight smooth tube revolves with angular velocity ω in a horizontal plane about one extremity which is fixed; if at zero time a particle inside it be at a distance a from the fixed end and moving with velocity V along the tube, shew that its distance at time t is $a\cosh\omega t+\frac{V}{\omega}\sinh\omega t$.

10. A thin straight smooth tube is made to revolve upwards with a constant angular velocity ω in a vertical plane about one extremity O; when it is in a horizontal position, a particle is at rest in it at a distance a from the fixed end O; if ω be very small, shew that it will reach O in a time $\left(\frac{6a}{g\omega}\right)^{\frac{1}{3}}$ nearly.

11. A particle is at rest on a smooth horizontal plane which commences to turn about a straight line lying in itself with constant angular velocity ω downwards; if a be the distance of the particle from the axis of rotation at zero time, shew that the body will leave the plane at time t given by the equation

$$a\sinh\omega t+\frac{g}{2\omega^2}\cosh\omega t=\frac{g}{\omega^2}\cos\omega t.$$

12. A particle falls from rest within a straight smooth tube which is revolving with uniform angular velocity ω about a point O in its length, being acted on by a force equal to $m\mu$ (distance) towards O. Shew that the equation to its path in space is

$$r=a\cosh\left[\sqrt{\frac{\omega^2-\mu}{\omega^2}}\theta\right] \text{ or } r=a\cos\left\{\sqrt{\frac{\mu-\omega^2}{\omega^2}}\theta\right\}, \text{ according as } \mu \lessgtr \omega^2.$$

If $\mu=\omega^2$, shew that the path is a circle.

13. A particle is placed at rest in a rough tube at a distance a from one end, and the tube starts rotating horizontally with a uniform angular velocity ω about this end. Shew that the distance of the particle at time t is

$$ae^{-\omega t.\tan\epsilon}[\cosh(\omega t.\sec\epsilon)+\sin\epsilon\sinh(\omega t\sec\epsilon)],$$

where $\tan\epsilon$ is the coefficient of friction.

14. One end A of a rod is made to revolve with uniform angular velocity ω in the circumference of a circle of radius a, whilst the rod itself revolves in the opposite direction about that end with the same angular velocity. Initially the rod coincides with a diameter and a smooth ring capable of sliding freely along the rod is placed at the centre of the circle. Shew that the distance of the ring from A at time t is

$$\frac{a}{5}[4\cosh(\omega t)+\cos 2\omega t].$$

[If O be the centre of the circle and P, where $AP=r$, is the position of the ring at time t when both OA and AP have revolved through an angle θ, $(=\omega t)$, in opposite directions, the acceleration of A is $a\omega^2$ along OA and the acceleration of P relative to A is $\ddot{r}-r\dot{\theta}^2$, by Art. 49, *i.e.* $\ddot{r}-r\omega^2$. Hence the total acceleration of P along AP is $\ddot{r}-r\omega^2+a\omega^2\cos 2\omega t$, and this is zero since the ring is smooth.]

15. PQ is a tangent at Q to a circle of radius a; PQ is equal to ρ and makes an angle θ with a fixed tangent to the circle; shew that the accelerations of P along and perpendicular to QP are respectively

$$\ddot{\rho}-\rho\dot{\theta}^2+a\ddot{\theta}, \text{ and } \frac{1}{\rho}\frac{d}{dt}(\rho^2\dot{\theta})+a\dot{\theta}^2.$$

[The accelerations of Q along and perpendicular to QP are $a\ddot{\theta}$ and $a\dot{\theta}^2$; the accelerations of P relative to Q in these same directions are

$$\ddot{\rho}-\rho\dot{\theta}^2 \text{ and } \frac{1}{\rho}\frac{d}{dt}(\rho^2\dot{\theta}).]$$

16. Two particles, of masses m and m', connected by an elastic string of natural length a, are placed in a smooth tube of small bore which is made to rotate about a fixed point in its length with angular velocity ω. The coefficient of elasticity of the string is $2mm'a\omega^2 \div (m+m')$. Shew that, if the particles are initially just at rest relative to the tube and the string is just taut, their distance apart at time t is $2a-a\cos\omega t$.

17. A weight can slide along the spoke of a horizontal wheel, whose mass may be neglected, and is connected to the centre of the wheel by means of a light spring; when the wheel is fixed, the period of oscillation of the weight is $2\pi/n$. If the wheel is started to rotate freely with angular velocity $6n\sqrt{11}/55$, prove that the greatest extension of the spring is one-fifth of its original length.

18. A uniform chain AB is placed in a straight tube OAB which revolves in a horizontal plane, about the fixed point O, with uniform angular velocity ω. Shew that the motion of the middle point of the chain is the same as would be the motion of a particle placed at this middle point, and that the tension of the chain at any point P is $\frac{1}{2}m\omega^2.AP.PB$, where m is the mass of a unit length of the chain.

53. *A particle moves in a plane with an acceleration which is always directed to a fixed point O in the plane; to obtain the differential equation of its path.*

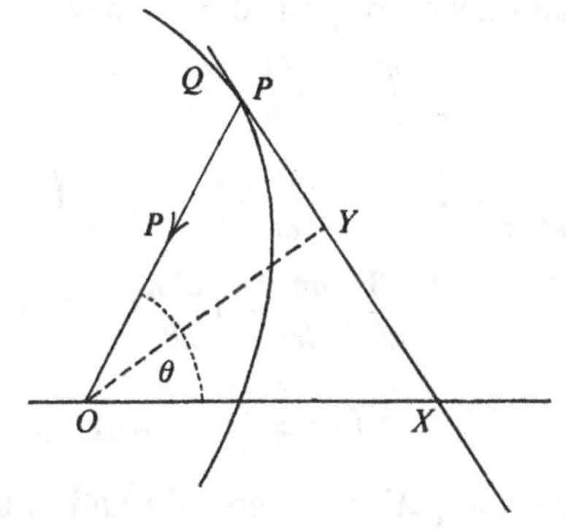

Referred to O as origin and a fixed straight line OX through O as initial line, let the polar coordinates of P be (r, θ). If P be the acceleration of the particle directed towards O, we have, by Art. 49,

$$\frac{d^2r}{dt^2}-r\left(\frac{d\theta}{dt}\right)^2=-P \quad \text{.......................(1).}$$

Also, since there is no acceleration perpendicular to OP, we have, by the same article,

$$\frac{1}{r}\frac{d}{dt}\left(r^2\frac{d\theta}{dt}\right)=0 \quad \text{.........................(2).}$$

(2) gives
$$r^2\frac{d\theta}{dt}=\text{const.}=h \text{ (say)} \quad \text{.....................(3).}$$

$$\therefore \frac{d\theta}{dt}=\frac{h}{r^2}=hu^2, \text{ if } u \text{ be equal to } \frac{1}{r}.$$

Then
$$\frac{dr}{dt}=\frac{d}{dt}\left(\frac{1}{u}\right)=-\frac{1}{u^2}\frac{du}{dt}=-\frac{1}{u^2}\frac{du}{d\theta}\frac{d\theta}{dt}=-h\frac{du}{d\theta},$$

and
$$\frac{d^2r}{dt^2}=\frac{d}{dt}\left(-h\frac{du}{d\theta}\right)=-h\frac{d}{d\theta}\left(\frac{du}{d\theta}\right)\frac{d\theta}{dt}=-h^2u^2\frac{d^2u}{d\theta^2}.$$

Hence equation (1) becomes

$$-h^2u^2\frac{d^2u}{d\theta^2}-\frac{1}{u}.h^2u^4=-P,$$

i.e.
$$\frac{d^2u}{d\theta^2}+u=\frac{P}{h^2u^2} \quad \text{........................(4).}$$

Again, if p be the perpendicular from the origin O upon the tangent at P, we have

$$\frac{1}{p^2}=\frac{1}{r^2}+\frac{1}{r^4}\left(\frac{dr}{d\theta}\right)^2=u^2+\left(\frac{du}{d\theta}\right)^2.$$

Hence, differentiating with respect to θ, we have

$$-\frac{2}{p^3}\frac{dp}{d\theta}=2u\frac{du}{d\theta}+2\frac{du}{d\theta}\frac{d^2u}{d\theta^2}.$$

$$\therefore -\frac{1}{p^3}\frac{dp}{dr}=\left[u+\frac{d^2u}{d\theta^2}\right]\frac{du}{dr}=\left(u+\frac{d^2u}{d\theta^2}\right)\left(-\frac{1}{r^2}\right).$$

$$\therefore \frac{1}{u^2p^3}\frac{dp}{dr}=u+\frac{d^2u}{d\theta^2}.$$

Hence (4) gives
$$P=\frac{h^2}{p^3}\frac{dp}{dr} \quad \text{...........................(5).}$$

Equation (4) gives the path in terms of r and θ, and (5) gives the (p, r) equation of the path.

54. *In every central orbit, the sectorial area traced out by the radius vector to the centre of force increases uniformly per unit of time, and the linear velocity varies inversely as the perpendicular from the centre upon the tangent to the path.*

Let Q be the position of the moving particle at time $t+\Delta t$, so that $\angle POQ=\Delta\theta$ and $OQ=r+\Delta r$.

The area
$$POQ=\tfrac{1}{2}OP.OQ\sin POQ$$
$$=\tfrac{1}{2}r(r+\Delta r)\sin\Delta\theta.$$

Hence the rate of description of sectorial area

$$=\underset{\Delta t=0}{\text{Lt}}\frac{\frac{1}{2}r(r+\Delta r)\sin\Delta\theta}{\Delta t}=\underset{\Delta t=0}{\text{Lt}}\left[\tfrac{1}{2}r(r+\Delta r).\frac{\sin\Delta\theta}{\Delta\theta}\cdot\frac{\Delta\theta}{\Delta t}\right]$$

$$=\tfrac{1}{2}r^2\frac{d\theta}{dt},\text{ in the limit,}$$

$$=\text{the constant }\tfrac{1}{2}h\text{ by equation (3) of the last article.}$$

The constant h is thus equal to twice the sectorial area described per unit of time.

Again, the sectorial area POQ=in the limit $\frac{1}{2}.PQ\times$perpendicular from O on PQ, and the rate of its description

$$=\underset{\Delta t=0}{\text{Lt}}\frac{1}{2}\cdot\frac{\Delta s}{\Delta t}\times\text{perpendicular from }O\text{ on }PQ.$$

Now, in the limit when Q is very close to P,

$$\frac{\Delta s}{\Delta t}=\text{the velocity }v,$$

and the perpendicular from O on PQ

=the perpendicular from O on the tangent at $P=p$.

$$\therefore\ h=v.p,\quad i.e.\ v=\frac{h}{p}.$$

Hence, when a particle moves under a force to a fixed centre, its velocity at any point P of its path varies inversely as the perpendicular from the centre upon the tangent to the path at P.

Since $v=\dfrac{h}{p}$, and in any curve

$$\frac{1}{p^2}=\frac{1}{r^2}+\frac{1}{r^4}\left(\frac{dr}{d\theta}\right)^2=u^2+\left(\frac{du}{d\theta}\right)^2,$$

$$\therefore\ v^2=h^2\left[u^2+\left(\frac{du}{d\theta}\right)^2\right].$$

55. *A particle moves in an ellipse under a force which is always directed towards its focus; to find the law of force, and the velocity at any point of its path.*

The equation to an ellipse referred to its focus is

$$r=\frac{l}{1+e\cos\theta},\quad i.e.\ u=\frac{1}{l}+\frac{e}{l}\cos\theta \quad \text{...............(1).}$$

$$\therefore\ \frac{d^2u}{d\theta^2}=-\frac{e}{l}\cos\theta.$$

Hence equation (4) of Art. 53 gives

$$P=h^2u^2\left[\frac{d^2u}{d\theta^2}+u\right]=\frac{h^2}{l}u^2 \quad \text{.....................(2).}$$

The acceleration therefore varies inversely as the square of the distance of the moving particle from the focus and, if it be $\frac{\mu}{(\text{distance})^2}$, then (2) gives

$$h=\sqrt{\mu l}=\sqrt{\mu\times\text{semi-latus-rectum}} \quad \text{...............(3).}$$

Also

$$v^2=h^2\left[u^2+\left(\frac{du}{d\theta}\right)^2\right]=h^2\left[\left(\frac{1}{l}+\frac{e}{l}\cos\theta\right)^2+\left(\frac{e}{l}\sin\theta\right)^2\right]$$

$$=\frac{\mu}{l}[1+2e\cos\theta+e^2]=\mu\left[2\frac{1+e\cos\theta}{l}-\frac{1-e^2}{l}\right]$$

$$=\mu\left[\frac{2}{r}-\frac{1}{a}\right],\ \text{by (1),} \quad \text{..(4),}$$

where $2a$ is the major axis of the ellipse.

It follows, since (4) depends only on the distance r, that the velocity at any point of the path depends only on the distance from the focus and that it is independent of the direction of the motion.

It also follows that the square of the velocity V of projection from any point whose distance from the focus is r_0, must be less than $\frac{2\mu}{r_0}$, and that the a of the corresponding ellipse is given by

$$V^2=\mu\left(\frac{2}{r_0}-\frac{1}{a}\right).$$

Periodic time. Since h is equal to twice the area described in a unit time, it follows, that it T be the time the particle takes to describe the whole arc of the ellipse, then

$$\tfrac{1}{2}h\times T=\text{area of the ellipse}=\pi ab.$$

Also $$h=\sqrt{\mu\times\text{semi-latus-rectum}}=\sqrt{\mu\frac{b^2}{a}}.$$

Hence $$T=\frac{2\pi ab}{h}=\frac{2\pi}{\sqrt{\mu}}a^{\frac{3}{2}}.$$

56. *Ex. Find the law of force towards the pole under which the curve* $r^n = a^n \cos n\theta$ *can be described.*

Here $$u^n a^n \cos n\theta = 1.$$

Hence, taking the logarithmic differential, we obtain

$$\frac{du}{d\theta} = u \tan n\theta.$$

$$\therefore \frac{d^2u}{d\theta^2} = \frac{du}{d\theta} \tan n\theta + nu \sec^2 n\theta = u[\tan^2 n\theta + n \sec^2 n\theta].$$

$$\therefore \frac{d^2u}{d\theta^2} + u = u(n+1) \sec^2 n\theta = (n+1)a^{2n}u^{2n+1}.$$

Hence equation (4) of Art. 53 gives

$$P = (n+1)h^2 a^{2n} u^{2n+3},$$

i.e. the curve can be described under a force to the pole varying inversely as the $(2n+3)$rd power of the distance.

Particular Cases. Let $n = -\frac{1}{2}$, so that the equation to the curve is

$$r = \frac{a}{\cos^2 \frac{\theta}{2}} = \frac{2a}{1+\cos\theta},$$

i.e. the curve is a parabola referred to its focus as pole.

Here $$P \propto \frac{1}{r^2}.$$

II. Let $n = \frac{1}{2}$, so that the equation is $r = \frac{a}{2}(1+\cos\theta)$, which is a cardioid.

Here $$P \propto \frac{1}{r^4}.$$

III. Let $n = 1$, so that the equation to the curve is $r = a\cos\theta$, *i.e.* a circle with a point on its circumference as pole.

Here $$P \propto \frac{1}{r^5}.$$

IV. Let $n = 2$, so that the curve is $r^2 = a^2 \cos 2\theta$, *i.e.* a lemniscate of Bernouilli, and $P \propto \frac{1}{r^7}$.

V. Let $n=-2$, so that the curve is the rectangular hyperbola $a^2=r^2\cos 2\theta$, the centre being pole, and $P\propto -r$, since in this case $(n+1)$ is negative. The force is therefore repulsive from the centre.

EXAMPLES

A particle describes the following curves under a force P to the pole, shew that the force is as stated:

1. Equiangular spiral; $P\propto\frac{1}{r^3}$.

2. Lemniscate of Bernouilli; $P\propto\frac{1}{r^7}$.

3. Circle, pole on its circumference; $P\propto\frac{1}{r^5}$.

4. $\frac{a}{r}=e^{n\theta}$, $n\theta$, $\cosh n\theta$, or $\sin n\theta$; $P\propto\frac{1}{r^3}$.

5. $r^n\cos n\theta=a^n$; $P\propto r^{2n-}$.

6. $r^n=A\cos n\theta+B\sin n\theta$; $P\propto\frac{1}{r^{2n+3}}$.

7. $r=a\sin n\theta$; $P\propto\frac{2n^2a^2}{r^5}-\frac{n^{\cdot}-1}{r^3}$.

8. $au=\tanh\left(\frac{\theta}{\sqrt{2}}\right)$ or $\coth\left(\frac{\theta}{\sqrt{2}}\right)$; $P\propto\frac{1}{r^5}$.

9. $au=\frac{\cosh\theta-2}{\cosh\theta+1}$ or $\frac{\cosh\theta+2}{\cosh\theta-1}$; $P\propto\frac{1}{r^4}$.

10. $a^2u^2=\frac{\cosh 2\theta-1}{\cosh 2\theta+2}$ or $\frac{\cosh 2\theta+1}{\cosh 2\theta-2}$; $P\propto\frac{1}{r^7}$.

11. Find the law of force to an internal point under which a body will describe a circle. Shew that the hodograph of such motion is an ellipse.

[Use formula (5) of Art. 53. The hodograph of the path of a moving point P is obtained thus: From a fixed point O draw a straight line OQ parallel to, and proportional to, the velocity of P; the locus of the point Q, for the different positions of P, is the hodograph of the path of P.]

12. A particle of unit mass describes an equiangular spiral, of angle α, under a force which is always in a direction perpendicular to the straight line joining the particle to the pole of the spiral; shew that the force is $\mu r^{2\sec^2\alpha-3}$, and that the rate of description of sectorial area about the pole is $\frac{1}{2}\sqrt{\mu\sin\alpha.\cos\alpha}.r^{\sec^2\alpha}$.

13. In an orbit described under a force to a centre the velocity at any point is inversely proportional to the distance of the point from the centre of force; shew that the path is an equiangular spiral.

14. The velocity at any point of a central orbit is $\frac{1}{n}$th of what it would be for a

circular orbit at the same distance; shew that the central force varies as $\frac{1}{r^{2n^2+1}}$ and that the equation to the orbit is

$$r^{n^2-1}=a^{n^2-1}\cos\{(n^2-1)\theta\}.$$

57. Apses. An apse is a point in a central orbit at which the radius vector drawn from the centre of force to the moving particle has a maximum or minimum value.

By the principles of the Differential Calculus u is a maximum or a minimum if $\frac{du}{d\theta}$ is zero, and if the first differential coefficient of u that does not vanish is of an even order.

If p be the perpendicular from the centre of force upon the tangent to the path at any point whose distance is r from the origin, then

$$\frac{1}{p^2}=u^2+\left(\frac{du}{d\theta}\right)^2.$$

When $\frac{du}{d\theta}$ is zero, $\frac{1}{p^2}=u^2=\frac{1}{r^2}$, so that the perpendicular in the case of the apse is equal to the radius vector. Hence at an apse the particle is moving at right angles to the radius vector.

58. *When the central acceleration is a single-valued function of the distance (i.e. when the acceleration is a function of the distance only and is always the same at the same distance), every apse-line divides the orbit into two equal and similar portions and thus there can only be two apse-distances.*

Let ABC be a portion of the path having three consecutive apses A, B, and C and let O be the centre of force.

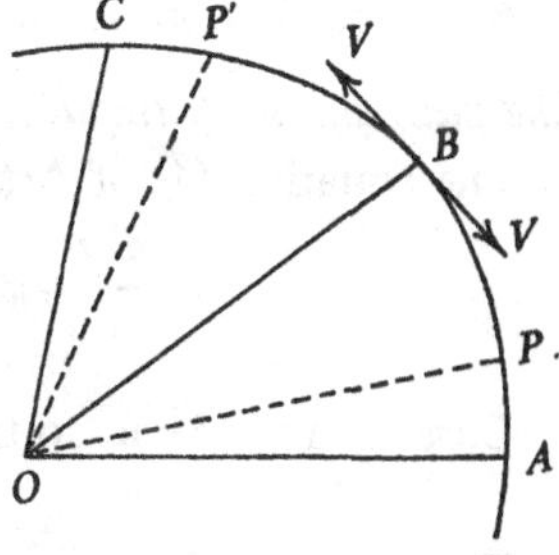

Let V be the velocity of the particle at B. Then, if the velocity of the particle were reversed at B, it would describe the path BPA. For, as the acceleration depends on the distance from O only, the velocity, by equations (1) and (3) of Art. 53, would depend only on the distance from O and not on the direction of the motion.

Again the original particle starting from B and the reversed particle, starting from B with equal velocity V, must describe similar paths. For the equations (1) and (3) of Art. 53, which do not depend on the direction of motion, shew that the value of r and θ at any time t for the first particle (*i.e.* OP' and $\angle BOP'$) are equal to the same quantities at the same time t for the second particle (*i.e.* OP and $\angle BOP$).

Hence the curves $BP'C$ and BPA are exactly the same; either, by being rotated about the line OB, would give the other. Hence, since A and C are the points where the radius vector is perpendicular to the tangent, we have $OA=OC$.

Similarly, if D were the next apse after C, we should have OB and OD equal, and so on.

Thus there are only two different apse-distances.

The angle between any two consecutive apsidal distances is called the **apsidal angle.**

59. When the central acceleration varies as some integral power of the distance, say μu^n, it is easily seen analytically that there are at most two apsidal distances.

For the equation of motion is

$$\frac{d^2u}{d\theta^2}+u=\frac{P}{h^2u^2}=\frac{\mu}{h^2}u^{n-2}.$$

$$\therefore \frac{h^2}{2}\left[\left(\frac{du}{d\theta}\right)^2+u^2\right]=\frac{\mu}{n-1}u^{n-1}+\text{const.}$$

The particle is at an apse when $\frac{du}{d\theta}=0$, and then this equation gives

$$u^{n-1}-\frac{n-1}{2}\frac{h^2}{\mu}u^2+C=0.$$

Whatever be the values of n or C this equation cannot have more than two changes of sign, and hence, by Descartes' Rule, it cannot have more than two positive roots.

60. *A particle moves with a central acceleration* $\frac{\mu}{(distance)^3}$; *to find the path and to distinguish the cases.*

The equation (4) of Art. 53 becomes

$$\frac{d^2u}{d\theta^2}+u=\frac{\mu}{h^2}u, \text{ i.e. } \frac{d^2u}{d\theta^2}=\left(\frac{\mu}{h^2}-1\right)u \quad \text{..............(1).}$$

Case I. Let $h^2<\mu$, so that $\frac{\mu}{h^2}-1$ is positive and equal to n^2, say.

The equation (1) is $\frac{d^2u}{d\theta^2}=n^2u$, the general solution of which is, as in Art. 29,

$$u=Ae^{n\theta}+Be^{-n\theta}=L\cosh n\theta+M\sinh n\theta,$$

where A, B, or L, M are arbitrary constants.

This is a spiral curve with an infinite number of convolutions about the pole. In the particular case when A or B vanishes, it is an equiangular spiral.

Case II. Let $h^2=\mu$, so that the equation (1) becomes

$$\frac{d^2u}{d\theta^2}=0.$$

$$\therefore\ u=A\theta+B=A(\theta-\alpha),$$

where A and α are arbitrary constants.

This represents a reciprocal spiral in general. In the particular case when A is zero, it is a circle.

Case III. Let $h^2>\mu$, so that $\frac{\mu}{h^2}-1$ is negative and equal to $-n^2$, say.

The equation (1) is therefore $\frac{d^2u}{d\theta^2}=-n^2u$, the solution of which is

$$u=A\cos(n\theta+B)=A\cos n(\theta-\alpha),$$

where A and α are arbitrary constants.

The apse is given by $\theta=\alpha$, $u=A$.

61. The equations (4) or (5) of Art. 53 will give the path when P is given and also the initial conditions of projection.

Ex. 1. *A particle moves with a central acceleration which varies inversely as the cube of the distance; if it be projected from an apse at a distance a from the origin with a velocity which is $\sqrt{2}$ times the velocity for a circle of radius a, shew that the equation to its path is*

$$r\cos\frac{\theta}{\sqrt{2}}=a.$$

Let the acceleration be μu^3.

If V_1 be the velocity in a circle of radius a with the same acceleration, then

$$\frac{V_1^2}{a}=\text{normal acceleration}=\frac{\mu}{a^3}.$$

$$\therefore\ V_1^2=\frac{\mu}{a^2}.$$

Hence, if V be the velocity of projection in the required path,

$$V=\sqrt{2}V_1=\frac{\sqrt{2\mu}}{a}.$$

The differential equation of the path is, from equation (4) of Art. 53,

$$\frac{d^2u}{d\theta^2}+u=\frac{\mu u^3}{h^2u^2}=\frac{\mu}{h^2}u.$$

Hence, multiplying by $\frac{du}{d\theta}$ and integrating, we have

$$\frac{1}{2}v^2=\frac{1}{2}h^2\left[\left(\frac{du}{d\theta}\right)^2+u^2\right]=\frac{\mu}{2}u^2+C \qquad \text{..............(1).}$$

The initial conditions give that when $u=\frac{1}{a}$, then $\frac{du}{d\theta}=0$, and $v=\frac{\sqrt{2\mu}}{a}$.

Hence (1) gives $$\frac{1}{2}\cdot\frac{2\mu}{a^2}=\frac{1}{2}h^2\left[\frac{1}{a^2}\right]=\frac{\mu}{2a^2}+C.$$

$$\therefore\ h^2=2\mu \text{ and } C=\frac{\mu}{2a^2}.$$

$\therefore$ from equation (1) we have

$$\left(\frac{du}{d\theta}\right)^2+u^2=\frac{u^2}{2}+\frac{1}{2a^2}.$$

$$\therefore\ \frac{du}{d\theta}=\sqrt{\frac{1}{2}\left(\frac{1}{a^2}-u^2\right)} \qquad \text{....................(2).}$$

$$\therefore\ \frac{\theta}{\sqrt{2}}=\int\frac{a\,du}{\sqrt{1-a^2u^2}}=\sin^{-1}au+\gamma.$$

If θ be measured from the initial radius vector, then $\theta=0$ when $u=\frac{1}{a}$, and therefore

$$\gamma=-\sin^{-1}(1)=-\frac{\pi}{2}.$$

$$\therefore\ au=\sin\left[\frac{\pi}{2}+\frac{\theta}{\sqrt{2}}\right]=\cos\frac{\theta}{\sqrt{2}}.$$

Hence the path is the curve $r\cos\frac{\theta}{\sqrt{2}}=a$.

If we take the negative sign on the right hand side of (2), we obtain the same result.

Ex. 2. *A particle, subject to a force producing an acceleration* $\mu\frac{r+2a}{r^5}$ *towards the origin, is projected from the point* $(a, 0)$ *with a velocity equal to the velocity from infinity at an angle* $\cot^{-1} 2$ *with the initial line. Shew that the equation to the path is*

$$r=a(1+2\sin\theta),$$

and find the apsidal angle and distances.

The "velocity from infinity" means the velocity that would be acquired by the particle in falling with the given acceleration from infinity to the point under consideration. Hence if this velocity be V we have, as in Art. 22,

$$\frac{1}{2}V^2=\int_\infty^a -\mu\left[\frac{x+2a}{x^5}\right]dx=\mu\left[\frac{1}{3}\frac{1}{x^3}+\frac{1}{2}\frac{a}{x^4}\right]_\infty^a=\mu\left[\frac{1}{3a^3}+\frac{1}{2a^3}\right],$$

so that $$V^2=\frac{5\mu}{3a^3} \quad\text{..............................(1).}$$

The equation of motion of the particle is

$$\frac{d^2u}{d\theta^2}+u=\frac{\mu}{h^2u^2}[u^4+2au^5]=\frac{\mu}{h^2}[u^2+2au^3],$$

$$\therefore\ \frac{1}{2}v^2=\frac{h^2}{2}\left[u^2+\left(\frac{du}{d\theta}\right)^2\right]=\mu\left[\frac{u^3}{3}+\frac{1}{2}au^4\right]+C \quad\text{............(2).}$$

If p_0 be the perpendicular from the origin upon the initial direction of projection, we have $p_0=a\sin\alpha$, where $\cot\alpha=2$, *i.e.* $p_0=\dfrac{a}{\sqrt{5}}$.

Hence, initially, we have

$$u^2+\left(\frac{du}{d\theta}\right)^2=\frac{1}{p_0^2}=\frac{5}{a^2} \quad\text{........................(3).}$$

Hence (2) gives, initially, from (1) and (3)

$$\frac{5\mu}{6a^3}=\frac{h^2}{2}\times\frac{5}{a^2}=\mu\left[\frac{1}{3a^3}+\frac{1}{2a^3}\right]+C,$$

so that $C=0$ and $h^2=\dfrac{\mu}{3a}$.

From (2) we then have

$$\frac{\mu}{6a}\left[u^2+\left(\frac{du}{d\theta}\right)^2\right]=\mu\left[\frac{u^3}{3}+\frac{1}{2}au^4\right],$$

i.e. $$\left(\frac{du}{d\theta}\right)^2=u^2[2au+3a^2u^2-1]=u^2[au+1][3au-1].$$

On putting $u=\frac{1}{r}$, this equation gives

$$\left(\frac{dr}{d\theta}\right)^2=(a+r)(3a-r),$$

and hence $$\theta=\int\frac{dr}{\sqrt{(a+r)(3a-r)}}.$$

Putting $r=a+y$, we have $\theta=\int\frac{dy}{\sqrt{4a^2-y^2}}=\sin^{-1}\frac{y}{2a}+\gamma.$

$$\therefore \sin(\theta-\gamma)=\frac{y}{2a}=\frac{r-a}{2a}.$$

If we measure θ from the initial radius vector, then $\theta=0$ when $r=a$, and hence $\gamma=0$.

Therefore the path is $\quad r=a(1+2\sin\theta)$.

Clearly $\frac{dr}{d\theta}=0$, *i.e.* we have an apse, when

$$\theta=\frac{\pi}{2},\ \frac{3\pi}{2},\ \frac{5\pi}{2},\ \text{etc.}$$

Hence the apsidal angle is π and the apsidal distances are equal to $3a$ and a, and the apses are both on the positive directions of the axis of y at distances $3a$ and a from the origin. The path is a limaçon and can be easily traced from its equation.

EXAMPLES

1. A particle moves under a central repulsive force $\left\{=\frac{m\mu}{(\text{distance})^3}\right\}$, and is projected from an apse at a distance a with velocity V. Shew that the equation to the path is $r\cos p\theta=a$, and that the angle θ described in time t is

$$\frac{1}{p}\tan^{-1}\left[\frac{pV}{a}t\right],\ \text{where } p^2=\frac{\mu+a^2V^2}{a^2V^2}.$$

2. A particle moves with a central acceleration, $=\frac{\mu}{(\text{distance})^5}$, and is projected from an apse at a distance a with a velocity equal to n times that which would be acquired in falling from infinity; shew that the other apsidal distance is $\frac{a}{\sqrt{n^2-1}}$.

If $n=1$, and the particle be projected in any direction, shew that the path is a circle passing through the centre of force.

3. A particle, moving with a central acceleration $\frac{\mu}{(\text{distance})^3}$, is projected from an apse at a distance a with a velocity V; shew that the path is

$$r \cosh \left[\frac{\sqrt{\mu - a^2V^2}}{aV}\theta\right] = a, \text{ or } r \cos \left[\frac{\sqrt{a^2V^2 - \mu}}{aV}\theta\right] = a,$$

according as V is $\lesseqgtr$ the velocity from infinity.

4. A particle moving under a constant force from a centre is projected in a direction perpendicular to the radius vector with the velocity acquired in falling to the point of projection from the centre. Shew that its path is

$$\left(\frac{a}{r}\right)^3 = \cos^2 \frac{3}{2}\theta,$$

and that the particle will ultimately move in a straight line through the origin in the same way as if its path had always been this line.

If the velocity of projection be double that in the previous case, shew that the path is

$$\frac{\theta}{2} = \tan^{-1}\sqrt{\frac{r-a}{a}} - \frac{1}{\sqrt{3}} \tan^{-1}\sqrt{\frac{r-a}{3a}}.$$

5. A particle moves with a central acceleration $\mu\left(r + \frac{2a^3}{r^2}\right)$, being projected from an apse at a distance a with twice the velocity for a circle at that distance; find the other apsidal distance, and shew that the equation to the path is

$$\frac{\theta}{2} = \tan^{-1}(t\sqrt{3}) - \frac{1}{\sqrt{5}} \tan^{-1}\left(\sqrt{\frac{5}{3}}t\right), \text{ where } t^2 = \frac{r-a}{3a-r}.$$

6. A particle moves with a central acceleration $\mu\left(r + \frac{a^4}{r^3}\right)$ being projected from an apse at distance a with a velocity $2\sqrt{\mu a}$; shew that it describes the curve $r^2[2 + \cos\sqrt{3}\theta] = 3a^2$.

7. A particle moves with a central acceleration $\mu(r^5 - c^4 r)$, being projected from an apse at distance c with a velocity $\sqrt{\frac{2\mu}{3}}c^3$; shew that its path is the curve $x^4 + y^4 = c^4$.

8. A particle moves under a central force $m\lambda[3a^2u^4 + 8au^2]$; it is projected from an apse at a distance a from the centre of force with velocity $\sqrt{10\lambda}$; shew that the second apsidal distance is half the first, and that the equation to the path is

$$2r = a\left[1 + \operatorname{sech}\frac{\theta}{\sqrt{5}}\right].$$

9. A particle describes an orbit with a central acceleration $\mu u^3 - \lambda u^5$, being projected from an apse at distance a with a velocity equal to that from infinity; shew that its path is

$$r = a \cosh\frac{\theta}{n}, \text{ where } n^2 + 1 = \frac{2\mu a^2}{\lambda}.$$

Prove also that it will be at distance r at the end of time

$$\sqrt{\frac{a^2}{2\lambda}}\left[a^2 \log\frac{r + \sqrt{r^2 - a^2}}{a} + r\sqrt{r^2 - a^2}\right].$$

10. In a central orbit the force is $\mu u^3(3+2a^2u^2)$; if the particle be projected at a distance a with a velocity $\sqrt{\frac{5\mu}{a^2}}$ in a direction making $\tan^{-1}\frac{1}{2}$ with the radius, shew that the equation to the path is $r=a\tan\theta$.

11. A particle moves under a force $m\mu\{3au^4-2(a^2-b^2)u^5\}$, $a>b$, and is projected from an apse at a distance $a+b$ with velocity $\sqrt{\mu}\div(a+b)$; shew that its orbit is $r=a+b\cos\theta$.

12. A particle moves with a central acceleration $\lambda^2(8au^2+a^4u^5)$; it is projected with velocity 9λ from an apse at a distance $\frac{a}{3}$ from the origin; shew that the equation to its path is

$$\frac{1}{\sqrt{3}}\sqrt{\frac{au+5}{au-3}}=\cot\frac{\theta}{\sqrt{6}}.$$

13. A particle, subject to a central force per unit of mass equal to

$$\mu\{2(a^2+b^2)u^5-3a^2b^2u^7\},$$

is projected at the distance a with a velocity $\frac{\sqrt{\mu}}{a}$ in a direction at right angles to the initial distance; shew that the path is the curve

$$r^2=a^2\cos^2\theta+b^2\sin^2\theta.$$

14. A particle moves with a central acceleration $\mu\left(u^5-\frac{a^2}{8}u^7\right)$; it is projected at a distance a with a velocity $\sqrt{\frac{25}{7}}$ times the velocity for a circle at that distance and at an inclination $\tan^{-1}\frac{4}{3}$ to the radius vector. Shew that its path is the curve

$$4r^2-a^2=\frac{3a^2}{(1-\theta)^2}.$$

15. A particle is acted on by a central repulsive force which varies as the nth power of the distance; if the velocity at any point of the path be equal to that which would be acquired in falling from the centre to the point, shew that the equation to the path is of the form

$$r^{\frac{n+3}{2}}\cos\frac{n+3}{2}\theta=\text{const.}$$

16. An elastic string, of natural length l, is tied to a particle at one end and is fixed at its other end to a point in a smooth horizontal table. The particle can move on the table and initially is at rest with the string straight and unstretched. A blow (which, if directed along the string would make the particle oscillate to a maximum distance $2l$ from the fixed end) is given to the particle in a direction inclined at an angle α to the string. Prove that the maximum length of the string during the ensuing motion is given by the greatest root of the equation

$$x^4-2lx^3+l^4\sin^2\alpha=0.$$

17. A particle of mass m is attached to a fixed point by an elastic string of natural length a, the coefficient of elasticity being nmg; it is projected from an apse at a distance a with velocity $\sqrt{2pgh}$; shew that the other apsidal distance is given by the equation

$$nr^2(r-a)-2pha(r+a)=0.$$

18. A particle acted on by a repulsive central force $\mu r \div (r^2-9c^2)^2$ is projected from an apse at a distance c with velocity $\sqrt{\frac{\mu}{8c^2}}$ shew that it will describe a three-cusped hypocycloid and that the time to the cusp is $\frac{4}{3}\pi c^2\sqrt{\frac{2}{\mu}}$.

[Use equation (5) of Art. 53, and we have $8p^2=9c^2-r^2$.

Also $hdt=p.ds=pdr.\frac{r}{\sqrt{r^2-p^2}}$, giving $ht=\int_c^{3c}\frac{rdr}{3}\sqrt{\frac{9c^2-r^2}{r^2-c^2}}$.

To integrate, put $r^2=c^2+8c^2\cos^2\phi$.]

19. Find the path described about a fixed centre of force by a particle, when the acceleration toward the centre is of the form $\frac{\mu}{r^2}+\frac{\mu'}{r^3}$, in terms of the velocity V at an apse whose distance is a from the centre of force.

20. Shew that the only law for a central attraction, for which the velocity in a circle at any distance is equal to the velocity acquired in falling from infinity to that distance, is that of the inverse cube.

21. A particle moves in a curve under a central attraction so that its velocity at any point is equal to that in a circle at the same distance and under the same attraction; shew that the law of force is that of the inverse cube, and that the path is an equiangular spiral.

22. A particle moves under a central force $m\mu \div (\text{distance})^n$ (where $n>1$ but not$=3$). If it be projected at a distance R in a direction making an angle β with the initial radius vector with a velocity equal to that due to a fall from infinity, shew that the equation to the path is

$$r^{\frac{n-3}{2}}\sin\beta=R^{\frac{n-3}{2}}\sin\left(\frac{n-3}{2}\theta+\beta\right).$$

If $n>3$ shew that the maximum distance from the centre is

$$R\operatorname{cosec}^{\frac{2}{n-3}}\beta,$$

and if $n \leqq 3$ then the particle goes to infinity.

23. A particle moves with central acceleration $\mu u^2+\nu u^3$ and the velocity of projection at distance R is V; shew that the particle will ultimately go off to infinity if $V^2>\frac{2\mu}{R}+\frac{\nu}{R^2}$.

24. A particle is projected from an apse at a distance a with a velocity $\frac{\sqrt{\mu+\lambda}}{a}$ and moves with a central attraction equal to

$$\frac{\mu}{2}(n-1)a^{n-3}r^{-n}+\lambda r^{-3}, \text{ where } n>3,$$

per unit of mass; shew that it will arrive at the centre of force in time

$$\frac{a^2}{2}\sqrt{\frac{\pi}{\mu}}\Gamma\left(\frac{n+1}{2n-6}\right)\Big/\Gamma\left(\frac{2}{n-3}\right).$$

25. In a central orbit if $P=\mu u^2(cu+\cos\theta)^{-3}$, shew that the path is one of the conics

$$(cu+\cos\theta)^2=a+b\cos(2\theta+a).$$

26. A particle, of mass m, moves under an attractive force to the pole equal to $\frac{m\mu}{r^2}\sin^2\theta$. It is projected with velocity $\sqrt{\frac{2\mu}{3a}}$ from an apse at a distance a. Shew that the equation to the orbit is $r(1+\cos^2\theta)=2a$, and that the time of a complete revolution is $(3a)^{\frac{3}{2}}\times\frac{\pi}{\sqrt{\mu}}$.

27. If a particle move with a central acceleration $\frac{\mu}{r^2}(1+k^2\sin^2\theta)^{-\frac{3}{2}}$, find the orbit and interpret the result geometrically.

[Multiplying the equation of motion, $h^2(\ddot{u}+u)=\mu(1+k^2\sin^2\theta)^{-\frac{3}{2}}$, by $\cos\theta$ and $\sin\theta$ in succession and integrating, we have

$$h^2(\dot{u}\cos\theta+u\sin\theta)=\mu\sin\theta(1+k^2\sin^2\theta)^{-\frac{1}{2}}+A,$$

and $$h^2(\dot{u}\sin\theta-u\cos\theta)=-\mu\cos\theta(1+k^2\sin^2\theta)^{-\frac{1}{2}}\div(1+k^2)+B.$$

Eliminating $\dot{u}$, we have

$$h^2u=\mu(1+k^2\sin^2\theta)^{\frac{1}{2}}\div(1+k^2)+A\sin\theta-B\cos\theta.]$$

28. A particle moves in a field of force whose potential is $\mu r^{-2}\cos\theta$ and it is projected at distance a perpendicular to the initial line with velocity $\frac{2}{a}\sqrt{\mu}$; shew that the orbit described is

$$r=a\sec\left[\sqrt{2}\log\tan\frac{\pi+\theta}{4}\right].$$

29. A particle is describing a circle of radius a under the action of a constant force λ to the centre when suddenly the force is altered to $\lambda+\mu\sin nt$, where μ is small compared with λ and t is reckoned from the instant of change. Shew that at any subsequent time t the distance of the particle from the centre of force is

$$a+\frac{\mu a}{3\lambda-an^2}\left[n\sqrt{\frac{a}{3\lambda}}\sin\left(t\sqrt{\frac{3\lambda}{a}}\right)-\sin nt\right].$$

What is the character of the motion if $3\lambda=an$?

[Use equations (1) and (2) of Art. 53; the second gives $r^2\dot{\theta}=\sqrt{\lambda a^3}$, and the first then becomes

$$\ddot{r}-\frac{\lambda a^3}{r^3}=-\lambda-\mu\sin nt.$$

Put $r=a+\xi$ where ξ is small, and neglect squares of ξ.]

62. *A particle describes a path which is nearly a circle about a centre of force $(=\mu u^n)$ at its centre; find the condition that this may be a stable motion.*

The equation of motion is

$$\frac{d^2u}{d\theta^2}+u=\frac{\mu}{h^2}u^{n-2} \quad \text{.........................(1).}$$

If the path is a circle of radius $\frac{1}{c}$, then $h^2=\mu c^{n-3}$ ....(2).

Suppose the particle to be slightly displaced from the circular path in such a way that h remains unaltered (for example, suppose it is given a small additional velocity in a direction away from the centre of force by means of a blow, the perpendicular velocity being unaltered).

In (1) put $u=c+x$, where x is small; then it gives

$$\frac{d^2x}{d\theta^2}+c+x=\frac{(c+x)^{n-2}}{c^{n-3}}=c+(n-2)x+ \quad\text{..............(3).}$$

Neglecting squares and higher powers of x, *i.e.* assuming that x is always small, we have

$$\frac{d^2x}{d\theta^2}=-(3-n)x.$$

If n be < 3, so that $3-n$ is positive, this gives

$$x=A\cos\left[\sqrt{3-n}\,\theta+B\right].$$

If n be > 3, so that $n-3$ is positive, the solution is

$$x=A_1e^{\sqrt{n-3}\,\theta}+B_1e^{-\sqrt{n-3}\,\theta},$$

so that x continually increases as θ increases; hence x is not always small and the orbit does not continue to be nearly circular.

If $n<3$, the approximation to the path is

$$u=c+A\cos\left[\sqrt{3-n}\,\theta+B\right] \quad\text{..................(4).}$$

The apsidal distances are given by the equation $\frac{du}{d\theta}=0$, *i.e.* by

$$0=\sin\left[\sqrt{3-n}\,\theta+B\right].$$

The solutions of this equation are a series of angles, the difference between their successive values being $\frac{\pi}{\sqrt{3-n}}$. This is therefore the apsidal angle of the path.

If $n=3$, this apsidal angle is infinite. In this case it would be found that the motion is unstable, the particle departing from the circular path altogether and describing a spiral curve.

The maximum and minimum values of u, in the case $n<3$, are $c+A$ and $c-A$, so that the motion is included between these values.

63. The general case may be considered in the same manner. Let the central acceleration be $\phi(u)$.

The equations (1) and (2) then become

$$\frac{d^2u}{d\theta^2}+u=\frac{\mu}{h^2}\cdot\frac{\phi(u)}{u^2} \quad\text{..........................(5),}$$

and

$$h^2c^3=\mu\phi(c) \quad\text{............................(6).}$$

Also (3) is now

$$\frac{d^2x}{d\theta^2}+c+x=\frac{c^3}{\phi(c)}\cdot\frac{\phi(c+x)}{(c+x)^2}$$

$$=\frac{c}{\phi(c)}\left[\phi(c)+x\phi'(c)+\ldots\right]\left[1-\frac{2x}{c}+\ldots\right]$$

$$=c-2x+x\frac{c\phi'(c)}{\phi(c)},\text{ neglecting squares of } x.$$

$$\therefore\ \frac{d^2x}{d\theta^2}=-\left\{3-\frac{c\phi'(c)}{\phi c}\right\}x,$$

and the motion is stable only if

$$\frac{c\phi'(c)}{\phi(c)}<3.$$

In this case the apsidal angle is

$$\pi\div\left\{3-\frac{c\phi'(c)}{\phi(c)}\right\}^{\frac{1}{2}}.$$

64. If, in addition to the central acceleration P, we have an acceleration T perpendicular to P, the equations of motion are

$$\frac{d^2r}{dt^2}-r\left(\frac{d\theta}{dt}\right)^2=-P \quad\ldots\ldots\ldots\ldots(1),$$

and

$$\frac{1}{r}\frac{d}{dt}\left(r^2\frac{d\theta}{dt}\right)=T \quad\ldots\ldots\ldots\ldots(2).$$

Let $r^2\frac{d\theta}{dt}=h$. In this case h is not a constant.

Then (2) gives

$$T=u\frac{dh}{dt}=u\frac{dh}{d\theta}\cdot\frac{d\theta}{dt}=hu^3\frac{dh}{d\theta} \quad\ldots\ldots\ldots\ldots(3).$$

$$\therefore\ \frac{dr}{dt}=\frac{dr}{d\theta}\frac{d\theta}{dt}=-\frac{1}{u^2}\frac{du}{d\theta}\cdot hu^2=-h\frac{du}{d\theta},$$

and

$$\frac{d^2r}{dt^2}=-\frac{d}{d\theta}\left(h\frac{du}{d\theta}\right)\frac{d\theta}{dt}=-hu^2\left[h\frac{d^2u}{d\theta^2}+\frac{dh}{d\theta}\frac{du}{d\theta}\right]$$

$$=-h^2u^2\frac{d^2u}{d\theta^2}-\frac{T}{u}\frac{du}{d\theta},$$

by equation (3).

Therefore (1) gives

$$-h^2u^2\frac{d^2u}{d\theta^2}-\frac{T}{u}\frac{du}{d\theta}-h^2u^3=-P,$$

i.e.

$$\frac{d^2u}{d\theta^2}+u=\frac{P-\frac{T}{u}\frac{du}{d\theta}}{h^2u^2} \qquad \text{.......................(4).}$$

This may also be written in the form

$$\frac{d}{d\theta}\left[\frac{\frac{P}{u^2}-\frac{T}{u^3}\frac{du}{d\theta}}{\frac{d^2u}{d\theta^2}+u}\right]=\frac{d}{d\theta}(h^2)=2h\frac{dh}{d\theta}=\frac{2T}{u^3},$$

from equation (3).

EXAMPLES

1. One end of an elastic string, of unstretched length a, is tied to a point on the top of a smooth table, and a particle attached to the other end can move freely on the table. If the path be nearly a circle of radius b, shew that its apsidal angle is approximately $\pi\sqrt{\frac{b-a}{4b-3a}}$.

2. If the nearly circular orbit of a particle be $p^2(a^{m-2}-r^{m-2})=b^m$, shew that the apsidal angle is $\frac{\pi}{\sqrt{m}}$ nearly.

[Using equation (5) of Art. 53 we see that P varies as r^{m-3}; the result then follows from Art. 62.]

3. A particle moves with a central acceleration $\frac{\mu}{r^2}-\frac{\lambda}{r^3}$; shew that the apsidal angle is $\pi\div\sqrt{1+\frac{\lambda}{h^2}}$, where $\frac{h}{2}$ is the constant areal velocity.

4. Find the apsidal angle in a nearly circular orbit under the central force ar^m+br^n.

5. Assuming that the moon is acted on by a force $\frac{\mu}{(\text{distance})^2}$ to the earth and that the effect of the sun's disturbing force is to cause a force $m^2\times$ distance from the earth to the moon, shew that, the orbit being nearly circular, the apsidal angle is $\pi\left(1+\frac{3}{2}\frac{m^2}{n^2}\right)$ nearly, where $\frac{2\pi}{n}$ is a mean lunar month, and cubes of m are neglected.

6. A particle is moving in an approximately circular orbit under the action of a central force $\frac{\mu}{r^2}$ and a small constant tangential retardation f; shew that, if the mean distance be a, then $\theta=nt+\frac{3}{2}\frac{f}{a}t^2$, where $\mu=a^3n^2$ and the squares of f are neglected.

7. *Two particles of masses M and m are attached to the ends of an inextensible string which passes through a smooth fixed ring, the whole resting on a horizontal table. The particle m being projected at right angles to the string, shew that its path is*

$$a=r\cos\left[\sqrt{\frac{m}{m+M}}\theta\right].$$

The tension of the string being T, the equations of motion are

$$\frac{d^2r}{dt^2}-r\left(\frac{d\theta}{dt}\right)^2=-\frac{T}{m} \quad \text{.....(1)},$$

$$\frac{1}{r}\frac{d}{dt}\left(r^2\frac{d\theta}{dt}\right)=0 \quad \text{.....(2)},$$

and

$$\frac{d^2}{dt^2}(l-r)=-\frac{T}{M} \quad \text{.....(3)}.$$

(2) gives

$$r^2\dot{\theta}=h \quad \text{.....(4)},$$

and then (1) and (3) give

$$\left(1+\frac{M}{m}\right)\ddot{r}=\frac{h^2}{r^3}.$$

$$\therefore \quad \left(1+\frac{M}{m}\right)\dot{r}^2=-\frac{h^2}{r^2}+A=h^2\left(\frac{1}{a^2}-\frac{1}{r^2}\right),$$

since $\dot{r}$ is zero initially, when $r=a$.

This equation and (4) give

$$\left(1+\frac{M}{m}\right)\left(\frac{dr}{d\theta}\right)^2=\left(1+\frac{M}{m}\right)\dot{r}^2\div\dot{\theta}^2=\frac{r^2-a^2}{a^2}r^2.$$

$$\therefore \quad \theta\times\sqrt{\frac{m}{m+M}}=\int\frac{adr}{r\sqrt{r^2-a^2}}=\cos^{-1}\frac{a}{r}+C,$$

and C vanishes if θ be measured from the initial radius vector.

$$\therefore \quad a=r\cos\left[\sqrt{\frac{m}{m+M}}\theta\right] \text{ is the path.}$$

8. Two masses M, m are connected by a string which passes through a hole in a smooth horizontal plane, the mass m hanging vertically. Shew that M describes on the plane a curve whose differential equation is

$$\left(1+\frac{m}{M}\right)\frac{d^2u}{d\theta^2}+u=\frac{mg}{M}\frac{1}{h^2u^2}.$$

Prove also that the tension of the string is

$$\frac{Mm}{M+m}(g+h^2u^3).$$

9. In the previous question if $m=M$, and the latter be projected on the plane with velocity $\sqrt{\frac{8ag}{3}}$ from an apse at a distance a, shew that the former will rise through a distance a.

10. Two particles, of masses M and m, are connected by a light string; the string passes through a small hole in the table, m hangs vertically, and M describes a curve on the table which is very nearly a circle whose centre is the hole; shew that the apsidal angle of the orbit of M is $\pi\sqrt{\frac{M+m}{3M}}$.

11. A particle of mass m can move on a smooth horizontal table. It is attached to a string which passes through a smooth hole in the table, goes under a small smooth pulley of mass M and is attached to a point in the under side of the table so that the parts of the string hang vertically. If the motion be slightly disturbed, when the mass m is describing a circle uniformly, so that the angular momentum is unchanged, shew that the apsidal angle is $\pi\sqrt{\dfrac{M+4m}{12m}}$.

12. Two particles on a smooth horizontal table are attached by an elastic string, of natural length a, and are initially at rest at a distance a apart. One particle is projected at right angles to the string. Shew that if the greatest length of the string during the subsequent motion be $2a$, then the velocity of projection is $\sqrt{\dfrac{8a\lambda}{3m}}$, where m is the harmonic mean between the masses of the particles and λ is the modulus of elasticity of the string.

[Let the two particles be A and B of masses M and M', of which B is the one that is projected. When the connecting string is of length r and therefore of tension T, such that $T=\lambda\dfrac{r-a}{a}$, the acceleration of A is $\dfrac{T}{M}$ along AB, and that of B is $\dfrac{T}{M'}$, along BA. To get the relative motion we give to both B and A an acceleration equal and opposite to that of A. The latter is then " reduced to rest " and the acceleration of B relative to A is along BA and

$$=\frac{T}{M}+\frac{T}{M'}=\frac{2}{m}\lambda\frac{r-a}{a}=\frac{2\lambda}{ma}\frac{1-au}{u}.$$

The equation to the relative path of B is now

$$\frac{d^2u}{d\theta^2}+u=\frac{2\lambda}{mah^2}\frac{1-au}{u^3}.$$

Integrate and introduce the conditions that the particle is projected from an apse at a distance a with velocity V. The fact that there is another apse at a distance $2a$ determines V.]

13. A particle is moving in a circular orbit, of radius a, under a force of intensity $\mu u^3(2a^2u^2-1)$ towards the centre. Shew that the orbit is unstable and that if a slight disturbance takes place, inward or outward, the path may be represented by either $r=a\tanh\theta$ or $r=a\coth\theta$.

14. Einstein's discussion of planetary motion suggests the following problem:

A particle moves in one plane subject to an acceleration to a fixed centre of magnitude $\mu\left(\dfrac{1}{r^2}+\dfrac{3h^2}{c^2r^4}\right)$, h being the moment of the velocity of the particle about the centre of acceleration, and c the velocity of light. Shew that the angle between successive apse-lines is $\pi\left(1+\dfrac{3h^2}{c^2l^2}\right)$, $\dfrac{h}{cl}$ being small, and l being the latus rectum of the ellipse which the particle would describe with the same moment of momentum, if the law were $\dfrac{\mu}{r^2}$.

Supposing the planet Mercury to be subject to an acceleration of this type directed towards the Sun, shew that its apse-line progresses at the rate of 42·9" per century, given that $l=5{\cdot}55\times10^7$ kilometres, $\dfrac{\mu}{c^2}=1{\cdot}47$ kilometres, and that the periodic time of Mercury is 87·97 days.

CHAPTER V

UNIPLANAR MOTION WHEN THE ACCELERATION IS CENTRAL AND VARYING AS THE INVERSE SQUARE OF THE DISTANCE

65. In the present chapter we shall consider the motion when the central acceleration follows the Newtonian Law of Attraction.

This law may be expressed as follows; between every two particles, of masses m_1 and m_2 placed at a distance r apart, the mutual attraction is

$$\gamma\frac{m_1 m_2}{r^2}$$

units of force, where γ is a constant, depending on the units of mass and length employed, and known as the constant of gravitation.

If the masses be measured in pounds, and the length in feet, the value of γ is $1{\cdot}05\times10^{-9}$ approximately, and the attraction is expressed in poundals.

If the masses be measured in grammes, and the length in centimetres, the value of γ is $6{\cdot}66\times10^{-8}$ approximately, and the attraction is expressed in dynes.

66. *A particle moves in a path so that its acceleration is always directed to a fixed point and is equal to* $\frac{\mu}{(distance)^2}$; *to shew that its path is a conic section and to distinguish between the three cases that arise.*

When $P=\frac{\mu}{r^2}$, the equation (5) of Art. 53 becomes

$$\frac{h^2}{p^3}\frac{dp}{dr}=\frac{\mu}{r^2} \quad \text{.............................(1).}$$

Integrating we have, by Art. 54,

$$v^2=\frac{h^2}{p^2}=\frac{2\mu}{r}+C \quad \text{..........................(2).}$$

Now the (p, r) equation of an ellipse and hyperbola, referred to a focus, are respectively

$$\frac{b^2}{p^2}=\frac{2a}{r}-1, \text{ and } \frac{b^2}{p^2}=\frac{2a}{r}+1 \quad \text{....................(3),}$$

where $2a$ and $2b$ are the transverse and conjugate axes.

Hence, when C is negative, (2) is an ellipse; when C is positive, it is a hyperbola.

Also when $C=0$, (2) becomes $\frac{p^2}{r}=$constant, and this is the (p, r) equation of a parabola referred to its focus.

Hence (2) always represents a conic section, whose focus is at the centre of force, and which is an

$$\left.\begin{matrix}\text{ellipse}\\ \text{parabola}\\ \text{or hyperbola}\end{matrix}\right\}\text{ according as } C \text{ is } \left\{\begin{matrix}\text{negative}\\ \text{zero}\\ \text{or positive}\end{matrix}\right\},$$

i.e. according as $v^2 \lesseqgtr \frac{2\mu}{r}$, *i.e.* according as the square of the velocity at any point P is $\lesseqgtr \frac{2\mu}{SP}$, where S is the focus.

Again, comparing equations (2) and (3), we have, in the case of the ellipse,

$$\frac{h^2}{b^2}=\frac{\mu}{a}=\frac{C}{-1}.$$

$$\therefore\ h=\sqrt{\mu\frac{b^2}{a}}=\sqrt{\mu\times\text{semi-latus-rectum}}, \text{ and } C=-\frac{\mu}{a}.$$

Hence, in the case of the ellipse, $v^2=\mu\left(\frac{2}{r}-\frac{1}{a}\right)$ ....(4).

So, for the hyperbola, $v^2=\mu\left(\frac{2}{r}+\frac{1}{a}\right)$,

and, for the parabola, $v^2=\frac{2\mu}{r}$.

It will be noted that in each case the velocity at any point does not depend on the direction of the velocity.

Since h is twice the area described in the unit of time (Art. 54), therefore, if T be the time of describing the ellipse, we have

$$T=\frac{\text{area of the ellipse}}{\frac{1}{2}h}=\frac{\pi ab}{\frac{1}{2}\sqrt{\mu\frac{b^2}{a}}}=\frac{2\pi}{\sqrt{\mu}}a^{\frac{3}{2}} \quad ....(5),$$

so that the square of the periodic time varies as the cube of the major axis.

COR. 1. If a particle be projected at a distance R with velocity V in any direction the path is an ellipse, parabola or hyperbola, according as $V^2 <=> \frac{2\mu}{R}$.

Now the square of the velocity that would be acquired in falling from infinity to the distance R, by Art. 31,

$$=2\int_{\infty}^{R}\left(-\frac{\mu}{r^2}\right)dr=\left[\frac{2\mu}{r}\right]_{\infty}^{R}=\frac{2\mu}{R}.$$

Hence the path is an ellipse, parabola or hyperbola according as the velocity at any point is $<=>$ that acquired in falling from infinity to the point.

COR. 2. The velocity V_1 for the description of a circle of radius R is given by

$$\frac{V_1^2}{R}=\text{normal acceleration}=\frac{\mu}{R^2}, \text{ so that } V_1^2=\frac{\mu}{R}$$

and
$$\therefore V_1=\frac{\text{velocity from infinity}}{\sqrt{2}}.$$

67. In the previous article the branch of the hyperbola described is the one nearest the centre of force.

If the central acceleration be from the centre and if it vary as the inverse square of the distance, the further branch is described. For in this case the equation of motion is

$$\frac{h^2}{p^3}\frac{dp}{dr}=-\frac{\mu}{r^2}. \quad \therefore \frac{h^2}{p^2}=-\frac{2\mu}{r}+C \quad \text{..................(1)}.$$

Now the (p, r) equation of the further branch of a hyperbola is

$$\frac{b^2}{p^2}=1-\frac{2a}{r},$$

and this always agrees with (1) provided that $\frac{h^2}{b^2}=\frac{\mu}{a}=C$, so that

$h=\sqrt{\mu\times\text{semi-latus-rectum}}$, and $v^2=\frac{h^2}{p^2}=\mu\left(\frac{1}{a}-\frac{2}{r}\right)$.

68. *Construction of the orbit given the point of projection and the direction and magnitude of the velocity of projection.*

Let S be the centre of attraction, P the point of projection, TPT' the direction of projection, and V the velocity of projection.

Case I. Let $V^2<\frac{2\mu}{SP}$; then, by Art. 66, the path is an ellipse whose major axis $2a$ is given by the equation

$$V^2=\mu\left(\frac{2}{R}-\frac{1}{a}\right), \text{ where } R=SP, \text{ so that } 2a=\frac{2R\mu}{2\mu-V^2R}.$$

Draw PS', so that PS' and PS are on the same side of TPT', making $\angle T'PS'=\angle TPS$, and take

$$PS'=2a-SP=2a-R=\frac{V^2R^2}{2\mu-V^2R}.$$

Then S' is the second focus and the elliptic path is therefore known.

Case II. Let $V^2=\frac{2\mu}{SP}$, so that the path is a parabola. Draw the direction PS' as in Case I; in this case this is the direction of the axis of the parabola. Draw SU parallel to PS' to meet TPT' in U; draw SY perpendicular to TPT' and YA perpendicular to SU. Then A is the vertex of the required parabola whose focus is S, and the curve can be constructed.

The semi-latus-rectum $=2SA=2.\frac{SY^2}{SP}=\frac{2p_0^2}{R}$, where p_0 is the perpendicular from S on the direction of projection.

Case III. Let $V^2>\frac{2\mu}{SP}$, so that the path is a hyperbola of transverse axis $2a$ given by the equation

$$V^2=\mu\left(\frac{2}{R}+\frac{1}{a}\right), \text{ and hence } 2a=\frac{2\mu R}{V^2R-2\mu}.$$

In this case PS' lies on the opposite side of TPT' from PS, such that $\angle TPS=\angle TPS'$, and $S'P-SP=2a$, so that

$$S'P=R+2a=\frac{V^2R^2}{V^2R-2\mu}.$$

The path can then be constructed, since S' is the second focus.

69. Kepler's Laws. The astronomer Kepler, after many years of patient labour, discovered three laws connecting the motions of the various planets about the sun. They are:

1. *Each planet describes an ellipse having the sun in one of its foci.*

2. *The areas described by the radii drawn from the planet to the sun are, in the same orbit, proportional to the times of describing them.*

3. *The squares of the periodic times of the various planets are proportional to the cubes of the major axes of their orbits.*

70. From the second law we conclude, by Art. 54, that the acceleration of each planet, and therefore the force on it, is directed towards the Sun.

From the first law it follows, by Art. 55 or Art. 66, that the acceleration of each planet varies inversely as the square of its distance from the Sun.

From the third law it follows, since from Art. 66 we have

$$T^2 = \frac{4\pi^2}{\mu} . a^3,$$

that the absolute acceleration μ (*i.e.* the acceleration at unit distance from the Sun) is the same for all planets.

Laws similar to those of Kepler have been found to hold for the planets and their satellites.

It follows from the foregoing considerations that we may assume Newton's Law of Gravitation to be true throughout the Solar System.

71. Kepler's Laws were obtained by him, by a process of continually trying hypotheses until he found one that was suitable; he started with the observations made and recorded for many years by Tycho Brahé, a Dane, who lived from A.D. 1546 to 1601.

The first and second laws were enunciated by Kepler in 1609 in his book on the motion of the planet Mars. The third law was announced ten years later in a book entitled *On the Harmonies of the World.* The explanation of these laws was given by Newton in his *Principia* published in the year 1687.

72. Kepler's third law, in the form given in Art. 69, is only true on the supposition that the Sun is fixed, or that the mass of the planet is neglected in comparison with that of the Sun.

A more accurate form is obtained in the following manner.

Let S be the mass of the Sun, P that of any of its planets, and γ the constant of gravitation. The force of attraction between the two is thus $\gamma . \frac{S.P}{r^2}$, where r is the distance between the Sun and planet at any instant.

The acceleration of the planet is then $\alpha \left(= \frac{\gamma S}{r^2}\right)$ towards the Sun, and that of the Sun is $\beta \left(= \frac{\gamma P}{r^2}\right)$ towards the planet.

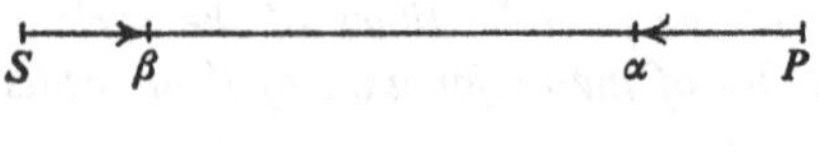

S β α P

To obtain the acceleration of the planet relative to the Sun we must give to both an acceleration β along the line PS. The acceleration of the Sun is then zero and that of the planet is $\alpha+\beta$ along PS. If, in addition, we give to each a velocity equal and opposite to that of the Sun we have the motion of P relative to the Sun supposed to be at rest.

The relative acceleration of the planet with respect to the Sun then $=\alpha+\beta=\dfrac{\gamma(S+P)}{r^2}$.

Hence the μ of Art. 66 is $\gamma\,(S+P)$, and, as in that article, we then have $T=\dfrac{2\pi}{\sqrt{\gamma(S+P)}}a^{\frac{3}{2}}$.

If T_1, be the time of revolution and a_1 the semi-major axis of the relative path of another planet P_1, we have similarly

$$T_1=\frac{2\pi}{\sqrt{\gamma(S+P_1)}}a_1^{\frac{3}{2}}. \quad \therefore \frac{S+P}{S+P_1}\cdot\frac{T^2}{T_1^2}=\frac{a^3}{a_1^3}.$$

Since Kepler's Law, that $\dfrac{T^2}{T_1^2}$ varies as $\dfrac{a^3}{a_1^3}$, is very approximately true, it follows that $\dfrac{S+P_1}{S+P}$ is very nearly unity, and hence that P and P_1 are either very nearly equal or very small compared with S. But it is known that the masses of the planets are very different; hence they must be very small compared with that of the Sun.

73. The corrected formula of the last article may be used to give an approximate value to the ratio of the mass of a planet to that of the Sun in the case where the planet has a small satellite, whose periodic time and mean distance from the planet are known.

In the case of the satellite the attraction of the planet is the force which for all practical purposes determines its path.

If P be the mass of the planet and D its mean distance from the Sun, then, as in the previous article,

$$T=\frac{2\pi}{\sqrt{\gamma(S+P)}}D^{\frac{3}{2}}.$$

Similarly, if p be the mass of the satellite, d its mean distance from the planet, and t its periodic time, then

$$t=\frac{2\pi}{\sqrt{\gamma(P+p)}}d^{\frac{3}{2}}. \quad \therefore \frac{S+P}{P+p}\frac{T^2}{t^2}=\frac{D^3}{d^3}.$$

The quantities T, t, D and d being known, this gives a value for $\dfrac{S+P}{P+p}$.

As a numerical example take the case of the Earth E and the Moon m.

Then
$$\frac{S+E}{E+m}=\frac{t^2}{T^2}\cdot\frac{D^3}{d^3}.$$

Now $T=365\frac{1}{4}$ days, $t=27\frac{1}{3}$ days, $D=93{,}000{,}000$ miles, and $d=240{,}000$ miles, all the values being approximate.

$$\therefore \frac{S+E}{E+m}=\left(\frac{27\frac{1}{3}}{365\frac{1}{4}}\right)^2\times\left(\frac{9300}{24}\right)^3=325900 \text{ nearly.}$$

Therefore $S+E=325{,}900$ times the sum of the masses of the Earth and Moon. Also $m=\frac{1}{81}E$ nearly.

$\therefore S=330{,}000\,E$ nearly.

This is a fairly close approximation to the accurate result.

If the Sun be assumed to be a sphere of radius 440,000 miles and mean density n times that of the Earth, assumed to be a sphere of radius 4000 miles, this gives

$$n\times(440{,}000)^3=330{,}000\times(4000)^3.$$

$$\therefore n=\frac{330{,}000}{110^3}=\frac{330}{1331}=\text{about }\frac{1}{4}.$$

Hence the mean density of the Sun

$$=\frac{1}{4}\text{ that of the Earth}=\frac{1}{4}\times5.527=\text{about }1{\cdot}4\text{ grammes per cub. cm.,}$$

so that the mean density of the Sun is nearly half as much again as that of water.

74. It is not necessary to know the mean distance and periodic time of the planet P in order to determine its mass, or rather the sum of its mass and that of its satellite.

For if E and m be the masses of the Earth and Moon, R the distance of the Earth from the Sun, r that of the Moon from the Earth, if Y denote a year and y the mean lunar month, then we have

$$Y=\frac{2\pi}{\sqrt{\gamma(S+E)}}R^{\frac{3}{2}} \quad\text{.........................(1),}$$

$$y=\frac{2\pi}{\sqrt{\gamma(E+m)}}.r^{\frac{3}{2}} \quad\text{.........................(2).}$$

Also, as in the last article,

$$t=\frac{2\pi}{\sqrt{\gamma(P+p)}}.d^{\frac{3}{2}} \quad\text{.........................(3).}$$

From (1) and (3),

$$(P+p)\frac{t^2}{d^3}=(S+E)\frac{Y^2}{R^3} \quad \ldots\ldots,\ldots\ldots\ldots\ldots\ldots(4).$$

From (2) and (3),

$$(P+p)\frac{t^2}{d^3}=(E+m).\frac{y^2}{r^3} \quad \ldots\ldots\ldots\ldots\ldots\ldots\ldots(5).$$

Equation (4) gives the ratio of $P+p$ to $S+E$.
Equation (5) gives the ratio of $P+p$ to $E+m$.

EXAMPLES

1. Shew that the velocity of a particle moving in an ellipse about a centre of force in the focus is compounded of two constant velocities, $\frac{\mu}{h}$ perpendicular to the radius and $\frac{\mu e}{h}$ perpendicular to the major axis.

2. A particle describes an ellipse about a centre of force at the focus; shew that, at any point of its path, the angular velocity about the other focus varies inversely as the square of the normal at the point.

3. A particle moves with a central acceleration $\left[=\frac{\mu}{(\text{distance})^2}\right]$; it is projected with velocity V at a distance R. Shew that its path is a rectangular hyperbola if the angle of projection is

$$\sin^{-1}\frac{\mu}{VR\left(V^2-\frac{2\mu}{R}\right)^{\frac{1}{2}}}.$$

4. A particle describes an ellipse under a force $\frac{\mu}{(\text{distance})^2}$ towards the focus; if it was projected with velocity V from a point distant r from the centre of force, shew that its periodic time is

$$\frac{2\pi}{\sqrt{\mu}}\left[\frac{2}{r}-\frac{V^2}{\mu}\right]^{-\frac{3}{2}}.$$

5. If the velocity of the Earth at any point of its orbit, assumed to be circular, were increased by about one-half, prove that it would describe a parabola about the Sun as focus.

Shew also that, if a body were projected from the Earth with a velocity exceeding 7 miles per second, it will not return to the Earth and may even leave the Solar System.

6. A particle is projected from the Earth's surface with velocity v; shew that, if the dimunution of gravity be taken into account, but the resistance of the air neglected, the path is an ellipse of major axis $\frac{2ga^2}{2gd-v^2}$, where a is the Earth's radius.

7. Shew that an unresisted particle falling to the Earth's surface from a great distance would acquire a velocity $\sqrt{2ga}$, where a is the Earth's radius.

Prove that the velocity acquired by a particle similarly falling into the Sun is to the Earth's velocity in the square root of the ratio of the diameter of the Earth's orbit to the radius of the Sun.

8. If a planet were suddenly stopped in its orbit, supposed circular, shew that it would fall into the Sun in a time which is $\frac{\sqrt{2}}{8}$ times the period of the planet's revolution.

9. The eccentricity of the Earth's orbit round the Sun is $\frac{1}{60}$; shew that the Earth's distance from the Sun exceeds the length of the semi-major axis of the orbit during about 2 days more than half the year.

10. The mean distance of Mars from the Sun being 1·524 times that of the Earth, find the time of revolution of Mars about the Sun.

11. The time of revolution of Mars about the Sun is 687 days and his mean distance 141½ millions of miles; the distance of the Satellite Deimos from Mars is 14,600 miles and his time of revolution 30 hrs. 18 mins.; shew that the mass of the Sun is little more than three million times that of Mars.

12. The time of revolution of Jupiter about the Sun is 11·86 years and his mean distance 483 millions of miles; the distance of his first satellite is 261,000 miles, and his time of revolution 1 day 18½ hrs.; shew that the mass of Jupiter is a little less than one-thousandth of that of the Sun.

13. The outer satellite of Jupiter revolves in 16¾ days approximately, and its distance from the planet's centre is 26½ radii of the latter. The last discovered satellite revolves in 12 hours nearly; find its distance from the planet's centre.

Find also the approximate ratio of Jupiter's mean density to that of the Earth, assuming that the Moon's distance is 60 times the Earth's radius and that her siderial period is 27⅓ days nearly.

[Use equations (2) and (3) of Art. 74, and neglect m in comparison with E, and p in comparison with P.]

14. A planet is describing an ellipse about the Sun as focus; shew that its velocity away from the Sun is greatest when the radius vector to the planet is at right angles to the major axis of the path, and that it then is $\frac{2\pi ae}{T\sqrt{1-e^2}}$, where $2a$ is the major axis, e the eccentricity, and T the periodic time.

75. *To find the time of description of a given arc of an elliptic orbit starting from the nearer end of the major axis.*

The equation $r^2\frac{d\theta}{d}=h$ of Art. 53 gives

$$ht=\int_0^0 r^2 d\theta=\int_0^0 \frac{l^2}{(1+e\cos\theta)^2}\cdot d\theta \quad \text{..............(1).}$$

If $e<1$, then by the well-known result in Integral Calculus

$$\int_0^\theta \frac{d\theta}{1+e\cos\theta}=\frac{2}{\sqrt{1-e^2}}\tan^{-1}\left[\sqrt{\frac{1-e}{1+e}}\tan\frac{\theta}{2}\right] \quad \text{......(2).}$$

Differentiating with respect to the constant e, we have

$$\int_0^\theta \frac{-\cos\theta}{(1+e\cos\theta)^2}d\theta = \frac{2e}{(1-e^2)^{\frac{3}{2}}}\tan^{-1}\left[\sqrt{\frac{1-e}{1+e}}\tan\frac{\theta}{2}\right] - \frac{1}{1-e^2}\frac{\sin\theta}{1+e\cos\theta} \quad \text{......(3).}$$

$$\therefore \int_0^\theta \frac{1}{(1+e\cos\theta)^2}d\theta = \int_0^\theta\left[\frac{1}{1+e\cos\theta} - \frac{e\cos\theta}{(1+e\cos\theta)^2}\right]d\theta$$

$$= \frac{2}{(1-e^2)}\tan^{-1}\left[\sqrt{\frac{1-e}{1+e}}\tan\frac{\theta}{2}\right] - \frac{e}{1-e^2}\frac{\sin\theta}{1+e\cos\theta} \quad \text{......(4).}$$

Hence, since $\frac{l^2}{h} = \frac{l^2}{\sqrt{\mu l}} = \frac{a^{\frac{3}{2}}(1-e^2)^{\frac{3}{2}}}{\sqrt{\mu}}$, we have, by (1),

$$t = \frac{a^{\frac{3}{2}}}{\sqrt{\mu}}\left[2\tan^{-1}\left(\sqrt{\frac{1-e}{1+e}}\tan\frac{\theta}{2}\right) - e\sqrt{1-e^2}\frac{\sin\theta}{1+e\cos\theta}\right] \quad \text{......(5).}$$

ALITER. If we change the variable θ into a new variable ϕ given by the relation $(1+e\cos\theta)(1-e\cos\phi) = 1-e^2$, so that

$$\cos\theta = \frac{\cos\phi - e}{1-e\cos\phi}, \quad \sin^2\theta = \frac{(1-e^2)\sin^2\phi}{(1-e\cos\phi)^2},$$

and $\sin\theta . d\theta = \frac{\sin\phi(1-e^2)}{(1-e\cos\phi)^2}d\phi$, we have $d\theta = \frac{\sqrt{1-e^2}}{1-e\cos\phi}d\phi$.

Hence

$$\int_0^\theta \frac{d\theta}{(1+e\cos\theta)^2} = \int_0^\phi \frac{1-e\cos\phi}{(1-e^2)^{\frac{3}{2}}}d\phi = \frac{1}{(1-e^2)^{\frac{3}{2}}}[\phi - e\sin\phi] \quad \text{...(6).}$$

Now $\quad \tan^2\frac{\phi}{2} = \frac{1-\cos\phi}{1+\cos\phi} = \frac{1-e}{1+e}\frac{1-\cos\theta}{1+\cos\theta} = \frac{1-e}{1+e}\tan^2\frac{\theta}{2},$

and $\quad \sin\phi = \frac{\sqrt{1-e^2}}{1+e\cos\theta}\sin\theta.$

Substituting in (6), we have result (4), and proceed as above.

76. *To find the time similarly for a hyperbolic orbit.*

If $e>1$, then $\int_0^\theta \frac{d\theta}{1+e\cos\theta} = \frac{1}{\sqrt{e^2-1}}\log\frac{\sqrt{e+1}+\sqrt{e-1}\tan\frac{\theta}{2}}{\sqrt{e+1}-\sqrt{e-1}\tan\frac{\theta}{2}}.$

Differentiating with respect to e, we have

$$\int_0^\theta \frac{-\cos\theta}{(1+e\cos\theta)^2}d\theta = \frac{-e}{(e^2-1)^{\frac{3}{2}}}\log\left[\frac{\sqrt{e+1}+\sqrt{e-1}\tan\frac{\theta}{2}}{\sqrt{e+1}-\sqrt{e-1}\tan\frac{\theta}{2}}\right] + \frac{1}{e^2-1}\frac{\sin\theta}{1+e\cos\theta}.$$

$$\therefore \int\frac{d\theta}{(1+e\cos\theta)^2} = \int_0^\theta\left[\frac{1}{1+e\cos\theta} - \frac{e\cos\theta}{(1+e\cos\theta)^2}\right]d\theta$$

$$= -\frac{1}{(e^2-1)^{\frac{3}{2}}}\log\left[\frac{\sqrt{e+1}+\sqrt{e-1}\tan\frac{\theta}{2}}{\sqrt{e+1}-\sqrt{e-1}\tan\frac{\theta}{2}}\right] + \frac{1}{e^2-1}\frac{\sin\theta}{1+e\cos\theta}.$$

Hence, since in this case $\dfrac{l^2}{h} = \dfrac{l^{\frac{3}{2}}}{\sqrt{\mu}} = \dfrac{a^{\frac{3}{2}}(e^2-1)^{\frac{3}{2}}}{\sqrt{\mu}}$,

the equation (1) of the last article gives

$$t = \frac{a^{\frac{3}{2}}}{\sqrt{\mu}}\left[e\sqrt{e^2-1}\frac{\sin\theta}{1+e\cos\theta} - \log\frac{\sqrt{e+1}+\sqrt{e-1}\tan\frac{\theta}{2}}{\sqrt{e+1}-\sqrt{e-1}\tan\frac{\theta}{2}}\right].$$

ALITER. Change the variable θ into a new variable ϕ, such that $(1+e\cos\theta)(e\cosh\phi-1) = e^2-1$, so that

$$\cos\theta = \frac{e-\cosh\phi}{e\cosh\phi-1}, \quad \sin^2\theta = \frac{(e^2-1)\sinh^2\phi}{(e\cosh\phi-1)^2},$$

and
$$d\theta = \frac{\sqrt{e^2-1}\,d\phi}{e\cosh\phi-1}.$$

Then $$\int_0^\theta\frac{d\theta}{(1+e\cos\theta)^2} = \frac{1}{(e^2-1)^{\frac{3}{2}}}\int_0^\phi(e\cosh\phi-1)d\phi = \frac{1}{(e^2-1)^{\frac{3}{2}}}[e\sinh\phi-\phi].$$

Now
$$\tanh^2\frac{\phi}{2} = \frac{\cosh\phi-1}{\cosh\phi+1} = \frac{e-1}{e+1}\tan^2\frac{\theta}{2},$$

and
$$\sinh\phi = 2\tanh\frac{\phi}{2}\Big/\left(1-\tanh^2\frac{\phi}{2}\right) = \sqrt{e^2-1}\frac{\sin\theta}{1+e\cos\theta}.$$

$$\therefore \int_0^\theta \frac{d\theta}{(1+e\cos\theta)^2} = \frac{e}{e^2-1}\,\frac{\sin\theta}{1+e\cos\theta} - \frac{2}{(e^2-1)}\tanh^{-1}\left[\sqrt{\frac{e-1}{e+1}}\tan\frac{\theta}{2}\right],$$

which is the same as before, since $\tanh^{-1} x = \frac{1}{2}\log\frac{1+x}{1-x}$.

77. *In the case of a parabolic orbit to find the corresponding time.*

The equation to the parabola is $r = \frac{d}{1+\cos\theta}$, where $2d$ is the latus-rectum and θ is measured from the axis. Hence the equation (3) of Art. 53 gives

$$h.t = \int r^2 d\theta = \int \frac{d^2}{(1+\cos\theta)^2} d\theta.$$

$$\therefore \frac{ht}{d^2} = \int_0^\theta \frac{d\theta}{4\cos^4\frac{\theta}{2}} = \frac{1}{4}\int_0^\theta \sec^2\frac{\theta}{2}.\sec^2\frac{\theta}{2}d\theta$$

$$= \frac{1}{2}\int_0^\theta \left(1+\tan^2\frac{\theta}{2}\right) d\left(\tan\frac{\theta}{2}\right) = \frac{1}{2}\left[\tan\frac{\theta}{2} + \frac{1}{3}\tan^3\frac{\theta}{2}\right].$$

But
$$\frac{h}{d^2} = \frac{\sqrt{\mu d}}{d^2} = \frac{\sqrt{\mu}}{d^{\frac{3}{2}}}.$$

$$\therefore t = \frac{d^{\frac{3}{2}}}{2\sqrt{\mu}}\left[\tan\frac{\theta}{2} + \frac{1}{3}\tan^3\frac{\theta}{2}\right] = \sqrt{\frac{2a^3}{\mu}}\left[\tan\frac{\theta}{2} + \frac{1}{3}\tan^3\frac{\theta}{2}\right],$$

if a be the apsidal distance.

78. *Motion of a projectile, variations of gravity being taken into consideration but the resistance of the air being neglected.*

The attraction of the Earth at a point outside it at a distance r from the centre is $\frac{\mu}{r^2}$. Hence the path of a projectile in vacuo is one of the cases of Art. 66, one of the foci of the path described being at the centre of the Earth.

If R be the radius of the Earth, then $\frac{\mu}{R^2}$ = the value of gravity at the surface of the Earth $= g$, so that $\mu = gR^2$.

The path of a projectile which starts from a point on the Earth's surface is therefore an ellipse, parabola, or hyperbola according as $V^2 \lesseqgtr \frac{2\mu}{R}$, *i.e.* according as $V^2 \lesseqgtr 2gR$.

79. The maximum range of a particle starting from the Earth's surface with a given velocity may be obtained as follows:

Let S be the centre of the Earth and P the point of projection. Let

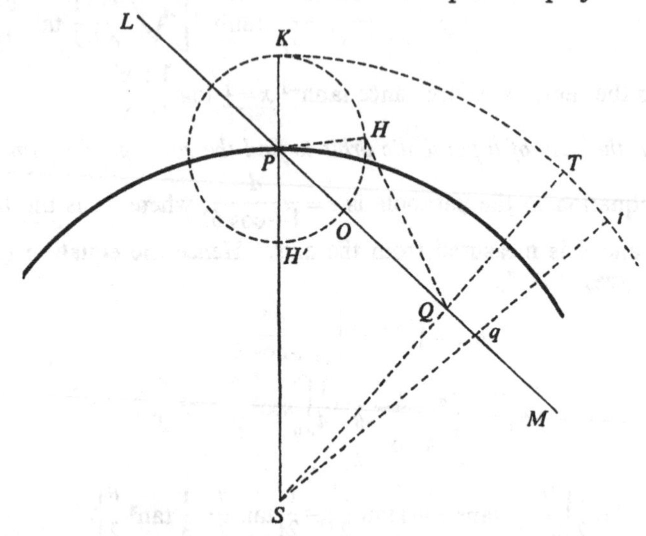

K be the point vertically above P to which the velocity, V, of projection is due, so that, by Art. 31, we have

$$V^2 = 2\mu\left[\frac{1}{SP} - \frac{1}{SK}\right] = 2gR^2\left[\frac{1}{R} - \frac{1}{R+h}\right] \quad \ldots\ldots\ldots\ldots(1),$$

where R is the radius of the Earth and PK is h.

If H be the second focus of the path, the semi-major axis is $\frac{1}{2}(R+PH)$. Hence, by equation (4) of Art. 66,

$$V^2 = \mu\left[\frac{2}{SP} - \frac{2}{R+PH}\right] = 2gR^2\left[\frac{1}{R} - \frac{1}{R+PH}\right].$$

By comparing this with equation (1) we have $PH = h$, so that the locus of the second focus is, for a constant velocity of projection, a circle whose centre is P and radius h. It follows that the major axis of the path is $SP+PH$ or SK.

The ellipse, whose foci are S and H, meet a plane LPM, passing through the point of projection, in a point Q, such that $SQ+QH=SK$. Hence, if SQ meet in T the circle whose centre is S and radius SK, we have $QT=QH$. Since there is, in general, another point, H', on the circle of foci equidistant with H from Q, we have, in general, two paths for a given range.

The greatest range on the plane LPM is clearly Pq where qt equals qO. Hence $Sq+qP=Sq+qO+OP=Sq+qt+PK=SK+PK$.

Therefore q lies on an ellipse, whose foci are the centre of the Earth and the point of projection, and which passes through K.

Hence we obtain the **maximum range.**

80. Suppose that the path described by a planet P about the Sun S is the ellipse of the figure. Draw PN perpendicular to the major axis and produce it to meet the auxiliary circle in Q. Let C be the centre.

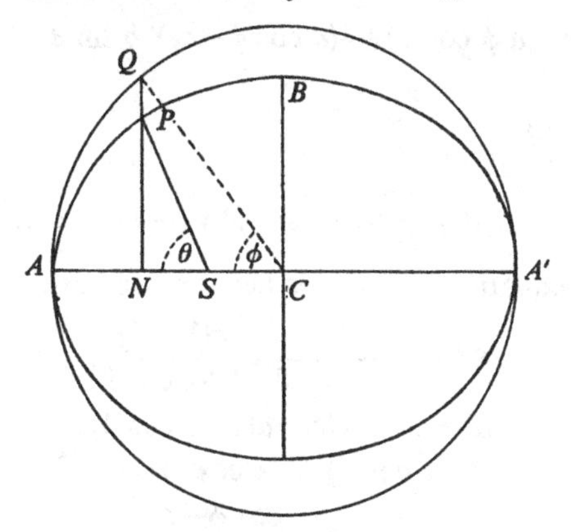

The points A and A' are called respectively the Perihelion and Aphelion of the path of the planet.

The angle ASP is called the True Anomaly and the angle ACQ the Eccentric Anomaly. In the case of any of the planets the eccentricity of the path is small, being never as large as ·1 except in the case of Mercury when it is ·2; the foci of the path are therefore very near C, the ellipse differs little in actual shape from the auxiliary circle, and hence the difference between the True and Eccentric Anomaly is a small quantity.

If $\frac{2\pi}{n}$ be the time of a complete revolution of the planet, so that n is its mean angular velocity, then nt is defined to be the Mean Anomaly and n is the Mean Motion. It is clear therefore that nt would be the Anomaly of an imaginary planet which moved so that its angular velocity was equal to the mean angular velocity of P.

Since $$\frac{2\pi}{n}=\frac{2\pi}{\sqrt{\mu}}a^{\frac{3}{2}} \text{ (Art. 66)}, \therefore n=\frac{\sqrt{\mu}}{a^{\frac{3}{2}}}.$$

Let θ be the True Anomaly ASP, and ϕ the Eccentric Anomaly ACQ.

If h be twice the area described in a unit of time, then

$$\frac{h}{2}t = \text{Sectorial area } ASP$$

$$= \text{Curvilinear area } ANP + \text{triangle } SNP$$

$$= \frac{b}{a} \times \text{Curvilinear area } ANQ + \text{triangle } SNP$$

$$= \frac{b}{a}(\text{Sector } ACQ - \text{triangle } CNQ) + \tfrac{1}{2}SN.NP$$

$$= \frac{b}{a}(\tfrac{1}{2}a^2\phi - \tfrac{1}{2}a^2 \sin\phi \cos\phi) + \tfrac{1}{2}(a\cos\phi - ae).b\sin\phi$$

$$= \frac{ab}{2}(\phi - e\sin\phi).$$

$$\therefore\ nt = \frac{nab}{h}(\phi - e\sin\phi) = \phi - e\sin\phi \quad \text{..............(1).}$$

By the polar equation to a Conic Section, we have

$$SP = \frac{l}{1+e\cos\theta} = \frac{a(1-e^2)}{1+e\cos\theta},$$

and
$$SP = a - e.CN = a(1 - e\cos\phi).$$

$$\therefore\ (1-e\cos\phi)(1+e\cos\theta) = 1 - e^2,$$

and
$$\therefore\ \cos\theta = \frac{\cos\phi - e}{1 - e\cos\phi} \quad \text{........................(2).}$$

81. If e be small, a first approximation from (1) to the value of ϕ is nt, and a second approximation is $nt + e\sin nt$.

From (2), a first approximation to the value of θ is ϕ, and a second approximation is $\phi + \lambda$ where

$$\cos\phi - \lambda\sin\phi = \frac{\cos\phi - e}{1 - e\cos\phi},$$

and
$$\therefore\ \lambda = \frac{e\sin\phi}{1 - e\cos\phi} = e\sin\phi \text{ approx.}$$

Hence, as far as the first power of e,

$$\theta = \phi + e\sin\phi = nt + e\sin nt + e\sin(nt + e\sin nt)$$
$$= nt + 2e\sin nt.$$

Also
$$SP = \frac{a(1-e^2)}{1+e\cos\theta} = a(1 - e\cos\theta),$$

to the same approximation,

$$= a - ae\cos(nt + 2e\sin nt) = a - ae\cos nt.$$

If an approximation be made as far as squares of e, the results are found to be

$$\phi = nt + e\sin nt + \frac{e^2}{2}\sin 2nt,$$

$$\theta = nt + 2e\sin nt + \frac{5e^2}{4}\sin 2nt,$$

and
$$r = a\left\{1 - e\cos nt + \frac{e^2}{2}(1-\cos 2nt)\right\}.$$

82. From equation (2) of Art. 80, we have

$$\tan^2\frac{\theta}{2} = \frac{1-\cos\theta}{1+\cos\theta} = \frac{(1+e)(1-\cos\phi)}{(1-e)(1+\cos\phi)} = \frac{1+e}{1-e}\tan^2\frac{\phi}{2},$$

so that
$$\phi = 2\tan^{-1}\left[\sqrt{\frac{1-e}{1+e}}\tan\frac{\theta}{2}\right],$$

and

$$\sin\phi = \frac{2\tan\frac{\phi}{2}}{1+\tan^2\frac{\phi}{2}} = \frac{2\sqrt{\frac{1-e}{1+e}}\tan\frac{\theta}{2}}{1+\frac{1-e}{1+e}\tan^2\frac{\theta}{2}} = \sqrt{1-e^2}\frac{\sin\theta}{1+e\cos\theta}.$$

Hence, from equation (1) of the same article, remembering that $n = \frac{\sqrt{\mu}}{a^{\frac{3}{2}}}$, we have

$$t = \frac{a^{\frac{3}{2}}}{\sqrt{\mu}}\left[2\tan^{-1}\left\{\sqrt{\frac{1-e}{1+e}}\tan\frac{\theta}{2}\right\} - e\sqrt{1-e^2}\frac{\sin\theta}{1+e\cos\theta}\right].$$

This is the result of Art. 75 and gives the time of describing any arc of the ellipse, starting from perihelion.

83. When a particle is describing an elliptic orbit, it may happen that at some point of the path it receives an impulse so that it describes another path; or the strength of the centre of force may be altered so that the path is altered. To obtain the new orbit we shall want to know how the major axis has been altered in magnitude and position, what is the new eccentricity, etc.

The new orbit will not necessarily be an ellipse and the student will find it a useful exercise to examine the various cases for himself. Similar considerations apply when the initial orbit is a circle, hyperbola, or parabola.

84. Tangential disturbing force.

Let APA' be the path of a particle moving about a centre of force at S, and let H be the other focus.

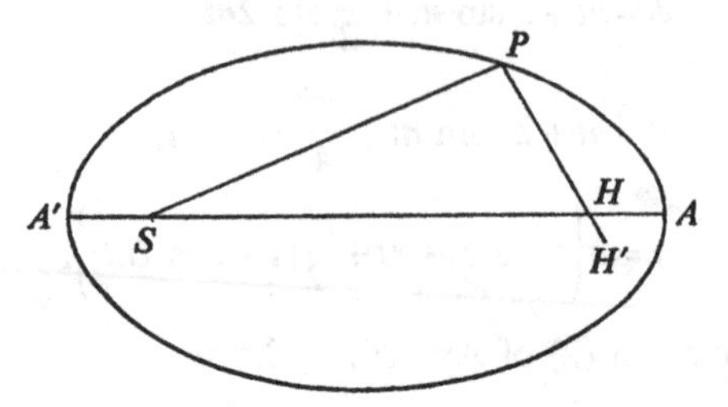

When the particle has arrived at P let its velocity be changed to $v+u$, the direction being unaltered; let $2a'$ be the new major axis.

Then we have

$$v^2=\mu\left[\frac{2}{SP}-\frac{1}{a}\right];\ (v+u)^2=\mu\left[\frac{2}{SP}-\frac{1}{a'}\right]\quad\text{.............(1).}$$

Hence, by subtraction, we have $\frac{1}{a'}$.

Since the direction of motion is unaltered at P, the new focus lies on PH; and, if H' be its position, we have

$$HH'=(H'P+SP)-(HP+SP)=2a'-2a.$$

If the change of velocity u be small and equal to δv, say, then by differentiating the first of equations (1) we have

$$2v\delta v=\frac{\mu}{a^2}\delta a.$$

[For SP is constant as far as these instantaneous changes are concerned.]

Hence δa, the increase in the semi-major axis,

$$=\frac{2v.\delta v.a^2}{\mu}\quad\text{.............................(2).}$$

Again, since HH' is now small, we have

$$\tan HSH'=\frac{HH'\sin H}{2ae+HH'\cos H}=\frac{2\delta a.\sin H}{2ae}.$$

Hence $\delta\psi$, the angle through which the major axis moves,

$$=HSH'=\frac{\delta a\sin H}{ae}=\frac{2va}{e\mu}.\sin H.\delta v\quad\text{...............(3).}$$

Since the direction of motion at P is unaltered by the blow, the value of h is altered in the ratio $\frac{v+\delta v}{v}$, so that $\delta h=\frac{\delta v}{v}h$.

But
$$h^2=\mu a(1-e^2).$$
$$\therefore 2hdh=\mu\delta a(1-e^2)-\mu a.2e\delta e.$$
$$\therefore \mu a.2e\delta e=2v\delta v.a^2(1-e^2)-2\frac{\delta v}{v}h^2,$$

so that
$$\delta e=\frac{\delta v}{v}.\frac{(1-e^2)}{e}.\frac{av^2-\mu}{\mu} \quad\text{.....................(4).}$$

This gives the increase in the value of the eccentricity.

Since the periodic time $T=\frac{2\pi}{\sqrt{\mu}}a^{\frac{3}{2}}$,

$$\therefore \frac{\delta T}{T}=\frac{3}{2}\frac{\delta a}{a}=\frac{3va\delta v}{\mu} \quad\text{.......................(5).}$$

85. If the disturbing force is not tangential, the velocity it produces must be compounded with the velocity in the orbit to give the new velocity and tangent at the point P. The equations (1) or (2) of the last article now give the magnitude, $2a'$, of the new major axis.

Also since the moment of the velocity of the point P about the focus S is equal to

$$\sqrt{\mu\times\text{semi-latus-rectum}},\ \textit{i.e.}\text{ to }\mu\sqrt{a'(1-e'^2)},$$

we obtain the new eccentricity.

Finally by drawing a line making with the new tangent at P an angle equal to that made by SP, and taking on it a point H', such that $SP+H'P$ is equal to the new major axis, we obtain the new second focus and hence the new position of the major axis of the orbit.

86. *Effect on the orbit of an instantaneous change in the value of the absolute acceleration* μ.

When the particle is at a distance r from the centre of force, let the value of μ be instantaneously changed to μ', and let the new values of the major axis and eccentricity be $2a'$ and e'.

Since the velocity is instantaneously unaltered in magnitude, we have

$$\mu\left(\frac{2}{r}-\frac{1}{a}\right)=v^2=\mu'\left(\frac{2}{r}-\frac{1}{a'}\right) \quad\text{.....................(1),}$$

an equation to give a'.

The moment of the velocity about S being unaltered, h remains the same, so that

$$\sqrt{\mu a(1-e^2)}=h=\sqrt{\mu' a'(1-e'^2)} \quad\text{..................(2)},$$

giving e'.

The direction of the velocity at distance r being unaltered, we obtain the new positions of the second focus and of the new major axis as in the previous article.

If the change $\delta\mu$ in μ be very small the change δa in a is obtained by differentiating the first equation in (1), where v and r are treated as constants, and we have

$$\frac{\delta a}{a^2}=-v^2.\frac{\delta\mu}{\mu^2}.$$

So, from (2), we have, on taking logarithmic differentials,

$$\frac{\delta\mu}{\mu}+\frac{\delta a}{a}-\frac{2e\delta e}{1-e^2}=0.$$

$$\therefore \frac{2e\delta e}{1-e^2}=\frac{\delta\mu}{\mu}-\frac{v^2 a}{\mu^2}\delta\mu=\frac{\delta\mu}{\mu}\left(1-\frac{v^2 a}{\mu}\right).$$

Again, since the periodic time $T=\dfrac{2\pi}{\sqrt{\mu}}a^{\frac{3}{2}}$,

$$\therefore \frac{dT}{T}=\frac{3}{2}\frac{\delta a}{a}-\frac{1}{2}\frac{\delta\mu}{\mu}=-\frac{1}{2}\frac{\delta\mu}{\mu}\left(1+\frac{3av^2}{\mu}\right).$$

EXAMPLES

1. If the period of a planet be 365 days and the eccentricity e is $\frac{1}{60}$, shew that the times of describing the two halves of the orbit, bounded by the latus rectum passing through the centre of force, are

$$\frac{365}{2}\left[1\pm\frac{1}{15\pi}\right] \text{ very nearly.}$$

2. The perihelion distance of a comet describing a parabolic path is $\frac{1}{n}$ of the radius of the Earth's path supposed circular; shew that the time that the comet will remain within the Earth's orbit is

$$\frac{2}{3\pi}.\frac{n+2}{n}.\sqrt{\frac{n-1}{2n}} \text{ of a year.}$$

[If S be the Sun, a the radius of the Earth's path, A the perihelion of the comet's path, and P the intersection of the paths of the earth and comet, then

$$a=SP=\frac{\frac{2a}{n}}{1+\cos\theta}, \text{ so that } \cos\theta=\frac{2}{n}-1, \text{ and therefore}$$

$$\tan\frac{\theta}{2}=\sqrt{n-1}.$$

Now use the formula of Art. 77, remembering that $\frac{2\pi}{\sqrt{\mu}}a^{\frac{3}{2}}=$ one year.]

3. The Earth's path about the Sun being assumed to be a circle, shew that the longest time that a comet, which describes a parabolic path, can remain within the Earth's orbit is $\frac{2}{3\pi}$ of a year.

4. A planet, of mass M and periodic time T, when at its greatest distance from the Sun comes into collision with a meteor of mass m, moving in the same orbit in the opposite direction with velocity v; if $\frac{m}{M}$ be small, shew that the major axis of the planet's path is reduced by

$$\frac{4m}{M}\cdot\frac{vT}{\pi}\sqrt{\frac{1-e}{1+e}}.$$

5. When a periodic comet is at its greatest distance from the Sun its velocity v is increased by a small quantity δv. Shew that the comet's least distance from the Sun is increased by the quantity $4\delta v.\left\{\frac{a^3(1-e)}{\mu(1+e)}\right\}^{\frac{1}{2}}$.

6. A small meteor, of mass m, falls into the Sun when the Earth is at the end of the minor axis of its orbit; if M be the mass of the Sun, shew that the major axis of the Earth's orbit is lessened by $2a.\frac{m}{M}$, that the periodic time is lessened by $\frac{2m}{M}$ of a year, and that the major axis of its orbit is turned through an angle $\frac{b}{ae}.\frac{m}{M}$.

7. The Earth's present orbit being taken to be circular, find what its path would be if the Sun's mass were suddenly reduced to $\frac{1}{n}$ of what it is now.

8. A comet is moving in a parabola about the Sun as focus; when at the end of its latus-rectum its velocity suddenly becomes altered in the ratio $n:1$, where $n<1$; shew that the comet will describe an ellipse whose eccentricity is $\sqrt{1-2n^2+2n^4}$, and whose major axis is $\frac{l}{1-n^2}$, where $2l$ was the latus-rectum of the parabolic path.

9. A body is moving in an ellipse about a centre of force in the focus; when it arrives at P the direction of motion is turned through a right-angle, the speed being unaltered; shew that the body will describe an ellipse whose eccentricity varies as the distance of P from the centre.

10. Two particles, of masses m_1 and m_2, moving in co-planar parabolas round the Sun, collide at right angles and coalesce when their common distance from the Sun is R. Shew that the subsequent path of the combined particles is an ellipse of major axis $\frac{(m_1+m_2)^2}{2m_1m_2}R$.

11. A particle is describing an ellipse under the action of a force to one of its foci. When the particle is at one extremity of the minor axis a blow is given to it and the subsequent orbit is a circle; find the magnitude and direction of the blow.

12. A particle m is describing an ellipse about the focus with angular momentum mh, and when at the end of the minor axis receives a small impulse mu along the radius vector to the focus. Shew that the major axis of the path is altered by $\frac{4abeu}{h}$, that the eccentricity is altered by $\frac{ua}{h}(1-e^2)^{\frac{3}{2}}$, and that the major axis is turned through the angle $\frac{au(1-e^2)}{h}$, where a, b are the semi-axes and e the eccentricity of the ellipse.

13. A particle is describing a parabolic orbit (latus-rectum $4a$) about a centre of force (μ) in the focus, and on its arriving at a distance r from the focus moving towards the vertex the centre of force ceases to act for a certain time τ. When the force begins again to operate prove that the new orbit will be an ellipse, parabola or hyperbola according as

$$\tau <=> 2r\sqrt{\frac{r-a}{2\mu}}.$$

14. Shew that the maximum range of a projectile on a horizontal plane through the point of projection is $2h.\frac{R+h}{R+2h}$, where R is the radius of the Earth, and h is the greatest height to which the projectile can be fired.

[Use the result of Art. 79.]

15. When variations of gravity and the spherical shape of the Earth are taken into account, shew that the maximum range attainable by a gun placed at the sea level is $2R\sin^{-1}\left(\frac{h}{R}\right)$, and that the necessary angle of elevation is $\frac{1}{2}\cos^{-1}\left(\frac{h}{R}\right)$, where R is the Earth's radius and h is the greatest height above the surface to which the gun can send the ball.

16. Shew that the least velocity with which a body must be projected from the Equator of the Earth so as to hit the surface again at the North Pole is about $4\frac{1}{2}$ miles per second, and that the corresponding direction of projection makes an angle of $67\frac{1}{2}°$ with the vertical at the point of projection.

CHAPTER VI

TANGENTIAL AND NORMAL ACCELERATIONS. UNIPLANAR CONSTRAINED MOTION

87. In the present chapter will be considered questions which chiefly involve motions where the particle is constrained to move in definite curves. In these cases the accelerations are often best measured along the tangent and normal to the curve. We must therefore first determine the tangential and normal accelerations in the case of any plane curve.

88. *To shew that the accelerations along the tangent and normal to the path of a particle are* $\frac{d^2s}{dt^2}\left(=v\frac{dv}{ds}\right)$ *and* $\frac{v^2}{\rho}$, *where* ρ *is the radius of curvature of the curve at the point considered.*

Let v be the velocity at time t along the tangent at any point P, whose arcual distance from a fixed point C on the path is s, and let $v+\Delta v$ be the velocity at time $t+\Delta t$ along the tangent at Q, where $PQ=\Delta s$.

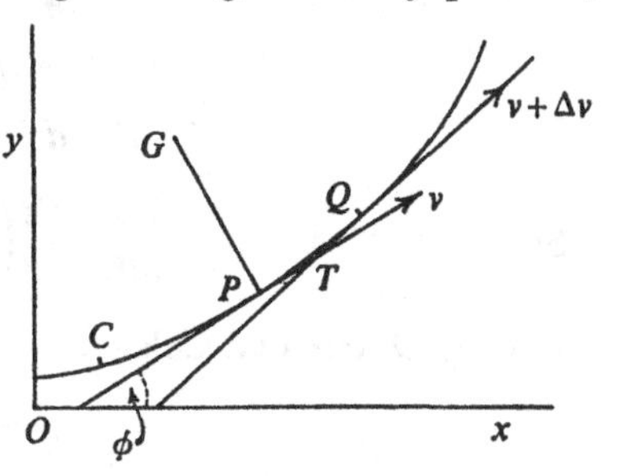

Let ϕ and $\phi+\Delta\phi$ be the angles that the tangents at P and Q make with a fixed line Ox, so that $\Delta\phi$ is the angle between the tangents at P and Q.

Then, by definition, the acceleration along the tangent at P

$$=\underset{\Delta=0}{\text{Lt}}\frac{\left[\begin{array}{c}\text{velocity along the tangent at time } t+\Delta t\\ -\text{the same at time } t\end{array}\right]}{\Delta t}$$

$$=\underset{\Delta t=0}{\text{Lt}}\frac{(v+\Delta v)\cos\Delta\phi-v}{\Delta t}$$

$$=\underset{\Delta t=0}{\text{Lt}}\frac{+\Delta v-v}{\Delta t},$$

on neglecting small quantities of the second order,

$$=\frac{dv}{dt}=\frac{d^2s}{dt^2}.$$

Also $$\frac{dv}{dt}=\frac{dv}{ds}\frac{ds}{dt}=v\frac{dv}{ds}.$$

Again the acceleration along the normal at P

$$=\underset{\Delta t=0}{\text{Lt}}\frac{\left[\begin{array}{c}\text{velocity along the normal at time } t+\Delta t\\ -\text{the same at time } t\end{array}\right]}{\Delta t}$$

$$=\underset{\Delta t=0}{\text{Lt}}\frac{(v+\Delta v)\sin\Delta\phi}{\Delta t}$$

$$=\underset{\Delta t=0}{\text{Lt}}(v+\Delta v).\frac{\sin\Delta\phi}{\Delta\phi}.\frac{\Delta\phi}{\Delta s}.\frac{\Delta s}{\Delta t}=v.1.\frac{1}{\rho}.v=\frac{v^2}{\rho}.$$

COR. In the case of a circle we have $\rho=a$, $s=a\theta$, $v=a\dot{\theta}$ and the accelerations are $a\ddot{\theta}$ and $a\dot{\theta}^2$.

89. The tangential and normal accelerations may also be directly obtained from the accelerations parallel to the axes.

For $$\frac{dx}{dt}=\frac{dx}{ds}.\frac{ds}{dt}.$$

$$\therefore\ \frac{d^2x}{dt^2}=\frac{d^2x}{ds^2}\left(\frac{ds}{dt}\right)^2+\frac{dx}{ds}\frac{d^2s}{dt^2}.$$

So $$\frac{d^2y}{dt^2}=\frac{d^2y}{ds^2}\left(\frac{ds}{dt}\right)^2+\frac{dy}{ds}\frac{d^2s}{dt^2}.$$

But, by Differential Calculus,

$$\frac{1}{\rho}=\frac{-\dfrac{d^2x}{ds^2}}{\dfrac{dy}{ds}}=\frac{\dfrac{d^2y}{ds^2}}{\dfrac{dx}{ds}},$$

$$\therefore\ \frac{d^2x}{dt^2}=-\frac{dy}{ds}.\frac{1}{\rho}.\left(\frac{ds}{dt}\right)^2+\frac{dx}{ds}\frac{d^2s}{dt^2}=-\frac{\sin\phi}{\rho}v^2+\frac{d^2s}{dt}.\cos\phi,$$

and $$\frac{d^2y}{dt^2}=\frac{dx}{ds}.\frac{1}{\rho}.\left(\frac{ds}{dt}\right)^2+\frac{dy}{ds}.\frac{d^2s}{dt^2}=\frac{\cos\phi}{\rho}.v^2+\frac{d^2s}{dt^2}\sin\phi.$$

Therefore the acceleration along the tangent

$$=\frac{d^2x}{dt^2}\cos\phi+\frac{d^2y}{dt^2}\sin\phi=\frac{d^2s}{dt^2}=\frac{dv}{dt}=\frac{dv}{ds}\frac{ds}{dt}=v\frac{dv}{ds},$$

and the acceleration along the normal $=-\dfrac{d^2x}{dt^2}\sin\phi+\dfrac{d^2y}{dt^2}\cos\phi=\dfrac{v^2}{\rho}.$

90. *Ex.* *A curve is described by a particle having a constant acceleration in a direction inclined at a constant angle to the tangent; shew that the curve is an equiangular spiral.*

Here $\frac{vdv}{ds} = f\cos\alpha$ and $\frac{v^2}{\rho} = f\sin\alpha$, where f and α are constants,

$$\therefore\ 2f\cos\alpha s + \text{const.} = v^2 = f\sin\alpha . \rho = f\sin\alpha\frac{ds}{d\psi}.$$

$$\therefore\ \frac{1}{2}\frac{ds}{d\psi} = s\cot\alpha + A, \text{ where } A \text{ is a constant.}$$

$$\therefore\ \log(s\cot\alpha + A) = 2\psi\cot\alpha + \text{const.}$$

$$\therefore\ s = -A\tan\alpha + Be^{2\psi\cot\alpha},$$

which is the intrinsic equation of an equiangular spiral.

EXAMPLES

1. Find the intrinsic equation to a curve such that, when a point moves on it with constant tangential acceleration, the magnitudes of the tangential velocity and the normal acceleration are in a constant ratio.

2. A point moves along the arc of a cycloid in such a manner that the tangent at it rotates with constant angular velocity; shew that the acceleration of the moving point is constant in magnitude.

3. A point moves in a curve so that its tangential and normal accelerations are equal and the tangent rotates with constant angular velocity; find the path.

4. If the relation between the velocity of a particle and the arc it has described be

$$2as = \log\frac{b+ac^2}{b+av^2},$$

find the tangential force acting on the particle and the time that must elapse from the beginning of the motion till the velocity has the value V.

5. Shew that a cycloid can be a free path for a particle acted on at each point by a constant force parallel to the corresponding radius of the generating circle, this circle being placed at the vertex.

6. A heavy particle lying in limiting equilibrium on a rough plane, inclined at an angle α to the horizontal, is projected with velocity V horizontally along the plane; shew that the limiting velocity is $\frac{1}{2}V$ and find the intrinsic equation to the path.

7. A circle rolls on a straight line, the velocity of its centre at any instant being v and its acceleration f; find the tangential and normal accelerations of a point on the edge of the circle whose angular distance from the point of contact is θ.

91. *A particle is compelled to move on a given smooth plane curve under the action of given forces in the plane; to find the motion.*

Let P be a point of the curve whose arcual distance from a fixed point C is s, and let v be the velocity at P. Let X, Y be the components parallel to two rectangular axes Ox, Oy of the forces acting on the particle when at P; since the curve is smooth the only reaction will be a force R along the normal at P.

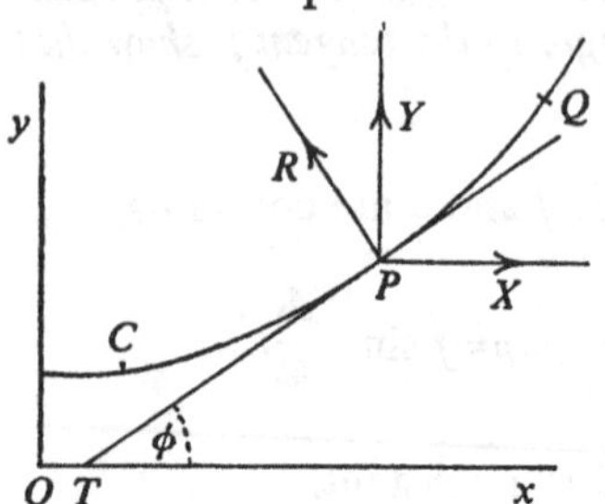

Resolving along the tangent and normal, we have

$$m\frac{vdv}{ds}=\text{force along } TP=X\cos\phi+Y\sin\phi$$

$$=X\frac{dx}{ds}+Y\frac{dy}{ds} \quad\text{..........(1)},$$

and

$$m.\frac{v^2}{\rho}=-X\sin\phi+Y\cos\phi+R$$

$$=-X\frac{dy}{ds}+Y\frac{dx}{ds}+R \quad\text{..........(2)}.$$

When v is known, equation (2) gives R at any point.

Equation (1) gives

$$\frac{1}{2}mv^2=\int(Xdx+Ydy) \quad\text{..........(3)}.$$

Suppose that $Xdx+Ydy$ is the complete differential of some function $\phi(x, y)$, so that $X=\frac{d\phi}{dx}$ and $Y=\frac{d\phi}{dy}$.

Then

$$\frac{1}{2}mv^2=\int\left(\frac{d\phi}{dx}dx+\frac{d\phi}{dy}dy\right)=\phi(x, y)+C \quad\text{..........(4)}.$$

Suppose that the particle started with a velocity V from a point whose coordinates are x_0, y_0. Then

$$\tfrac{1}{2}mV^2=\phi(x_0, y_0)+C.$$

Hence, by subtraction,

$$\tfrac{1}{2}mv^2-\tfrac{1}{2}mV^2=\phi(x, y)-\phi(x_0, y_0) \quad\text{..........(5)}.$$

This result is quite independent of the path pursued between the initial point and P, and would therefore be the same whatever be the form of the restraining curve.

From the definition of Work it is clear that $Xdx+Ydy$ represents the work done by the forces X, Y during a small displacement ds along the curve. Hence the right-hand side of (3) or of (4) represents the total work done on the particle by the external forces, during its motion from the point of projection to P, added to an arbitrary constant.

Hence, when the components of the forces are equal to the differentials with respect to x and y of some function $\phi(x, y)$, it follows from (5) that

The change in the Kinetic Energy of the particle

= the Work done by the External Forces.

Forces of this kind are called Conservative Forces.

The quantity $\phi(x, y)$ is known as the Work-Function of the system of forces. From the ordinary definition of a Potential Function, it is clear that $\phi(x, y)$ is equal to the Potential of the given system of forces added to some constant.

If the motion be in three dimensions we have, similarly, that the forces are Conservative when $\int(Xdx+Ydy+Zdz)$ is a perfect differential, and an equation similar to (5) will also be true. [See Art. 131.]

92. The Potential Energy of the particle, due to the given system of forces, when it is in the position P

= the work done by the forces as the particle moves to some standard position.

Let the latter position be the point (x_1, y_1). Then the potential energy of the particle at P

$$=\int_{(x,\,y)}^{(x_1,\,y_1)}(Xdx+Ydy)=\int_{(x,\,y)}^{(x_1,\,y_1)}\left(\frac{d\phi}{dx}dx+\frac{d\phi}{dy}dy\right)$$

$$=\Big[\phi(x, y)\Big]_{(x,\,y)}^{(x_1,\,y_1)}=\phi(x_1, y_1)-\phi(x, y).$$

Hence, from equation (4) of the last article,

(Kinetic Energy + Potential Energy) of the particle when at P

$$=\phi(x; y)+C+\phi(x_1, y_1)-\phi(x, y)$$

$$=C+\phi(x_1, y_1)=\text{a constant.}$$

Hence, when a particle moves under the action of a Conservative System of Forces, the sum of its Kinetic and Potential Energies is constant throughout the motion.

93. In the particular case when gravity is the only force acting we have, if the axis of y be vertical, $X=0$ and $Y=-mg$.

Equation (3) then gives $\frac{1}{2}mv^2=-mgy+C$.

Hence, if Q be a point of the path, this gives

kinetic energy at P − kinetic energy at Q

= mg × difference of the ordinates at P and Q

= the work done by gravity as the particle passes from Q to P.

This result is important; from it, given the kinetic energy at any known point of the curve, we have the kinetic energy at any other point of the path, if the curve be smooth.

94. If the only forces acting on a particle be perpendicular to its direction of motion (as in the case of a particle tethered by an inextensible string, or moving on a smooth surface) its velocity is constant; for the work done by the string or reaction is zero.

95. *All forces which are one-valued functions of distances from fixed points are Conservative Forces.*

Let a force acting on a particle at the point (x, y) be a function $\psi(r)$ of the distance r from a fixed point (a, b) so that

$$r^2 = (x-a)^2 + (y-b)^2.$$

Also let the force act towards the point (a, b).

Then
$$r\frac{dr}{dx} = (x-a), \text{ and } r\frac{dr}{dy} = y-b.$$

The component X of this force parallel to the axis of x

$$= -\psi(r) \times \frac{x-a}{r},$$

if the force be an attraction, and the component Y parallel to y

$$= -\psi(r) \times \frac{y-b}{r}.$$

Hence
$$Xdx + Ydy = -\psi(r) \times \frac{(x-a)dx + (y-b)dy}{r}$$

$$= -\psi(r)\frac{rdr}{r} = -\psi(r)dr.$$

Hence, if $F(r)$ be such that $\frac{d}{dr}F(r) = -\psi(r)$ ....(1),

we have

$$\int (Xdx + Ydy) = \int \frac{d}{dr}F(r)dr = F(r) + \text{const.}$$

Such a force therefore satisfies the condition of being a Conservative Force.

If the force be a central one and follow the law of the inverse square, so that $\psi(r)=\frac{\mu}{r^2}$, then $F(r)=-\int\psi(r)dr=\frac{\mu}{r}$ and hence

$$\int(Xdx+Ydy)=\frac{\mu}{r}+\text{constant}.$$

96. *The work done in stretching an elastic string is equal to the extension produced multiplied by the mean of the initial and final tensions.*

Let a be the unstretched length of the string, and λ its modulus of elasticity, so that, when its length is x, its tension

$$=\lambda.\frac{x-a}{a}, \text{ by Hooke's Law.}$$

The work done in stretching it from a length b to a length c

$$=\int_b^c T.dx=\int_b^c \lambda\frac{x-a}{a}dx=\frac{\lambda}{2a}\Big[(x-a)^2\Big]_b^c$$

$$=\frac{\lambda}{2a}[(c-a)^2-(b-a)^2]=(c-b)\left[\lambda\frac{b-a}{a}+\lambda\frac{c-a}{a}\right]\times\tfrac{1}{2}$$

$$=(c-b)\times\text{mean of the initial and final tensions.}$$

Ex. A and B are two points in the same horizontal plane at a distance 2a apart; AB is an elastic string whose unstretched length is 2a. To O, the middle point of AB, is attached a particle of mass m which is allowed to fall under gravity; find its velocity when it has fallen a distance x and the greatest vertical distance through which it moves.

When the particle is at P, where $OP=x$, let its velocity be v, so that its kinetic energy then is $\frac{1}{2}mv^2$.

The work done by gravity $=mg.x$.

The work done against the tension of the string

$$=2\times(BP-BO)\times\frac{1}{2}\lambda\frac{BP-BO}{a}=\frac{\lambda}{a}(BP-a)^2=\frac{\lambda}{a}[\sqrt{x^2+a^2}-a]^2.$$

Hence, by the Principle of Energy,

$$\tfrac{1}{2}mv^2=mgx-\frac{\lambda}{a}[\sqrt{x^2+a^2}-a]^2.$$

The particle comes to rest when $v=0$, and then x is given by the equation

$$mgxa=\lambda[\sqrt{x^2+a^2}-a]^2.$$

EXAMPLES

1. If an elastic string, whose natural length is that of a uniform rod, be attached to the rod at both ends and suspended by the middle point, shew by means of the Principle of Energy, that the rod will sink until the strings are inclined to the horizon at an angle θ given by the equation

$$\cot^3\frac{\theta}{2}-\cot\frac{\theta}{2}=2n,$$

given that the modulus of elasticity of the string is n times the weight of the rod.

2. A heavy ring, of mass m, slides on a smooth vertical rod and is attached to a light string which passes over a small pulley distant a from the rod and has a mass M ($>m$) fastened to its other end. Shew that, if the ring be dropped from a point in the rod in the same horizontal plane as the pulley, it will descend a distance $\frac{2Mma}{M^2-m^2}$ before coming to rest.

Find the velocity of m when it has fallen through any distance x.

3. A shell of mass M is moving with velocity V. An internal explosion generates an amount of energy E and breaks the shell into two portions whose masses are in the ratio $m_1 : m_2$. The fragments continue to move in the original line of motion of the shell. Shew that their velocities are $V+\sqrt{\frac{2m_2E}{m_1M}}$ and $V-\sqrt{\frac{2m_1E}{m_2M}}$.

4. An endless elastic string, of natual length $2\pi a$, lies on a smooth horizontal table in a circle of radius a. The string is suddenly set in motion about its centre with angular velcoity ω. Shew that if left to itself the string will expand and that, when its radius is r, its angular velocity is $\frac{a^2}{r^2}\omega$, and the square of its radial velocity from the centre is $\frac{a^2\omega^2}{r^2}(r^2-a^2)-\frac{2\pi\lambda(r-a)^2}{ma}$, where m is the mass and λ the modulus of elasticity of the string.

5. Four equal particles are connected by strings, which form the sides of a square, and repel one another with a force equal to $\mu\times$distance; if one string be cut, shew that, when either string makes an angle θ with its original position, its angular velocity is $\sqrt{\frac{4\mu\sin\theta\,(2+\sin\theta)}{2-\sin^2\theta}}$.

[As in Art 47 the centre of mass of the whole system remains at rest; also the repulsion, by the well-known property, on each particle is the same as if the whole of the four particles were collected at the centre and $=4\mu\times$distance from the fixed centre of mass. Equate the total kinetic energy to the total work done by the repulsion.]

6. A uniform string, of mass M and length $2a$, is placed symmetrically over a smooth peg and has particles of masses m and m' attached to its extremities; shew that when the string runs off the peg its velocity is

$$\sqrt{\frac{M+2(m-m')}{M+m+m'}ag}.$$

7. A heavy uniform chain, of length $2l$, hangs over a small smooth fixed pulley, the length $l+c$ being at one side and $l-c$ at the other; if the end of the shorter portion be held, and then let go, shew that the chain will slip off the pulley in time $\left(\frac{l}{g}\right)^{\frac{1}{2}}\log\frac{l+\sqrt{l^2-c^2}}{c}$.

8. A uniform chain, of length l and weight W, is placed on a line of greatest slope of a smooth plane, whose inclination to the horizontal is α, and just reaches the bottom of the plane where there is a small smooth pulley over which it can run. Shew that, when a length x has run off, the tension at the bottom of the plane is $W(1-\sin\alpha)\frac{x(l-x)}{l^2}$.

9. Over a small smooth pulley is placed a uniform flexible cord; the latter is initially at rest and lengths $l-a$ and $l+a$ hang down on the two sides. The pulley is now made to move with constant upward acceleration f. Shew that the string will leave the pulley after a time

$$\sqrt{\frac{l}{f+g}}\cosh^{-1}\frac{l}{a}.$$

97. Oscillations of a Simple Pendulum.

A particle m is attached by a light string, of length l, to a fixed point and oscillates under gravity through a small angle; to find the period of its motion.

When the string makes an angle θ with the vertical, the equation of motion is

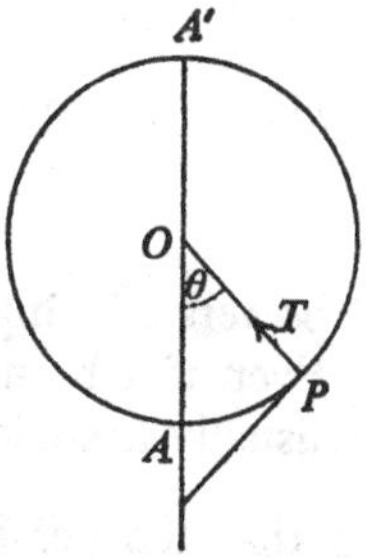

$$m\frac{d^2s}{dt^2}=-mg\sin\theta \quad\text{.......................(1).}$$

But $s=l\theta$.

$$\therefore\ \ddot{\theta}=-\frac{g}{l}\sin\theta=-\frac{g}{l}\theta,\ \text{to a first approximation.}$$

If the pendulum swings through a small angle α on each side of the vertical, so that $\theta=\alpha$ and $\dot{\theta}=0$ when $t=0$, this equation gives

$$\theta=\alpha\cos\left[\sqrt{\frac{g}{l}}t\right],$$

so that the motion is simple harmonic and the time, T_1, of a very small oscillation $=2\pi\sqrt{\frac{l}{g}}$, as in Art. 22.

For a higher approximation we have, from equation (1),

$$l\dot{\theta}^2=2g(\cos\theta-\cos\alpha)\quad\text{.......................(2),}$$

since $\dot{\theta}$ is zero when $\theta=\alpha$.

[This equation follows at once from the Principle of Energy.]

$$\therefore\ \sqrt{\frac{2g}{l}}.t=\int_0^\alpha\frac{d\theta}{\sqrt{\cos\theta-\cos\alpha}},$$

where t is the time of a quarter-swing.

$$\therefore\ 2\sqrt{\frac{g}{l}}t=\int_0^\alpha\frac{d\theta}{\sqrt{\sin^2\frac{\alpha}{2}-\sin^2\frac{\theta}{2}}}.$$

A′ O θ T P A

Put $$\sin\frac{\theta}{2}=\sin\frac{\alpha}{2}.\sin\phi.$$

$$\therefore\ 2\sqrt{\frac{g}{l}}.t=\int_0^{\frac{\pi}{2}}\frac{2\sin\frac{\alpha}{2}\cos\phi d\phi}{\cos\frac{\theta}{2}.\sin\frac{\alpha}{2}\cos\phi}.$$

$$\therefore\quad t=\sqrt{\frac{l}{g}}.\int_0^{\frac{\pi}{2}}\frac{d\phi}{\left(1-\sin^2\frac{\alpha}{2}\sin^2\phi\right)^{\frac{1}{2}}}\quad\text{.............................(3)}$$

$$=\sqrt{\frac{l}{g}}\int_0^{\frac{\pi}{2}}\left[1+\frac{1}{2}\sin^2\frac{\alpha}{2}.\sin^2\phi+\frac{1.3}{2.4}\sin^4\frac{\alpha}{2}\sin^4\phi+\ldots\right]d\phi$$

$$=\sqrt{\frac{l}{g}}.\frac{\pi}{2}\left[1+\frac{1}{2^2}\sin^2\frac{\alpha}{2}+\left(\frac{1.3}{2.4}\right)^2\sin^4\frac{\alpha}{2}\right.$$
$$\left.+\left(\frac{1.3.5}{2.4.6}\right)^2\sin^6\frac{\alpha}{2}+\ldots\right]\quad\text{......(4).}$$

Hence a second approximation to the required period, T_2,

$$=T_1\left[1+.\frac{1}{4}.\sin^2\frac{\alpha}{2}\right]=T_1\left[1+\frac{\alpha^2}{16}\right],$$

if powers of α higher than the second are neglected.

Even if α be not very small, the second term in the bracket of (4) is usually a sufficient approximation. For example, suppose $\alpha=30°$, so that the pendulum swings through an angle of 60°; then $\sin^2\frac{\alpha}{2}=\sin^2 15°=\cdot067$, and (4) gives

$$t=\frac{\pi}{2}\sqrt{\frac{l}{g}}[1+\cdot017+\cdot00063+\ldots].$$

[The student who is acquainted with Elliptic Functions will see that (3) gives

$$\sin\phi=\text{sn}\left(t\sqrt{\frac{g}{l}}\right),\ \left(\text{mod. }\sin\frac{\alpha}{2}\right),$$

so that $$\sin\frac{\theta}{2}=\sin\frac{\alpha}{2}\ \text{sn}\left(t\sqrt{\frac{g}{l}}\right),\ \left(\text{mod. }\sin\frac{\alpha}{2}\right).$$

The time of a complete oscillation is also, by (3), equal to $\sqrt{\frac{l}{g}}$ multiplied by the real period of the elliptic function with modulus $\sin\frac{\alpha}{2}$.]

98. The equations (1) and (2) of the previous article give the motion in a circle in any case, when α is not necessarily small. If ω be the angular velocity of the particle when passing through the lowest point A, we have

$$l\dot{\theta}^2 = 2g\cos\theta + \text{const.} = l\omega^2 - 2g(1-\cos\theta) \quad \text{.........(5).}$$

This equation cannot in general be integrated without the use of Elliptic Functions, which are beyond the scope of this book.

If T be the tension of the string, we have

$$T - mg\cos\theta = \text{force along the normal } PO$$
$$= ml\dot{\theta}^2 = ml\omega^2 - 2mg(1-\cos\theta),$$
$$\therefore\ T = m\{l\omega^2 - g(2-3\cos\theta)\} \quad \text{..................(6).}$$

Hence T vanishes and becomes negative, and hence circular motion ceases, when $\cos\theta = \dfrac{2g - l\omega^2}{3g}$.

Particular Case. Let the angular velocity at A be that due to a fall from the highest point A', so that

$$l^2\omega^2 = 2g.2l, \text{ i.e. } \omega^2 = \frac{4g}{l}.$$

Then (5) gives $\qquad \dot{\theta}^2 = \dfrac{2g}{l}(1+\cos\theta).$

$$\therefore\ t\sqrt{\frac{2g}{l}} = \int \frac{d\theta}{\sqrt{1+\cos\theta}} = \frac{1}{\sqrt{2}}\int \frac{d\theta}{\cos\frac{\theta}{2}}.$$

$$\therefore\ t = \frac{1}{2}\sqrt{\frac{l}{g}}\left[2\log\tan\left(\frac{\pi}{4}+\frac{\theta}{4}\right)\right]_0^\theta$$

$$= \sqrt{\frac{l}{g}}\log\frac{\cos\frac{\theta}{4}+\sin\frac{\theta}{4}}{\cos\frac{\theta}{4}-\sin\frac{\theta}{4}} = \sqrt{\frac{l}{g}}\log\frac{1+\sin\frac{\theta}{2}}{\cos\frac{\theta}{2}}$$

$$= \sqrt{\frac{l}{g}}\log\left[\sec\frac{\theta}{2}+\tan\frac{\theta}{2}\right],$$

giving the time t of describing an angle θ from the lowest point.

Also in this case

$$T = m\{4g - 2g + 3g\cos\theta\} = mg[2 + 3\cos\theta].$$

Circular motion therefore ceases when $\cos\theta=-\frac{2}{3}$, and then $\sec\frac{\theta}{2}=\sqrt{6}$ and $\tan\frac{\theta}{2}=\sqrt{5}$. Therefore the time during which the circular motion lasts

$$=\sqrt{\frac{l}{g}}\log_e(\sqrt{5}+\sqrt{6}).$$

99. *Ex.* 1. Shew that a pendulum, which beats seconds when it swings through 3° on each side of the vertical, will lose about 12 secs. per day if the angle be 4° and about 27 secs. per day if the angle be 5°.

Ex. 2. A heavy bead slides on a smooth fixed vertical circular wire of radius a; if it be projected from the lowest point with velocity just sufficient to carry it to the highest point, shew that the radius to the bead is at time t inclined to the vertical at an angle $2\tan^{-1}\left[\sinh\sqrt{\frac{g}{a}}t\right]$, and that the bead will be an infinite time in arriving at the highest point.

100. *Motion on a smooth cycloid whose axis is vertical and vertex lowest.*

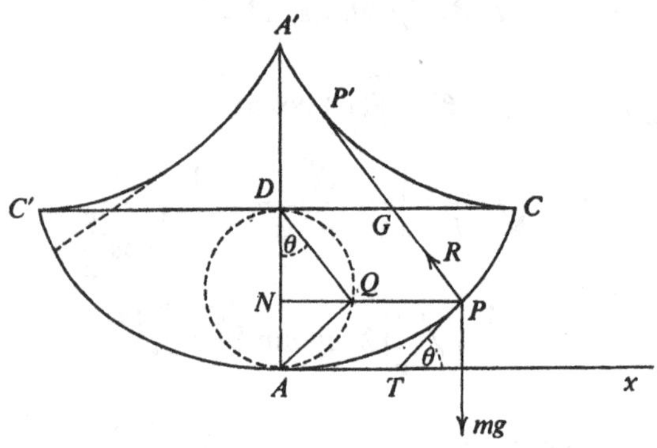

Let AQD be the generating circle of the cycloid $CPAC'$, P being any point on it; let PT be the tangent at P and PQN perpendicular to the axis meeting the generating circle in Q. The two principal properties of the cycloid are that the tangent TP is parallel to AQ, and that the arc AP is equal to twice the line AQ.

Hence, if PTx be θ, we have

$$\theta=\angle QAx=ADQ,$$

and
$$s=\text{arc } AP=2.AQ=4a\sin\theta \qquad (1),$$

if a be the radius of the generating circle.

If R be the reaction of the curve along the normal, and m the particle at P, the equations of motion are then

$$m\frac{d^2s}{dt^2}=\text{force along } PT=-mg\sin\theta \quad \text{..............(2)},$$

and $$m.\frac{v^2}{\rho}=\text{force along the normal}=R-mg\cos\theta \quad \text{......(3)}.$$

From (1) and (2), we then have

$$\frac{d^2s}{dt^2}=-\frac{g}{4a}s \quad \text{..............................(4)},$$

so that the motion is simple harmonic, and hence, as in Art. 22, the time to the lowest point

$$=\frac{\frac{\pi}{2}}{\sqrt{\frac{g}{4a}}}=\pi\sqrt{\frac{a}{g}},$$

and is therefore always the same whatever be the point of the curve at which the particle started from rest.

Integrating equation (4), we have

$$v^2=\left(\frac{ds}{dt}\right)^2=-\frac{g}{4a}s^2+C=-g.4a\sin^2\theta+C$$

$$=4ag(\sin^2\theta_0-\sin^2\theta),$$

if the particle started from rest at the point where $\theta=\theta_0$.

[This equation can be written down at once by the Principle of Energy.]

Also $$\rho=\frac{ds}{d\theta}=4a\cos\theta.$$

Therefore (3) gives

$$R=mg\cos\theta+mg\frac{\sin^2\theta_0-\sin^2\theta}{\cos\theta}=mg\frac{\cos 2\theta+\sin^2\theta_0}{\cos\theta},$$

giving the reaction of the curve at any point of the path.

On passing the lowest point the particle ascends the other side until it is at the height from which it started, and thus it oscillates backwards and forwards.

101. The property proved in the previous article will be still true if, instead of the material curve, we substitute a string tied to the particle in such a way that the particle describes a cycloid and the string is always normal to the curve. This will be the case if the string unwraps and wraps itself on the evolute of the cycloid. It can be easily shewn that the evolute of a cycloid is two halves of an equal cycloid.

For, since $\rho=4a\cos\theta$, the points on the evolute corresponding to A and C are A', where $AD=DA'$, and C itself. Let the normal PG meet this evolute in P', and let the arc CP' be σ. By the property of the evolute

$$\sigma=\text{arc } P'C=P'P, \text{ the radius of curvature at } P,$$
$$=4a\cos\theta=4a\sin P'GD.$$

Hence, by (1) of the last article, the curve is a similar cycloid whose vertex is at C and whose axis is vertical. This holds for the arc CA. The evolute for the arc $C'A$ is the similar semi-cycloid $C'A'$.

Hence, if a string, or flexible wire, of length equal to the arc CA', *i.e.* $4a$, be attached at A' and allowed to wind and unwind itself upon fixed metal cheeks in the form of the curve $CA'C'$, a particle P attached to its other end will describe the cycloid CAC', and the string will always be normal to the curve CAC'; the times of oscillation will therefore be always isochronous, whatever be the angle through which the string oscillates. In actual practice, a pendulum is only required to swing through a small angle, so that only small portions of the two arcs near A' are required. This arrangement is often adopted in the case of the pendulum of a small clock, the upper end of the supporting wire consisting of a thin flat spring which coils and uncoils itself from the two metal cheeks at A'.

102. Motion on a rough curve under gravity.

Whatever be the curve described under gravity with friction, we have, if ϕ be the angle measured from the horizontal made by the tangent, and if s increases with ϕ

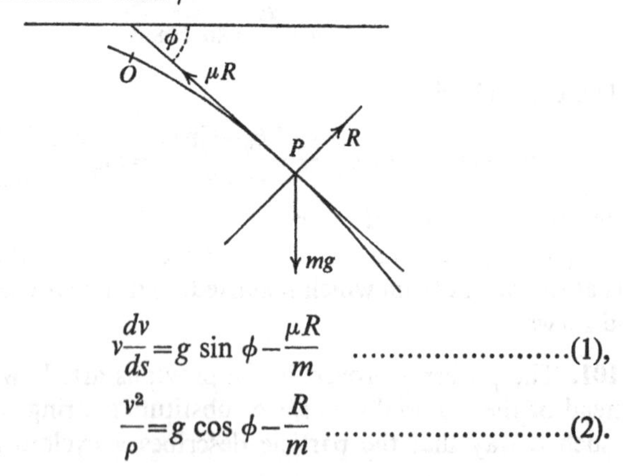

$$v\frac{dv}{ds}=g\sin\phi-\frac{\mu R}{m} \quad \text{.......................(1),}$$

and

$$\frac{v^2}{\rho}=g\cos\phi-\frac{R}{m} \quad \text{..........................(2).}$$

$$\therefore \frac{1}{2}\frac{d.v^2}{ds} - \mu\frac{v^2}{\rho} = g(\sin\phi - \mu\cos\phi).$$

$$\therefore \frac{dv^2}{d\phi} - 2\mu v^2 = 2g\rho(\sin\phi - \mu\cos\phi).$$

Multiplying by $e^{-2\mu\phi}$ and integrating, we have

$$v^2 e^{-2\mu\phi} = 2g\int \rho e^{-2\mu\phi}(\sin\phi - \mu\cos\phi) + \text{constant}.$$

When the curve is given, so that ρ is known in terms of ϕ, this gives v^2, and hence $\left(\frac{ds}{d\phi}\right)^2\left(\frac{d\phi}{dt}\right)^2$. Hence $\frac{d\phi}{dt}$ is known, and therefore theoretically t in terms of ϕ.

103. *If the cycloid of Art. 100 be rough with a coefficient of friction μ, to find the motion, the particle sliding downwards.*

In this case the friction, μR, acts in the direction TP produced. Since $s = 4a\sin\theta$, we have $\rho = 4a\cos\theta$, and $v = 4a\cos\theta.\dot\theta$, so that the equations of motion are

$$m\frac{d}{dt}(4a\cos\theta.\dot\theta) = \mu R - mg\sin\theta \quad \text{..............(1)},$$

and

$$mv^2/\rho = m.4a\cos\theta.\dot\theta^2 = R - mg\cos\theta \quad \text{............(2)}.$$

$$\therefore \frac{d}{dt}(\dot\theta\cos\theta) - \mu\cos\theta.\dot\theta^2 = -\frac{g}{4a}(\sin\theta - \mu\cos\theta),$$

i.e.

$$\frac{d}{dt}[\dot\theta\cos\theta e^{-\mu\theta}] = -\frac{g}{4a}(\sin\theta - \mu\cos\theta)e^{-\mu\theta} \text{...........(3)}.$$

Now

$$\frac{d}{dt}[e^{-\mu\theta}(\sin\theta - \mu\cos\theta)] = (1+\mu^2)e^{-\mu\theta}\cos\theta.\dot\theta.$$

Hence (3) gives

$$\frac{d^2}{dt^2}[e^{-\mu\theta}(\sin\theta - \mu\cos\theta)] = -(1+\mu^2)\frac{g}{4a}[e^{-\mu\theta}(\sin\theta - \mu\cos\theta)].$$

$$\therefore e^{-\mu\theta}(\sin\theta - \mu\cos\theta) = A\cos\left[\sqrt{\frac{g(1+\mu^2)}{4a}}t + B\right] \quad \text{...(4)},$$

where A and B are constants depending on the initial conditions.

Differentiating (4), we obtain

$$v^2 = 16a^2\cos^2\theta.\dot\theta^2 = \frac{4ag}{1+\mu^2}[A^2e^{2\mu\theta} - (\sin\theta - \mu\cos\theta)^2].$$

EXAMPLES ON CHAPTER VI

1. A particle slides down the smooth curve $y=a\sinh\frac{x}{a}$, the axis of x being horizontal and the axis of y downwards, starting from rest at the point where the tangent is inclined at α to the horizon; shew that it will leave the curve when it has fallen through a vertical distance $a\sec\alpha$.

2. A particle descends a smooth curve under the action of gravity, describing equal vertical distances in equal times, and starting in a vertical direction. Shew that the curve is a semi-cubical parabola, the tangent at the cusp of which is vertical.

3. A particle is projected with velocity V from the cusp of a smooth inverted cycloid down the arc; shew that the time of reaching the vertex is $2\sqrt{\frac{a}{g}}\tan^{-1}\left[\frac{\sqrt{4ag}}{V}\right]$.

4. A particle slides down the arc of a smooth cycloid whose axis is vertical and vertex lowest; prove that the time occupied in falling down the first half of the vertical height is equal to the time of falling down the second half.

5. A particle is placed very close to the vertex of a smooth cycloid whose axis is vertical and vertex upwards, and is allowed to run down the curve. Shew that it leaves the curve when it is moving in a direction making with the horizontal an angle of 45°.

6. A ring is strung on a smooth closed wire which is in the shape of two equal cycloids joined cusp to cusp, in the same plane and symmetrically situated with respect to the line of cusps. The plane of the wire is vertical, the line of cusps horizontal, and the radius of the generating circle is a. The ring starts from the highest point with velocity v. Prove that the times from the upper vertex to the cusp, and from the cusp to the lower vertex are respectively

$$2\sqrt{\frac{a}{g}}\sinh^{-1}\left(\frac{\sqrt{4ag}}{v}\right) \text{ and } 2\sqrt{\frac{a}{g}}\sin^{-1}\sqrt{\frac{4ag}{v^2+8ag}}.$$

7. A particle moves in a smooth tube in the form of a catenary, being attracted to the directrix by a force proportional to the distance from it. Shew that the motion is simple harmonic.

8. A particle, of mass m, moves in a smooth circular tube, of radius a, under the action of a force, equal to $m\mu\times$distance, to a point inside the tube at a distance c from its centre; if the particle be placed very nearly at its greatest distance from the centre of force, shew that it will describe the quadrant ending at its least distance in time

$$\sqrt{\frac{a}{\mu c}}\log(\sqrt{2}+1).$$

9. A bead is constrained to move on a smooth wire in the form of an equiangular spiral. It is attracted to the pole of the spiral by a force, $=m\mu(\text{distance})^{-2}$, and starts from rest at a distance b from the pole. Shew that, if the equation to the spiral be $r=ae^{\theta\cot\alpha}$, the time of arriving at the pole is $\frac{\pi}{2}\sqrt{\frac{b^3}{2\mu}}.\sec\alpha$.

Find also the reaction of the curve at any instant.

10. A smooth parabolic tube is placed, vertex downwards, in a vertical plane; a particle slides down the tube from rest under the influence of gravity; prove that in any position the reaction of the tube is $2w\frac{h+a}{\rho}$, where w is the weight of the particle, ρ the radius of curvature, $4a$ the latus rectum, and h the original vertical height of the particle above the vertex.

11. From the lowest point of a smooth hollow cylinder whose cross-section is an ellipse, of major axis $2a$ and minor axis $2b$, and whose minor axis is vertical, a particle is projected from the lowest point in a vertical plane perpendicular to the axis of the cylinder; shew that it will leave the cylinder if the velocity of projection lie between

$$\sqrt{2gb} \quad \text{and} \quad \sqrt{g\frac{a^2+4b^2}{b}}.$$

12. A small bead, of mass m, moves on a smooth circular wire, being acted upon by a central attraction $\frac{m\mu}{(\text{distance})^2}$ to a point within the circle situated at a distance b from its centre. Shew that, in order that the bead may move completely round the circle, its velocity at the point of the wire nearest the centre of force must not be less than $\sqrt{\frac{4\mu b}{a^2-b^2}}$.

13. A small bead moves on a thin elliptic wire under a force to the focus equal to $\frac{\mu}{r^2}+\frac{\lambda}{r^3}$. It is projected from a point on the wire distant R from the focus with the velocity which would cause it to describe the ellipse freely under a force $\frac{\mu}{r^2}$. Shew that the reaction of the wire is

$$\frac{\lambda}{\rho}\left[\frac{1}{r^2}-\frac{1}{ar}+\frac{1}{R^2}\right],$$

where ρ is the radius of curvature.

14. If a particle is made to describe a curve in the form of the four-cusped hypocycloid $x^{\frac{2}{3}}+y^{\frac{2}{3}}=a^{\frac{2}{3}}$ under the action of an attraction perpendicular to the axis and varying as the cube root of the distance from it, shew that the time of descent from any point to the axis of x is the same, *i.e.* that the curve is a Tautochrone for this law of force.

15. A small bead moves on a smooth wire in the form of an epicycloid, being acted upon by a force, varying as the distance, from the centre of the epicycloid; shew that its oscillations are always isochronous. Shew that the same is true if the curve be a hypocycloid and the force always towards, instead of from, the centre.

16. A curve in a vertical plane is such that the time of describing any arc, measured from a fixed point O, is equal to the time of sliding down the chord of the arc; shew that the curve is a lemniscate of Bernouilli, whose node is at O and whose axis is inclined at 45° to the vertical.

17. A particle is projected along the inner surface of a rough sphere and is acted on by no forces; shew that it will return to the point of projection at the end of time $\frac{a}{\mu V}(e^{2\mu\pi}-1)$, where a is the radius of the sphere, V is the velocity of projection and μ is the coefficient of friction.

18. A bead slides down a rough circular wire, which is in a vertical plane, starting from rest at the end of a horizontal diameter. When it has described an angle θ about the centre, shew that the square of its angular velocity is

$$\frac{2g}{a(1+4\mu^2)}[(1-2\mu^2)\sin\theta+3\mu(\cos\theta-e^{-2\mu\theta})],$$

where μ is the coefficient of friction and a the radius of the rod.

19. A particle falls from a position of limiting equilibrium near the top of a nearly smooth glass sphere. Shew that it will leave the sphere at the point whose radius is inclined to the vertical at an angle

$$\alpha+\mu\left\{2-\frac{4}{3}\frac{\alpha}{\sin\alpha}\right\},$$

where $\cos\alpha=\frac{2}{3}$, and μ is the small coefficient of friction.

20. A particle is projected horizontally from the lowest point of a rough sphere of radius a. After describing an arc less than a quadrant it returns and comes to rest at the lowest point. Shew that the initial velocity must be $\sin\alpha\sqrt{2ga\frac{1+\mu^2}{1-2\mu^2}}$, where μ is the coefficient of friction and $a\alpha$ is the arc through which the particle moves.

21. The base of a rough cycloidal arc is horizontal and its vertex downwards; a bead slides along it starting from rest at the cusp and coming to rest at the vertex. Shew that $\mu^2 e^{\mu\pi}=1$.

22. A particle slides in a vertical plane down a rough cycloidal arc whose axis is vertical and vertex downwards, starting from a point where the tangent makes an angle θ with the horizon and coming to rest at the vertex. Shew that $\mu e^{\mu\theta}=\sin\theta-\mu\cos\theta$.

23. A rough cycloid has its plane vertical and the line joining its cusps horizontal. A heavy particle slides down the curve from rest at a cusp and comes to rest again at the point on the other side of the vertex where the tangent is inclined at 45° to the vertical. Shew that the coefficient of friction satisfies the equation

$$3\mu\pi+4\log_e(1+\mu)=2\log_e 2.$$

24. A bead moves along a rough curved wire which is such that it changes its direction of motion with constant angular velocity. Shew that a possible form of the wire is an equiangular spiral.

25. *A particle is held at the lowest point of a catenary, whose axis is vertical, and is attached to a string which lies along the catenary but is free to unwind from it. If the particle be released, shew that the time that elapses before it is moving at an angle ϕ to the vertical is*

$$\sqrt{\frac{c}{2g}}\log\left(\frac{1+\sqrt{2}\sin\frac{\phi}{2}}{1-\sqrt{2}\sin\frac{\phi}{2}}\right),$$

and that its velocity then is $2\sqrt{gc}\sin\frac{\phi}{2}$, where c is the parameter of the catenary. Find also the tension of the string in terms of ϕ.

At time t, let the string PQ be inclined at an angle ϕ to the horizontal, where P is the particle and Q the point where the string touches the catenary. A being the lowest point, let

$$s = \text{arc } AQ = \text{line } PQ.$$

The velocity of P along QP = vel. of Q along the tangent + the vel. of P relative to Q

$$= (-\dot{s}) + \dot{s} = 0 \quad \text{.............................(1).}$$

The velocity of P perpendicular to QP similarly

$$= s.\dot{\phi} \quad \text{.............................(2).}$$

The acceleration of P along QP (by Arts. 4 and 49)

= acc. of Q along the tangent QP + the acc. of P relative to Q

$$= -+\ddot{s}(\ddot{s} - s\dot{\phi}^2) = -s\dot{\phi}^2 \quad \text{.............................(3).}$$

The acceleration of P perpendicular to QP

= acc. of Q in this direction + acc. of P relative to Q

$$= -\frac{\dot{s}^2}{\rho} + \frac{1}{s}\frac{d}{dt}(s^2\dot{\phi}) = -\dot{s}\dot{\phi} + [s\ddot{\phi} + 2\dot{s}\dot{\phi}]$$

$$= s\ddot{\phi} + \dot{s}\dot{\phi} \quad \text{.............................(4).}$$

These are the component velocities and accelerations for any curve, whether a catenary or not.

The equation of energy gives for the catenary

$$\tfrac{1}{2}m.(c \tan \phi\dot{\phi})^2 = mg(c - c \cos \phi) \quad \text{.............................(5).}$$

Resolving along the line PQ we have

$$mc \tan \phi . \dot{\phi}^2 = T - mg \sin \phi \quad \text{.............................(6).}$$

(5) and (6) give the results required.

26. A particle is attached to the end of a light string wrapped round a vertical circular hoop and is initially at rest on the outside of the hoop at its lowest point. When a length $a\theta$ of the string has become unwound, shew that the velocity v of the particle then is

$$\sqrt{2ag(\theta \sin \theta + \cos \theta - 1)},$$

and that the tension of the string is $\left(3 \sin \theta + \frac{2 \cos \theta}{\theta} - \frac{2}{\theta}\right)$ times the weight of the particle.

27. A particle is attached to the end of a fine thread which just winds round the circumference of a circle from the centre of which acts a repulsive force $m\mu$(distance); shew that the time of unwinding is $\frac{2\pi}{\sqrt{\mu}}$, and that the tension of the thread at any time t is $2m\mu^{\frac{3}{2}}at$, where a is the radius of the circle.

28. A particle is suspended by a light string from the circumference of a cylinder, of radius a, whose axis is horizontal, the string being tangential to the cylinder and its unwound length being $a\beta$. The particle is projected horizontally in a plane perpendicular to the axis of the cylinder so as to pass under it; shew that the least velocity it can have so that the string may wind itself completely up is

$$\sqrt{2ga(\beta - \sin \beta)}.$$

29. From the lowest point of a smooth hollow cylinder whose cross-section is one-half of the lemniscate $r^2 = a^2 \cos 2\theta$, with axis vertical and node downwards, a particle is projected with velocity V along the inner surface in the plane of a cross-section; shew that it will make a complete revolution if $3V^2 > 7ag$

30. *If a particle can describe a certain plane curve freely under one set of forces and can also describe it freely under a second set, then it can describe it freely when both sets act, provided that the initial kinetic energy in the last case is equal to the sum of the initial kinetic energies in the first two cases.*

Let the arc s be measured from the point of projection, and let the initial velocities of projection in the first two cases be U_1 and U_2.

Let the tangential and normal forces in the first case be T_1 and N_1, when an arc s has been described, and T_2 and N_2 similarly in the second case; let the velocities at this point be v_1 and v_2. Then

$$mv_1\frac{dv_1}{ds}=T_1; \qquad m\frac{v_1^2}{\rho}=N_1;$$

$$mv_2\frac{dv_2}{ds}=T_2; \text{ and } m\frac{v_2^2}{\rho}=N_2.$$

$$\therefore \quad \tfrac{1}{2}mv_1^2=\int_0^s T_1 ds+\tfrac{1}{2}mU_1^2,$$

and

$$\tfrac{1}{2}mv_2^2=\int_0^s T_2 ds+\tfrac{1}{2}mU_2^2.$$

$$\therefore \quad \tfrac{1}{2}m(v_1^2+v_2^2)=\int_0^s T_1 ds+\int_0^s T_2 ds+\tfrac{1}{2}mU_1^2+\tfrac{1}{2}mU_2^2 \quad \text{......(1)},$$

and

$$m\frac{v_1^2+v_2^2}{\rho}=N_1+N_2 \quad \text{......(2)}.$$

If the same curve be described freely when both sets of forces are acting, and the velocity be v at arcual distance s, and U be the initial velocity, we must have similarly

$$\tfrac{1}{2}mv^2=\int_0^s (T_1+T_2)ds+\tfrac{1}{2}mU^2 \quad \text{......(3)},$$

and

$$m\frac{v^2}{\rho}=N_1+N_2 \quad \text{......(4)}.$$

Provided that $\tfrac{1}{2}mU^2=\tfrac{1}{2}mU_1^2+\tfrac{1}{2}mU_2^2$ equations (1) and (3) give

$$v^2=v_1^2+v_2^2.$$

and then (4) is the same as (2), which is true.

Hence the conditions of motion are satisfied for the last case, if the initial kinetic energy for it is equal to the sum of the kinetic energies in the first two cases.

The same proof would clearly hold for more than two sets of forces.

COR. The Theorem may be extended as follows.

If particles of masses m_1, m_2, m_3... all describe one path under forces F_1, F_2, F_3...; then the same path can be described by a particle of mass M under all the forces acting simultaneously, provided its kinetic energy at the point of projection is equal to the sum of the kinetic energies of the particles m_1, m_2, m_3... at the same point of projection.

31. A particle moves under the influence of two forces $\frac{\mu}{r^5}$ to one point and $\frac{\mu'}{r'^5}$ to another point; shew that it is possible for the particle to describe a circle, and find the circle.

32. Shew that a particle can be made to describe an ellipse freely under the action of forces $\lambda r+\frac{\mu}{r^2}$, $\lambda r'+\frac{\mu'}{r'^2}$ directed towards its foci.

33. A circle, of radius a, is described by a particle under a force $\frac{\mu}{(\text{distance})^5}$ to a point on its circumference. If, in addition, there be a constant normal repulsive force $\frac{\mu'}{a^5}$, shew that the circle will still be described freely if the particle start from rest at a point where

$$r = a\sqrt[4]{\frac{\mu}{2\mu'}}.$$

34. Shew that a particle can describe a circle under two forces

$$\frac{uf^2}{r_1^5} \text{ and } \frac{uf'^2}{r_2^5}$$

directed to two centres of force, which are inverse points for the circle at distances f and f' from the centre, and that the velocity at any point is

$$\frac{\sqrt{uf}}{r_1^2}\left(\text{or } \frac{\sqrt{uf'}}{r_2^2}\right).$$

35. A ring, of mass m, is strung on a smooth circular wire, of mass M and radius a; if the system rests on a smooth table, and the ring be started with velocity v in the direction of the tangent to the wire, shew that the reaction of the wire is always $\frac{Mm}{M+m}\frac{v^2}{a}$.

36. O, A and B are three collinear points on a smooth table, such that $OA=a$ and $AB=b$. A string is laid along AB and to B is attached a particle. If the end A be made to describe a circle, whose centre is O, with uniform velocity v, shew that the motion of the string relative to the revolving radius OA is the same as that of a pendulum of length $\frac{gab}{v^2}$, and further that the string will not remain taut unless $a > 4b$.

37. A particle slides under gravity down a rough cycloid, whose axis is vertical and vertex downwards, starting from rest at the cusp. Shew that it will come to rest before reaching the lowest point if $\mu e^{\frac{\mu\pi}{2}} > 1$, where μ is the coefficient of friction.

Prove that this inequality is satisfied if $\mu = \frac{1}{2}$.

38. A smooth parabolic tube is fixed in a vertical plane with its vertex downwards. A particle starts from rest at the extremity of the latus rectum and slides down the tube; express as a definite integral the time taken to reach the vertex, and shew that this time is approximately $2{\cdot}7 \times \sqrt{\frac{a}{g}}$ seconds, where $4a$ is the length of the latus rectum.

CHAPTER VII

MOTION IN A RESISTING MEDIUM. MOTION OF PARTICLES OF VARYING MASS

104. When a body moves in a medium like air, it experiences a resistance to its motion which increases as its velocity increases, and which may therefore be assumed to be equal to some function of the velocity, such as $k\rho f(v)$, where ρ is the density of the medium and k is some constant depending on the shape of the body.

Many efforts have been made to discover the law of resistance, but without much success. It appears, however, that for projectiles moving with velocities under about 800 feet per second the resistance approximately varies as the square of the velocity, that for velocities between this value and about 1350 feet per second the resistance varies as the cube, or even a higher power, of the velocity, whilst for higher velocities the resistance seems to again follow the law of the square of the velocity.

For other motions it is found that other assumptions of the law for the resistance are more suitable. Thus in the case of the motion of an ordinary pendulum the assumption that the resistance varies as the velocity is the best approximation.

In any case the law assumed is more or less empiric, and its truth can only be tested by enquiring how far the results, which are theoretically obtained by its use, fit with the actually observed facts of the motion.

Whatever be the law of resistance, the forces are non-conservative, and the Principle of Conservation of Energy cannot be applied.

105. In the case of a particle falling under gravity in a resisting medium the velocity will never exceed some definite quantity.

For suppose the law of resistance to be $kv^n.m$. Then the downward acceleration is $g-kv^n$, and this vanishes when $kv^n=g$, *i.e.* when the velcoity $=\left(\frac{g}{k}\right)^{\frac{1}{n}}$. This therefore will be the maximum velocity possible, and it is called the limiting or terminal velocity.

It follows from this that we cannot tell the height from which drops of rain fall by observing their velocity on reaching the ground. For soon after they have started they will have approximately reached their terminal velocity, and will then continue to move with a velocity which is sensibly constant and very little differing from the terminal velocity.

In the case of a ship which is under steam there is a full speed beyond which it cannot travel. This full speed will depend on the dimensions of the ship and the size and power of its engines, etc.

But whatever the latter may be, there will be some velocity at which the work that must be done in overcoming the resistance of the water, which varies as some function of the velocity, will be just equivalent to the maximum amount of work that can be done by the engines of the ship, and then further increase of the speed of the ship is impossible.

106. *A particle falls under gravity (supposed constant) in a resisting medium whose resistance varies as the square of the velocity; to find the motion if the particle starts from rest.*

Let v be the velocity when the particle has fallen a distance x in time t from rest. The equation of motion is

$$\frac{d^2x}{dt^2}=g-\mu v^2.$$

Let $\mu=\frac{g}{k^2}$, so that
$$\frac{d^2x}{dt^2}=g\left(1-\frac{v^2}{k^2}\right) \quad \text{.........................(1).}$$

From (1) it follows that if v equalled k, the acceleration would be zero; the motion would then be unresisted and the velocity of the particle would continue to be k. For this reason k is called the "terminal velocity."

From (1),
$$v\frac{dv}{dx}=g\left(1-\frac{v^2}{k^2}\right),$$

so that
$$\frac{2g}{k^2}x=\int\frac{2v\,dv}{k^2-v^2}=-\log(k^2-v^2)+A.$$

Since v and x are both zero initially, $\therefore A=\log k^2$.

$$\therefore k^2-v^2=k^2e^{-\frac{2gx}{k^2}}.$$

$$\therefore v^2=k^2\left[1-e^{-\frac{2gx}{k^2}}\right] \quad \text{.......................(2).}$$

It follows that $x=\infty$ when $v=k$. Hence the particle would not actually acquire the "terminal velocity" until it had fallen an infinite distance.

Again (1) can be written

$$\frac{dv}{dt}=g\left(1-\frac{v^2}{k^2}\right).$$

$$\therefore \frac{gt}{k^2}=\int\frac{dv}{k^2-v^2}=\frac{1}{2k}\log\frac{k+v}{k-v}+B.$$

Since v and t were zero initially, $\therefore B=0$.

Hence $$\frac{k+v}{k-v}=e^{\frac{2gt}{k}}.$$

$$\therefore\ v=k\frac{e^{\frac{2gt}{k}}-1}{e^{\frac{2gt}{k}}+1}=k\tanh\left(\frac{gt}{k}\right) \quad \text{..................(3).}$$

From (2) and (3), we have

$$e^{-\frac{2gx}{k^2}}=1-\frac{v^2}{k^2}=1-\tanh^2\frac{gt}{k}=\frac{1}{\cosh^2\frac{gt}{k}},$$

so that $$e^{\frac{gx}{k^2}}=\cosh\frac{gt}{k},\quad\text{and}\quad x=\frac{k^2}{g}\log\cosh\frac{gt}{k} \quad \text{...........(4).}$$

107. *If the particle were projected upwards instead of downwards, to find the motion.*

Let V be the velocity of projection.

The equation of motion now is

$$\frac{d^2x}{dt^2}=-g-\mu v^2=-g\left(1+\frac{v^2}{k^2}\right) \quad \text{..................(5),}$$

where x is measured upwards.

Hence $$v\frac{dv}{dx}=-g\left(1+\frac{v^2}{k^2}\right).$$

$$\therefore\ \frac{2g}{k^2}x=-\int\frac{2v\,dv}{v^2+k^2}=-\log\left(v^2+k^2\right)+A,$$

where $$0=-\log\left(V^2+k^2\right)+A.$$

$$\therefore\ \frac{2gx}{k^2}=\log\frac{V^2+k^2}{v^2+k^2} \quad \text{.......................(6).}$$

Again (5) gives

$$\frac{dv}{dt}=-g\left(1+\frac{v^2}{k^2}\right).$$

$$\therefore\ -\frac{gt}{k^2}=\int\frac{dv}{k^2+v^2}=\frac{1}{k}\tan^{-1}\frac{v}{k}+B,$$

where $$0=\frac{1}{k}\tan^{-1}\frac{V}{k}+B.$$

$$\therefore\ \frac{gt}{k}=\tan^{-1}\frac{V}{k}-\tan^{-1}\frac{v}{k} \quad \text{.....................(7).}$$

Equation (6) gives the velocity when the particle has described any distance, and (7) gives the velocity at the end of any time.

108. *Ex. A person falls by means of a parachute from a height of* 800 *yards in* $2\frac{1}{2}$ *minutes. Assuming the resistance to vary as the square of the velocity, shew that in a second and a half his velocity differs by less than one per cent. from its value when he reaches the ground and find an approximate value for the limiting velocity.*

When the parachute has fallen a space x in time t, we have, by Art. 106, if

$$\mu=\frac{g}{k^2},$$

$$v^2=k^2\left[1-e^{-\frac{2gx}{k^2}}\right] \quad \text{.......................(1),}$$

$$v=k\tanh\left(\frac{gt}{k}\right) \quad \text{..........................(2),}$$

and

$$x=\frac{k^2}{g}\log\cosh\left(\frac{gt}{k}\right) \quad \text{.....................(3).}$$

Here

$$2400\frac{g}{k^2}=\log\cosh\left(\frac{150g}{k}\right).$$

$$\therefore e^{2400\frac{g}{k^2}}=\frac{e^{\frac{150g}{k}}+e^{-\frac{150g}{k}}}{2} \quad \text{.......................(4).}$$

The second term on the right hand is very small, since k is positive.

Hence (4) is approximately equivalent to $e^{2400\frac{g}{k^2}}=\frac{1}{2}e^{\frac{150g}{k}}$.

$$\therefore 2400\frac{g}{k^2}=\frac{150g}{k}-\log 2=\frac{150g}{k}, \text{ nearly.}$$

Hence $k=16$ is a first approximation.
Putting $k=16(1+y)$, (4) gives, for a second approximation,

$$e^{300(1-2y)}=\frac{e^{300(1-y)}+e^{-300(1-y)}}{2}=\frac{e^{300(1-y)}}{2}, \text{ very approx.}$$

$$\therefore e^{-300y}=\tfrac{1}{2}.$$

$$\therefore y=\frac{1}{300}\log_e 2=\frac{\cdot 693}{300}=0\cdot 0023.$$

Therefore a second approx. is $k=16\,(1+0\cdot 0023)$, giving the terminal velocity.

Also the velocity v_1, when the particle reaches the ground, is, by (1), given by

$$v_1^2=k^2\left[1-e^{-\frac{2.32.2400}{16^2}}\right]=k^2[1-e^{-600}]$$

$$=k^2, \text{ for all practical purposes.}$$

When v is 99% of the terminal velocity, (2) gives

$$\tanh\frac{gt}{k}=\frac{99}{100}=\cdot 99.$$

$$\therefore\ e^{\frac{2gt}{k}}=199=e^{5\cdot 3}, \text{ from the Tables.}$$

$$\therefore\ t=\frac{k}{2g}\times 5\cdot 3=\frac{16}{64}\times 5\cdot 3=1\cdot 325 \text{ approx.}$$

i.e. t is less than $1\frac{1}{2}$ secs.

EXAMPLES

1. A particle, of mass m, is falling under the influence of gravity through a medium whose resistance equals μ times the velocity.

If the particle is released from rest, shew that the distance fallen through in time t is $g\frac{m^2}{\mu^2}\left\{e^{-\frac{\mu t}{m}}-1+\frac{\mu t}{m}\right\}$.

2. A particle, of mass m, is projected vertically under gravity, the resistance of the air being mk times the velocity; shew that the greatest height attained by the particle is $\frac{V^2}{g}[\lambda-\log(1+\lambda)]$, where V is the terminal velocity of the particle and λV is its initial vertical velocity.

3. A heavy particle is projected vertically upwards with velocity u in a medium, the resistance of which is $gu^{-2}\tan^2\alpha$ times the square of the velocity, α being a constant. Shew that the particle will return to the point of projection with velocity $u\cos\alpha$, after a time

$$ug^{-1}\cot\alpha\left(\alpha+\log\frac{\cos\alpha}{1-\sin\alpha}\right).$$

4. A particle falls from rest under gravity through a distance x in a medium whose resistance varies as the square of the velocity; if v be the velocity actually acquired by it, v_0 the velocity it would have acquired had there been no resisting medium, and V the terminal velocity, shew that

$$\frac{v^2}{v_0^2}=1-\frac{1}{2}\frac{v_0^2}{V^2}+\frac{1}{2.3}\frac{v_0^4}{V^4}-\frac{1}{2.3.4}\frac{v_0^6}{V^6}+\dots$$

5. A particle is projected with velocity V along a smooth horizontal plane in a medium whose resistance per unit of mass is μ times the cube of the velocity. Shew that the distance it has described in time t is

$$\frac{1}{\mu V}[\sqrt{1+2\mu V^2 t}-1],$$

and that its velocity then is $\frac{V}{\sqrt{1+2\mu V^2 t}}$.

6. A heavy particle is projected vertically upwards with a velocity u in a medium the resistance of which varies as the cube of the particle's velocity. Determine the height to which the particle will ascend.

7. If the resistance vary as the fourth power of the velocity, the energy of m lbs. at a depth x below the highest point when moving in a vertical line under gravity will be $E \tan \dfrac{mgx}{E}$ when rising and,

$$E \tanh \frac{mgx}{E}$$

when falling, where E is the terminal energy in the medium.

8. A particle is projected in a resisting medium whose resistance varies as (velocity)^n, and it comes to rest after describing a distance s in time t. Find the values of s and t and shew that s is finite if $n<2$, but infinite if $n=$ or >2, whilst t is finite if $n<1$, but infinite if $n=$ or >1.

9. In the previous question if the resistance be k (velocity) and the initial velocity be V, shew that $v=Ve^{-kt}$ and $s=\dfrac{V}{k}(1-e^{-kt})$.

10. A heavy particle is projected vertically upwards in a medium the resistance of which varies as the square of the velocity. It has a kinetic energy K in its upward path at a given point; when it passes the same point on the way down, shew that its loss of energy is $\dfrac{K^2}{K+K'}$, where K' is the limit to which its energy approaches in its downward course.

11. If the resistance to the motion of a railway train vary as its mass and the square of its velocity, and the engine work at constant H.P., shew that full speed will never be attained, and that the distance traversed from rest when half-speed is attained is $\dfrac{1}{3\mu} \log_e \dfrac{8}{7}$, where μ is the resistance per unit mass per unit velocity.

Find also the time of describing this distance.

12. A ship, with engines stopped, is gradually brought to rest by the resistance of the water. At one instant the velocity is 10 ft. per sec. and one minute later the speed has fallen to 6 ft. per sec. For speeds below 2 ft. per sec. the resistance may be taken to vary as the speed, and for higher speeds to vary as the square of the speed. Shew that, before coming to rest, the ship will move through $900[1+\log_e 5]$ feet, from the point when the first velocity was observed.

13. A particle moves from rest at a distance a from a fixed point O under the action of a force to O equal to μ times the distance per unit of mass; if the resistance of the medium in which it moves be k times the square of the velocity per unit of mass, shew that the square of the velocity when it is at a distance x from O is

$\dfrac{\mu x}{k}-\dfrac{\mu a}{k}e^{2k(x-a)}+\dfrac{\mu}{2k^2}[1-e^{2k(x-a)}]$.

Shew also that when it first comes to rest it will be at a distance b given by $(1-2kb)e^{2kb}=(1+2ka)e^{-2ak}$.

14. A particle falls from rest at a distance a from the centre of the Earth towards the Earth, the motion meeting with a small resistance proportional to the square of the velocity v and the retardation being μ for unit velocity; shew that the kinetic energy at distance x from the centre is $mgr^2\left\{\dfrac{1}{x}-\dfrac{1}{a}+2\mu\left(1-\dfrac{x}{a}\right)-2\mu \log_e \dfrac{a}{x}\right\}$, the square of μ being neglected, and r being the radius of the Earth.

15. An attracting force, varying as the distance, acts on a particle initially at rest at a distance a. Shew that, if V be the velocity when the particle is at a distance x, and V' the velocity of the same particle when the resistance of the air is taken into account, then

$$V'=V\left[1-\frac{1}{3}k\frac{(2a+x)(a-x)}{a+x}\right]$$

nearly, the resistance of the air being given to be k times the square of the velocity per unit of mass, where k is very small.

109. *A particle is projected under gravity and a resistance equal to mk (velocity) with a velocity u at an angle α to the horizon; to find the motion.*

Let the axes of x and y be respectively horizontal and vertical, and the origin at the point of projection. Then the equations of motion are

$$\ddot{x}=-k\frac{ds}{dt}\cdot\frac{dx}{ds}=-k\frac{dx}{dt},$$

and
$$\ddot{y}=-k\frac{ds}{dt}\cdot\frac{dy}{ds}-g=-k\frac{dy}{dt}-g.$$

Integrating, we have

$$\log\dot{x}=-kt+\text{const.}=-kt+\log(u\cos\alpha),$$

and
$$\log(k\dot{y}+g)=-kt+\text{const.}=-kt+\log(ku\sin\alpha+g);$$

$$\therefore\ \dot{x}=u\cos\alpha e^{-kt}\quad\text{.......................(1)},$$

and
$$k\dot{y}+g=(ku\sin\alpha+g)e^{-kt}\quad\text{....................(2)}.$$

$$\therefore\ x=-\frac{u\cos\alpha}{k}e^{-kt}+\text{const.}=\frac{u\cos\alpha}{k}(1-e^{-kt})\quad\text{......(3)},$$

and
$$ky+gt=-\frac{ku\sin\alpha+g}{k}e^{-kt}+\text{const.}$$

$$=\frac{ku\sin\alpha+g}{k}(1-e^{-kt})\quad\text{....................(4)}.$$

Eliminating t, we have

$$y=\frac{g}{k^2}\log\left(1-\frac{kx}{u\cos\alpha}\right)+\frac{x}{u\cos\alpha}\left(u\sin\alpha+\frac{g}{k}\right)\quad\text{......(5)},$$

which is the equation to the path.

The greatest height is attained when $\dot{y}=0$, *i.e.* when

$$e^{-kt}=\frac{g}{ku\sin\alpha+g},\ \textit{i.e.}\text{ at time }\frac{1}{k}\log\left(1+\frac{ku\sin\alpha}{g}\right),$$

and then
$$y=\frac{u\sin\alpha}{k}-\frac{g}{k^2}\log\left(1+\frac{ku\sin\alpha}{g}\right).$$

It is clear from equations (3) and (4) that when $t=\infty$, $x=\dfrac{u\cos\alpha}{k}$ and $y=-\infty$. Hence the path has a vertical asymptote at a horizontal distance $\dfrac{u\cos\alpha}{k}$ from the point of projection. Also, then, $\dot{x}=0$ and $\dot{y}=-\dfrac{g}{k}$, *i.e.* the particle will then have just attained the limiting velocity.

Cor. If the right-hand side of (5) be expanded in powers of k, it becomes

$$y=\frac{g}{k^2}\left[-\frac{kx}{u\cos\alpha}-\frac{1}{2}\frac{k^2x^2}{u^2\cos^2\alpha}-\frac{1}{3}\frac{k^3x^3}{u^3\cos^3\alpha}-\dots\right]+\frac{x}{u\cos\alpha}\left(u\sin\alpha+\frac{g}{k}\right),$$

i.e. $$y=x\tan\alpha-\frac{gx^2}{2u^2\cos^2\alpha}-\frac{1}{3}\frac{gkx^3}{u^3\cos^3\alpha}-\frac{1}{4}\frac{gk^2x^4}{u^4\cos^4\alpha}-\dots$$

On putting k equal to zero, we have the ordinary equation to the trajectory for unresisted motion.

110. *A particle is moving under gravity in a medium whose resistance* $=m\mu$ *(velocity)*^{2}; *to find the motion.*

When the particle has described a distance s, let its tangent make an angle ϕ with the upward drawn vertical, and let v be its velocity.

The equations of motion are then

$$v\frac{dv}{ds}=-g\cos\phi-\mu v^2 \quad \dots\dots\dots(1),$$

and

$$\frac{v^2}{\rho}=g\sin\phi \quad \dots\dots\dots(2).$$

(1) gives $$\frac{d.v^2}{d\phi}\frac{d\phi}{ds}=-2g\cos\phi-2\mu v^2,$$

i.e., from (2), $$\frac{1}{\rho}\frac{d}{d\phi}(\rho\sin\phi)=-2\cos\phi-2\mu\rho\sin\phi.$$

$$\therefore\ \frac{1}{\rho}\frac{d\rho}{d\phi}\sin\phi+3\cos\phi=-2\mu\rho\sin\phi.$$

$$\therefore\ \frac{d}{d\phi}\left(\frac{1}{\rho}\right).\frac{1}{\sin^3\phi}-\frac{3\cos\phi}{\sin^4\phi}.\frac{1}{\rho}=\frac{2\mu}{\sin^3\phi}.$$

$$\therefore\ \frac{1}{\rho\sin^3\phi}=2\mu\int\frac{1}{\sin^3\phi}\,d\phi=-\mu\frac{\cos\phi}{\sin^2\phi}-\mu\log\frac{1+\cos\phi}{\sin\phi}+A \quad \dots(3).$$

(2) then gives

$$v^2\left[A-\mu\frac{\cos\phi}{\sin^2\phi}-\mu\log\frac{1+\cos\phi}{\sin\phi}\right]=\frac{g}{\sin^2\phi}.$$

Equation (3) gives the intrinsic equation of the path, but cannot be integrated further.

111. *A bead moves on a smooth wire in a vertical plane under a resistance* $\{=k(\textit{velocity})^2\}$; *to find the motion.*

When the bead has described an arcual distance s, let the velocity be v at an angle ϕ to the horizon (Fig., Art. 102), and let the reaction of the wire be R.

The equations of motion are

$$\frac{vdv}{ds}=g\sin\phi-kv^2 \quad\text{.......................(1)},$$

and

$$\frac{v^2}{\rho}=g\cos\phi-R \quad\text{.........................(2)}.$$

Let the curve be $s=f(\phi)$.

Then (1) gives

$$\frac{d}{d\phi}\left(\frac{v^2}{2}\right)=f'(\phi)\,[g\sin\phi-kv^2],$$

i.e.

$$\frac{d}{d\phi}(v^2)+2kf'(\phi).v^2=2g\sin\phi.f'(\phi),$$

a linear equation to give v^2.

Particular case. Let the curve be a circle so that $s=a\phi$, if s and ϕ be measured from the highest point.

(1) then gives

$$\frac{d}{d\phi}(v^2)+2akv^2=2ag\sin\phi.$$

$$\therefore\ v^2e^{2ak\phi}=2ag\int\sin\phi.e^{2ak\phi}$$

$$=\frac{2ag}{1+4a^2k^2}e^{2ak\phi}(2ak\sin\phi-\cos\phi)+C.$$

$$\therefore\ v^2=\frac{2ag}{1+4a^2k^2}(2ak\sin\phi-\cos\phi)+Ce^{-2ak\phi}.$$

EXAMPLES

1. A particle of unit mass is projected with velocity u at an inclination α above the horizon in a medium resistance is k times the velocity. Shew that its direction will again make an angle α with the horizon after a time $\frac{1}{k}\log\left\{1+\frac{2ku}{g}\sin\alpha\right\}$.

2. If the resistance vary as the velocity and the range on the horizontal plane through the point of projection is a maximum, shew that the angle α which the direction of projection makes with the vertical is given by

$$\frac{\lambda(1+\lambda\cos\alpha)}{\cos\alpha+\lambda}=\log\,[1+\lambda\sec\alpha],$$

where λ is the ratio of the velocity of projection to the terminal velocity.

3. A particle acted on by gravity is projected in a medium of which the resistance varies as the velocity. Shew that its acceleration retains a fixed direction and diminishes without limit to zero.

4. Shew that in the motion of a heavy particle in a medium, the resistance of which varies as the velocity, the greatest height above the level of the point of projection is reached in less than half the total time of the flight above that level.

5. If a particle be moving in a medium whose resistance varies as the velocity of the particle, shew that the equation of the trajectory can, by a proper choice of axes, be put into the form

$$y+ax=b\log x.$$

6. If the resistance of the air to a particle's motion be n times its weight, and the particle be projected horizontally with velocity V, shew that the velocity of the particle, when it is moving at an inclination ϕ to the horizontal, is $V(1-\sin\phi)^{\frac{n-1}{2}}(1+\sin\phi)^{-\frac{n+1}{2}}$.

7. A heavy bead, of mass m, slides on a smooth wire in the shape of a cycloid, whose axis is vertical and vertex upwards, in a medium whose resistance is $m\frac{v^2}{2c}$ and the distance of the starting point from the vertex is c; shew that the time of descent to the cusp is $\sqrt{\frac{8a(4a-c)}{gc}}$, where $2a$ is the length of the axis of the cycloid.

8. A heavy bead slides down a smooth wire in the form of a cycloid, whose axis is vertical and vertex downwards, from rest at a cusp, and is acted on besides its weight by a tangential resistance proportional to the square of the velocity. Determine the velocity after a fall through the height x.

9. If a point travel on an equiangular spiral towards the pole with uniform angular velocity about the pole, shew that the projection of the point on a straight line represents a resisted simple vibration.

10. A particle, moving in a resisting medium, is acted on by a central force $\frac{\mu}{r^n}$; if the path be an equiangular spiral of angle α, whose pole is at the centre of force, shew that the resistance is $\frac{n-3}{2}\frac{\mu\cos\alpha}{r^n}$.

11. A particle, of mass m, is projected in a medium whose resistance is mk (velocity), and is acted on by a force to a fixed point ($=m.\mu.$distance). Find the equation to the path, and, in the case when $2k^2=9\mu$, shew that it is a parabola and that the particle would ultimately come to rest at the origin, but that the time taken would be infinite.

12. If a high throw is made with a diabolo spool the vertical resistance may be neglected, but the spin and the vertical motion together account for a horizontal drifting force which may be taken as proportional to the vertical velocity. Shew that if the spool is thrown so as to rise to the height h and return to the point of projection, the spool is at its greatest distance c from the vertical through that point when it is at a height $\frac{2h}{3}$; and shew that the equation to the trajectory is of the form $4h^3x^2=27c^2y^2(h-y)$.

13. If a body move under a central force in a medium which exerts a resistance equal to k times the velocity per unit of mass, prove that $\frac{d^2u}{d\theta^2}+u=\frac{P}{h^2u^2}.e^{2kt}$, where h is twice the initial moment of momentum about the centre of force.

14. A particle moves with a central acceleration P in a medium of which the resistance is $k.(\text{velocity})^2$; shew that the equation to its path is $\frac{d^2u}{d\theta^2}+u=\frac{P}{h^2u^2}e^{2ks}$, where s is the length of the arc described, and h is twice the initial moment of momentum about the centre of force.

15. A particle moves in a resisting medium with a given central acceleration P; the path of the particle being given, shew that the resistance is $-\frac{1}{2p^2}\frac{d}{ds}\left(p^3\frac{dr}{dp}P\right)$.

112. *Motion where the mass moving varies.*

The equation $P=mf$ is only true when the mass m is constant. Newton's second law in its more fundamental form is

$$P=\frac{d}{dt}(mv) \quad \text{.............................(1)}.$$

Suppose that a particle gains in time δt an increment δm of mass and that this increment δm was moving with a velocity u.

Then in time δt the increment in the momentum of the particle

$$=m.\delta v+\delta m(v+\delta v-u),$$

and the impulse of the force in this time is $P\delta t$.

Equating these we have, on proceeding to the limit,

$$m\frac{dv}{dt}+v\frac{dm}{dt}-u\frac{dm}{dt}=P,$$

i.e.
$$\frac{d}{dt}(mv)=P+u\frac{dm}{dt} \quad \text{.........................(2)}.$$

When u is zero we have the result (1).

113. *Ex.* 1. *A spherical raindrop, falling freely, receives in each instant an increase of volume equal to λ times its surface at that instant; find the velocity at the end of time t, and the distance fallen through in that time.*

When the raindrop has fallen through a distance x in time t, let its radius be r and its mass M. Then

$$\frac{d}{dt}\left[M\frac{dx}{dt}\right]=Mg \quad \text{.........................(1)}.$$

Now $M = \frac{4}{3}\pi\rho r^3$, so that $4\pi r^2 \rho \frac{dr}{dt} = \frac{dM}{dt} = \rho . 4\lambda\pi r^2$, by the question.

$$\therefore \frac{dr}{dt} = \lambda, \text{ and } r = a + \lambda t,$$

where a is the initial radius.

Hence (1) gives $$\frac{a}{dt}\left[(a+\lambda t)^3 \frac{dx}{dt}\right] = (a+\lambda t)^3 g.$$

$$\therefore (a+\lambda t)^3 \frac{dx}{dt} = \frac{(a+\lambda t)^4}{4\lambda} g - \frac{a^4}{4\lambda} g,$$

since the velocity was zero to start with.

$$\therefore \frac{dx}{dt} = \frac{g}{4\lambda}\left[a + \lambda t - \frac{a^4}{(a+\lambda t)^3}\right],$$

and $$x = \frac{g}{4\lambda^2}\left[\frac{(a+\lambda t)^2}{2} + \frac{a^4}{2(a+\lambda t)^2}\right] - \frac{g}{4\lambda^2} a^2,$$

since x and t vanish together.

$$\therefore x = \frac{g}{8\lambda^2}\left[(a+\lambda t)^2 - 2a^2 + \frac{a^4}{(a+\lambda t)^2}\right]$$

$$= \frac{g}{8\lambda^2}\left[a + \lambda t - \frac{a^2}{a+\lambda t}\right]^2 = \frac{gt^2}{8}\left[\frac{2a+\lambda t}{a+\lambda t}\right]^2.$$

Ex. 2. *A mass in the form of a solid cylinder, the area of whose cross-section is A, moves parallel to its axis, being acted on by a constant force F, through a uniform cloud of fine dust of volume density ρ which is moving in a direction opposite to that of the cylinder with constant velocity V. If all the dust that meets the cylinder clings to it, find the velocity and distance described in any time t, the cylinder being originally at rest, and its initial mass m.*

Let M be the mass at time t and v the velocity. Then

$M . \delta v + \delta M (v + \delta v + V) =$ increase in the momentum in time $\delta t = F \delta t$.

$$\therefore M\frac{dv}{dt} + v\frac{dM}{dt} + V\frac{dM}{dt} = F \quad \text{.....................(1)}$$

in the limit.

Also $$\frac{dM}{dt} = A\rho(v+V) \quad \text{..........................(2).}$$

(1) gives $$Mv + MV = Ft + \text{const.} = Ft + mV.$$

Therefore (2) gives $$M\frac{dM}{dt} = A\rho(Ft + mV).$$

$$\therefore M^2 = A\rho(Ft^2 + 2mVt) + m^2.$$

Therefore (2) gives

$$v=-V+\frac{Ft+mV}{M}=-V+\frac{Ft+mV}{\sqrt{m^2+2mA\rho Vt+AF\rho t^2}} \quad \text{......(3).}$$

Also if the hinder end of the cylinder has described a distance x from rest, so that $v=\frac{dx}{dt}$, then $x=-Vt+\frac{1}{A\rho}\sqrt{m^2+2mA\rho Vt+AF\rho t^2}-\frac{m}{A\rho}$.

From (3) we have that the acceleration

$$\frac{dv}{dt}=\frac{m^2(F-A\rho V^2)}{(m^2+2mA\rho Vt+AF\rho t^2)^{\frac{3}{2}}},$$

so that the motion is always in the direction of the force, or opposite, according as $F \gtrless A\rho V^2$.

Ex. 3. *A uniform chain is coiled up on a horizontal plane and one end passes over a small light pulley at a height a above the plane; initially a length b, > a, hangs freely on the other side; find the motion.*

When the length b has increased to x, let v be the velocity; then in the time δt next ensuing the momentum of the part $(x+a)$ has increased by $m(x+a)\delta v$, where m is the mass per unit length. Also a length $m\delta x$ has been jerked into motion, and given a velocity $v+\delta v$. Hence

$$m(x+a)\delta v+m\delta x(v+\delta v)=\text{change in the momentum}$$
$$=\text{impulse of the acting force}=mg(x-a).\delta t.$$

Hence, dividing by δt and proceeding to the limit, we have

$$(x+a)\frac{dv}{dt}+v^2=(x-a)g.$$

$$\therefore\ v\frac{dv}{dx}.(x+a)+v^2=(x-a)g.$$

$$\therefore\ v^2(x+a)^2=\int_b^x 2(x^2-a^2)g=2\left\{\frac{x^3-b^3}{3}-a^2(x-b)\right\}g,$$

so that $$v^2=\frac{2g}{3}\frac{(x-b)(x^2+bx+b^2-3a^2)}{(x+a)^2}.$$

This equation cannot be integrated further.

In the particular case when $b=2a$, this gives $v^2=\frac{2g}{3}(x-b)$, so that the end descends with constant acceleration $\frac{g}{3}$.

The tension T of the chain at the coil is clearly given by $T\delta t=m\delta x.v$, so that $T=mv^2$.

EXAMPLES

1. A spherical raindrop of radius a cms. falls from rest through a vertical height h, receiving throughout the motion an accumulation of condensed vapour at the rate of k grammes per square cm. per second, no vertical force but gravity acting; shew that when it reaches the ground its radius will be $k\sqrt{\frac{2h}{g}}\left[1+\sqrt{1+\frac{ga^2}{2hk^2}}\right]$.

2. A mass in the form of a solid cylinder, of radius c, acted upon by no forces, moves parallel to its axis through a uniform cloud of fine dust, of volume density ρ, which is at rest. If the particles of dust which meet the mass adhere to it, and if M and u be the mass and velocity at the beginning of the motion, prove that the distance x traversed in time t is given by the equation $(M+\rho\pi c^2x)^2=M^2+2\rho\pi uc^2Mt$.

3. A particle of mass M is at rest and begins to move under the action of a constant force F in a fixed direction. It encounters the resistance of a stream of fine dust moving in the opposite direction with velocity V, which deposits matter on it at a constant rate ρ. Shew that its mass will be m when it has travelled a distance

$$\frac{k}{\rho^2}\left[m-M\left\{1+\log\frac{m}{M}\right\}\right] \text{ where } k=F-\rho V.$$

4. A spherical raindrop, whose radius is ·04 inches, begins to fall from a height of 6400 feet, and during the fall its radius grows, by precipitation of moisture, at the rate of 10^{-4} inches per second. If its motion be unresisted, shew that its radius when it reaches the ground is ·0420 inches and that it will have taken about 20 seconds to fall.

5. Snow slides off a roof clearing away a part of uniform breadth; shew that, if it all slide at once, the time in which the roof will be cleared is $\sqrt{\frac{6\pi a}{g\sin\alpha}\frac{\Gamma(\frac{2}{3})}{\Gamma(\frac{1}{6})}}$, but that, if the top move first and gradually set the rest in motion, the acceleration is $\frac{1}{3}g\sin\alpha$ and the time will be $\sqrt{\frac{6a}{g\sin\alpha}}$, where α is the inclination of the roof and a the length originally covered with snow.

6. A ball, of mass m, is moving under gravity in a medium which deposits matter on the ball at a uniform rate μ. Shew that the equation to the trajectory, referred to horizontal and vertical axes through a point on itself, may be written in the form

$$k^2uy=kx(g+kv)+gu(1-e^{\frac{kx}{u}}),$$

where u, v are the horizontal and vertical velocities at the origin and $mk=2\mu$.

7. A falling raindrop has its radius uniformly increased by access of moisture. If it have given to it a horizontal velocity, shew that it will then describe a hyperbola, one of whose asymptotes is vertical.

8. If a rocket, originally of mass M, throw off every unit of time a mass eM with relative velocity V, and if M' be the mass of the case, etc., shew that it cannot rise at once unless $eV>g$, nor at all unless $\frac{eMV}{M'}>g$. If it just rises vertically at once, shew that its greatest velocity is

$$V\log\frac{M}{M'}-\frac{g}{e}\left(1-\frac{M'}{M}\right),$$

and that the greatest height it reaches is

$$\frac{V^2}{2g}\left(\log\frac{M}{M'}\right)^2+\frac{V}{e}\left(1-\frac{M'}{M}-\log\frac{M}{M'}\right).$$

9. A heavy chain, of length l, is held by its upper end so that its lower end is at a height l above a horizontal plane; if the upper end is let go, shew that at the instant when half the chain is coiled up on the plane the pressure on the plane is to the weight of the chain in the ratio of 7 : 2.

10. A chain, of great length a, is suspended from the top of a tower so that its lower end touches the Earth; if it be then let fall, shew that the square of its velocity, when its upper end has fallen a distance x, is $2gr \log \dfrac{a+r}{a+r-x}$, where r is the radius of the Earth.

11. A chain, of length l, is coiled at the edge of a table. One end is fastened to a particle, whose mass is equal to that of the whole chain, and the other end is put over the edge. Shew that, immediately after leaving the table, the particle is moving with velocity $\dfrac{1}{2}\sqrt{\dfrac{5gl}{6}}$.

12. A uniform string, whose length is l and whose weight is W, rests over a small smooth pulley with its end just reaching to a horizontal plane; if the string be slightly displaced, shew that when a length x has been deposited on the plane the pressure on it is

$$W\left[2 \log \frac{l}{l-x} - \frac{x}{l}\right],$$

and that the resultant pressure on the pulley is $W\dfrac{l-2x}{l-x}$.

13. A mass M is attached to one end of a chain whose mass per unit of length is m. The whole is placed with the chain coiled up on a smooth table and M is projected horizontally with velocity V. When a length x of the chain has become straight, shew that the velocity of M is $\dfrac{MV}{M+mx}$, and that its motion is the same as if there were no chain and it were acted on by a force varying inversely as the cube of its distance from a point in its line of motion.

Shew also that the rate at which kinetic energy is dissipated is at any instant proportional to the cube of the velocity of the mass.

14. A weightless string passes over a smooth pulley. One end is attached to a coil of chain lying on a horizontal table, and the other to a length l of the same chain hanging vertically with its lower end just touching the table. Shew that after motion ensues the system will first be at rest when a length x of chain has been lifted from the table, such that $(l-x)e^{\frac{2x}{l}} = l$. Why cannot the Principle of Energy be directly applied to find the motion of such a system?

15. A ship's cable passes through a hole in the deck at a height a above the the coil in which the cable is heaped, then passes along the deck for a distance b, and out at a hole in the side of the ship, immediately outside of which it is attached to the anchor. If the latter be loosed find the resulting motion, and, if the anchor be of weight equal to $2a+\frac{1}{2}b$ of the cable, shew that it descends with uniform acceleration $\frac{1}{3}g$.

16. A mass M is fastened to a chain of mass m per unit length coiled up on a rough horizontal plane (coefficient of friction $=\mu$). The mass is projected from the coil with velocity V; shew that it will be brought to rest in a distance $\dfrac{M}{m}\left\{\left(1+\dfrac{3m}{2M\mu g}\dfrac{V^2}{}\right)^{\frac{1}{3}}-1\right\}$.

17. A uniform chain, of mass M and length l, is coiled up at the top of a rough plane inclined at an angle α to the horizon and has a mass M fastened to one end. This mass is projected down the plane with velocity V. If the system comes to rest when the whole of the chain is just straight, shew that $V^2=\frac{14gl}{3}\sec\epsilon\sin(\epsilon-\alpha)$, where ϵ is the angle of friction.

18. A uniform chain, of length l and mass ml, is coiled on the floor, and a mass mc is attached to one end and projected vertically upwards with velocity $\sqrt{2gh}$. Shew that, according as the chain does or does not completely leave the floor, the velocity of the mass on finally reaching the floor again is the velocity due to a fall through a height.

$$\frac{1}{3}\left[2l-c+\frac{a^3}{(l+c)^2}\right] \text{ or } a-c,$$

where $a^3=c^2(c+3h)$.

19. A uniform chain is partly coiled on a table, one end of it being just carried over a smooth pulley at a height h immediately above the coil and attached there to a weight equal to that of a length $2h$ of the chain. Shew that until the weight strikes the table, the chain uncoils with uniform acceleration $\frac{1}{3}g$, and that, after it strikes the table, the velocity at any moment is $\sqrt{\frac{2}{3}gh}\,e^{-\frac{x-h}{2h}}$, where x is the length of the chain uncoiled.

20. A string, of length l, hangs over a smooth peg so as to be at rest. One end is ignited and burns away at a uniform rate v. Shew that the other end will at time t be at a depth x below the peg, where x is given by the equation

$$(l-vt)\left(\frac{d^2x}{dt^2}+g\right)-v\frac{dx}{dt}-2gx=0.$$

[At time t let x be the longer, and y the shorter part of the string, so that $x+y=l-vt$. Also let V, $(=\dot{x})$, be the velocity of the string then. On equating the change of momentum in the ensuing time δt to the impulse of the acting force, we have

$$(x+y-v\delta t)(V+\delta V)-(x+y)V=(x-y)g\delta t,$$

giving
$$(x+y)\frac{dV}{dt}-vV=(x-y)g=(2x-l+vt)g, \text{ etc.}]$$

21. A chain, of mass m and length $2l$, hangs in equilibrium over a smooth pulley when an insect of mass M alights gently at one end and begins crawling up with uniform velocity V relative to the chain; shew that the velocity with which the chain leaves the pulley will be

$$\left[\frac{M^2}{(M+m)^2}V^2+\frac{2M+m}{M+m}gl\right]^{\frac{1}{2}}.$$

[Let V_0 be the velocity with which the chain starts, so that $V-V_0$ is the velocity with which the insect starts. Then $M(V-V_0)=$ the initial impulsive action between the insect and chain $=mV_0$, so that

$$V_0=\frac{M}{M+m}V.$$

At any subsequent time t let x be the longer, and y the shorter part of the chain, z the depth of the insect below the pulley, and P the force exerted by the insect on the chain. We then have

$$m\frac{d^2x}{dt^2}=P+\frac{mg}{2l}(x-y);\ M\frac{d^2z}{dt^2}=Mg-P;\ \text{and } \frac{dx}{dt}-\frac{dz}{dt}=V.$$

Also $$x+y=2l.$$
These equations give
$$(M+m)\dot{x}^2=2(M-m)gx+\frac{mg}{l}x^2+A.$$
Also, when $x=l$, $\dot{x}=V_0$, etc.]

22. A uniform cord, of length l, hangs over a smooth pulley and a monkey, whose weight is that of the length k of the cord, clings to one end and the system remains in equilibrium. If he start suddenly, and continue to climb with uniform relative velocity along the cord, shew that he will cease to ascend in space at the end of time
$$\left(\frac{l+k}{2g}\right)^{\frac{1}{2}} \cosh^{-1}\left(1+\frac{l}{k}\right).$$

23. One end of a heavy uniform chain, of length $5a$ and mass $5ma$, is fixed at a point O and the other passes over a small smooth peg at a distance a above O; the whole hangs in equilibrium with the free end at a depth $2a$ below the peg. The free end is slightly displaced downwards; prove that its velocity V, when the length of the free portion is x, is given by
$$V^2=\frac{2(x-2a)^2(x+10a)}{(x+6a)^2}g,$$
and find the impulsive tension at O at the instant when the part of the chain between O and the peg becomes tight.

24 A machine gun, of mass M, stands on a horizontal plane and contains shot, of mass M'. The shot is fired at the rate of mass m per unit of time with velocity u relative to the ground. If the coefficient of sliding friction between the gun and the plane is μ, shew that the velocity of the gun backward by the time the mass M' is fired is
$$\frac{M'}{M}u-\frac{(M+M')^2-M^2}{2mM}\mu g.$$

CHAPTER VIII

OSCILLATORY MOTION AND SMALL OSCILLATIONS

114. In the previous chapters we have had several examples of oscillatory motion. We have seen that wherever the equation of motion can be reduced to the form $\ddot{x} = -n^2 x$, or $\ddot{\theta} = -n^2 \theta$, the motion is simple harmonic with a period of oscillation equal to $\frac{2\pi}{n}$. We shall give in this chapter a few examples of a more difficult character.

115. *Small oscillations.* The general method of finding the small oscillations about a position of equilibrium is to write down the general equations of motion of the body. If there is only one variable, x say, find the value of x which makes $\dot{x}$, $\ddot{x}$ ... etc. zero, *i.e.* which gives the position of equilibrium. Let this value be a.

In the equation of motion put $x = a + \xi$, where ξ is small. For a small oscillation ξ will be small so that we may neglect its square. The equation of motion then generally reduces to the form $\ddot{\xi} = -\lambda \xi$, in which case the time of a small oscillation is $\frac{2\pi}{\sqrt{\lambda}}$.

For example, suppose the general equation of motion is

$$\frac{d^2x}{dt^2} + f(x)\left(\frac{dx}{dt}\right)^2 = F(x).$$

For the position of equilibrium we have

$$F(x) = 0, \text{ giving } x = a.$$

Put $x = a + \xi$ and neglect ξ^2.

The equation becomes

$$\frac{d^2\xi}{dt^2} = F(a+\xi) = F(a) + \xi F'(a) + \ldots,$$

by Taylor's theorem.

Since $F(a) = 0$ this gives $\frac{d^2\xi}{dt^2} = \xi . F'(a)$.

If $F'(a)$ be negative, we have a small oscillation and the position of equilibrium given by $x = a$ is stable.

If $F'(a)$ be positive, the corresponding motion is not oscillatory and the position of equilibrium is unstable.

116. *Ex.* 1. *A uniform rod, of length 2a, is supported in a horizontal position by two strings attached to its ends whose other extremities are tied to a fixed point; if the unstretched length of each string be l and the modulus of elasticity be n times the weight of the rod, shew that in the position of equilibrium the strings are inclined to the vertical at an angle α such that*

$$a \cot \alpha - l \cos \alpha = \frac{l}{2n},$$

and that the time of a small oscillation about the position of equilibrium is

$$2\pi \sqrt{\frac{a}{g} \frac{\cot \alpha}{1+2n \cos^3 \alpha}}.$$

When the rod is at depth x below the fixed point, let θ be the inclination of each string to the vertical, so that $x = a \cot \theta$ and the tension

$$= nmg \frac{\dfrac{a}{\sin \theta} - l}{l} = \frac{nmg}{l} \frac{a - l \sin \theta}{\sin \theta}.$$

The equation of motion is then

$$m\ddot{x} = mg - 2 . \frac{nmg}{l} . \frac{a - l \sin \theta}{\sin \theta} \cos \theta,$$

i.e.
$$-\frac{a}{\sin^2 \theta} \ddot{\theta} + \frac{2a \cos \theta}{\sin^3 \theta} \dot{\theta}^2 = g - 2\frac{ng}{l} \frac{a - l \sin \theta}{\sin \theta} \cos \theta,$$

i.e.
$$\ddot{\theta} - 2 \cot \theta \dot{\theta}^2 = -\frac{g}{a} \sin^2 \theta + \frac{2ng}{al} \sin \theta \cos \theta (a - l \sin \theta) \quad \text{...(1)}.$$

In the position of equilibrium when $\theta = \alpha$, we have $\dot{\theta} = 0$ and $\ddot{\theta} = 0$, and

$$\therefore a \cot \alpha - l \cos \alpha = \frac{l}{2n} \quad \text{.....................(2)}.$$

For a small oscillation put $\theta = \alpha + \psi$, where ψ is small, and

$$\therefore \sin \theta = \sin \alpha + \psi \cos \alpha, \text{ and } \cos \theta = \cos \alpha - \psi \sin \alpha.$$

In this case $\dot{\theta}^2$ is the square of a small quantity and is negligible, and (1) gives

$$\ddot{\psi} = -\frac{g}{a} (\sin \alpha + \psi \cos \alpha)^2 + \frac{2ng}{al} (\sin \alpha + \psi \cos \alpha)(\cos \alpha - \psi \sin \alpha)$$
$$[a - l(\sin \alpha + \psi \cos \alpha)]$$

$$= -\frac{g}{a} (\sin^2 \alpha + 2\psi \sin \alpha \cos \alpha) + \frac{2ng}{al} [\sin \alpha \cos \alpha + \psi (\cos^2 \alpha - \sin^2 \alpha)]$$
$$\left(\frac{l}{2n} \tan \alpha - l\psi \cos \alpha\right) \text{ by equation (2)}$$

$$= -\psi.\frac{g}{a}[2n \sin \alpha \cos^2 \alpha + \tan \alpha]$$

$$= -\psi.\frac{g}{a}\tan \alpha(1+2n \cos^3 \alpha).$$

Hence the required time $= 2\pi\sqrt{\dfrac{a}{g}\,\dfrac{\cot \alpha}{1+2n \cos^3 \alpha}}$.

Making use of the principle of the last article, if the right-hand side of (1) be $f(\theta)$, the equation for small oscillations is

$$\ddot{\psi} = \psi.f'(\alpha),$$

and

$$f'(\alpha) = -\frac{2g}{a}\sin \alpha \cos \alpha + \frac{2ng}{al}(\cos^2 \alpha - \sin^2 \alpha)(a - l \sin \alpha) - \frac{2ng}{a}\sin \alpha \cos^2 \alpha$$

$$= \text{etc., as before.}$$

Ex. 2. *A heavy particle is placed at the centre of a smooth circular table ; n strings are attached to it and, after passing over small pulleys symmetrically arranged at the circumference of the table, each is attached to a mass equal to that of the particle on the table. If the particle be slightly displaced, shew that the time of an oscillation is* $2\pi\sqrt{\dfrac{a}{g}\left(1+\dfrac{2}{n}\right)}$.

Let O be the centre of the board, $A_1, A_2, \ldots, A_n$ the pulleys, and let the particle be displaced along a line OA lying between OA_n and OA_1. When its distance $OP = x$, let $PA_r = y_r$ and $\angle POA_r = \alpha_r$. Also, let a be the radius of the table and l the length of a string.

Then $y_r = \sqrt{a^2 + x^2 - 2ax \cos \alpha_r} = a\left(1 - \dfrac{x}{a}\cos \alpha_r\right)$, since x is very small.

Let T_r be the tension of the string PA_r.

Then $mg - T_r = m\dfrac{d^2}{dt^2}(l - y_r) = m\ddot{x}\cos \alpha_r$.

$$\therefore\ T_r = m(g - \ddot{x}\cos \alpha_r).$$

Also $T_r.\cos APA_r = m(g - \ddot{x}\cos \alpha_r).\dfrac{a \cos \alpha_r - x}{y_r}$

$$= m(g - \ddot{x}\cos \alpha_r)\,\frac{a \cos \alpha_r - x}{a}\left(1 + \frac{x}{a}\cos \alpha_r\right)$$

$$= \frac{m}{a^2}(g - \ddot{x}\cos \alpha_r)\,[a^2 \cos \alpha_r - ax + ax \cos^2 \alpha_r].$$

Now if $POA_1 = \alpha$, then

$$\Sigma \cos \alpha_r = \cos \alpha + \cos\left(\alpha + \frac{2\pi}{n}\right) + \ldots \text{ to } n \text{ terms} = 0,$$

$$\Sigma \cos^2 \alpha_r = \frac{1}{2}\left[1 + \cos 2\alpha + 1 + \cos\left(2\alpha + \frac{4\pi}{n}\right) + \ldots\right] = \frac{n}{2},$$

and $$\Sigma \cos^3 \alpha_r = \frac{1}{4}\,\Sigma[3\cos \alpha_r + \cos 3\alpha_r] = 0.$$

Therefore the equation of motion of P is

$$m\ddot{x} = \Sigma T_r \cos APA_r = \frac{m}{a^2}\left[-agx.n + gax\frac{n}{2} - a^2\ddot{x}\frac{n}{2}\right].$$

$$\therefore\ \ddot{x}\left(1 + \frac{n}{2}\right) = -\frac{ng}{2a}x.$$

$$\therefore\ \ddot{x} = -\frac{ng}{a(2+n)}x,$$

and the time of a complete oscillation $= 2\pi\sqrt{\dfrac{a(2+n)}{ng}}$.

It can easily be shewn that the sum of the resolved parts of the tensions perpendicular to OP vanishes if squares of x be neglected.

Ex. 3. *Two particles, of masses m and m', are connected by an elastic string of natural length a and modulus of elasticity λ; m is on a smooth table and describes a circle of radius c with uniform angular velocity; the string passes through a hole in the table at the centre of the circle and m' hangs at rest at a distance c' below the table. Shew that, if m be slightly disturbed, the periods $\frac{2\pi}{p}$ of small oscillations about this state of steady motion are given by the equation*

$$a^2cmm'p^4 - \{mc + (4c + 3c' - 3a)m'\}a\lambda p^2 (c + c' - a)\lambda^2 = 0.$$

At any time during the motion let x and y be the distances of m and m' from the hole and T the tension, so that the equations of motion are

$$m(\ddot{x} - x\dot{\theta}^2) = -T = -\lambda\frac{x+y-a}{a} \quad \text{.................(1)},$$

$$\frac{1}{x}\frac{d}{dt}(x^2\dot{\theta}) = 0 \quad \text{............................(2)},$$

and $$m'\ddot{y} = m'g - T = m'g - \lambda\frac{x+y-a}{a} \quad \text{.................(3)}.$$

(2) gives $x^2\theta = \text{const.} = h$,

so that (1) gives $$\ddot{x} = \frac{h^2}{x^3} - \frac{\lambda}{ma}(x+y-a) \quad \text{.......................(4).}$$

When $x=c$, $y=c'$ we have equilibrium, so that $\ddot{x}=\ddot{y}=0$ then, and hence from (3) and (4)

$$m'g = m\frac{h^2}{c^3} = \frac{\lambda(c+c'-a)}{a} \quad \text{.......................(5).}$$

Hence (4) and (3) give, on putting $x=c+\xi$ and $y=c'+\eta$ where ξ and η are small,

$$\ddot{\epsilon} = \frac{h^2}{c^3}\left(1-\frac{3\xi}{c}\right) - \frac{\lambda}{ma}(c+c'-a+\xi+\eta) = -\frac{\lambda}{am}\left[\frac{4c+3c'-3a}{c}\xi+\eta\right],$$

and $$\ddot{\eta} = -\frac{\lambda}{am'}(\xi+\eta).$$

To solve these equations, put

$$\xi = A\cos(pt+\beta) \text{ and } \eta = B\cos(pt+\beta).$$

On substituting we have

$$A\left[-p^2+\frac{\lambda}{am}\,\frac{4c+3c'-3a}{c}\right]+\frac{\lambda}{am}B=0,$$

and $$A\,.\frac{\lambda}{am'}+B\left[-p^2+\frac{\lambda}{am'}\right]=0.$$

Equating the two values of $\frac{A}{B}$ thus obtained, we have, on reduction,

$$a^2cmm'p^4 - \{mc+m'(4c+3c'-3a)\}\,a\lambda p^2 + 3(c+c'-a)\,\lambda^2 = 0.$$

This equation gives two values, p_1^2 and p_2^2, for p^2, both values being positive.

The solution is thus of the form

$$\xi = A_1\cos(p_1t+\beta_1)+A_2\cos(p_2t+\beta_2)$$

with a similar expression for η.

Hence the oscillations are compounded of two simple harmonic motions whose periods are $\frac{2\pi}{p_1}$ and $\frac{2\pi}{p_2}$.

EXAMPLES

1. Two equal centres of repulsive force are at a distance $2a$, and the law of force is $\frac{\mu}{r^2}+\frac{\mu a}{r^3}$; find the time of the small oscillation of a particle on the line joining the centres.

If the centres be attractive, instead of repulsive, find the corresponding time for a small oscillation on a straight line perpendicular to it.

2. A heavy particle is attached by two equal light extensible strings to two fixed points in the same horizontal line distant $2a$ apart; the length of each string when unstretched was b and the modulus of elasticity is λ. The particle is at rest when the strings are inclined at an angle α to the vertical, and is then slightly displaced in a vertical direction; shew that the time of a complete small oscillation is $2\pi\sqrt{\dfrac{a\cot\alpha}{g}\cdot\dfrac{a-b\sin\alpha}{a-b\sin^3\alpha}}$.

3. Two equal heavy particles are fastened to the ends of a weightless rod, of length $2c$, and oscillate in a vertical plane in a smooth sphere of radius a; shew that the time of the oscillation is the same as that of a simple pendulum of length $\dfrac{a^2}{\sqrt{a^2-c^2}}$.

4. A heavy rectangular board is symmetrically suspended in a horizontal position by four light elastic strings attached to the corners of the board and to a fixed point vertically above its centre. Shew that the period of the small vertical oscillations is $2\pi\left(\dfrac{g}{c}+\dfrac{4c^2\lambda}{k^3M}\right)^{-\frac{1}{2}}$ where c is the equilibrium-distance of the board below the fixed point, a is the length of a semi-diagonal, $k=\sqrt{a^2+c^2}$ and λ is the modulus.

5. A rod of mass m hangs in a horizontal position supported by two equal vertical elastic strings, each of modulus λ and natural length a. Shew that, if the rod receives a small displacement parallel to itself, the period of a horizontal oscillation is $2\pi\sqrt{a\left(\dfrac{1}{g}+\dfrac{m}{2\lambda}\right)}$.

6. A light string has one end attached to a fixed point A, and, after passing over a smooth peg B at the same height as A and distant $2a$ from A, carries a mass P at the other end. A ring, of mass M, can slide on the portion of the string between A and B. Shew that the time of its small oscillation about its position of equilibrium is

$$4\pi[aMP(M+P)\div g(4P^2-M^2)^{\frac{3}{2}}]^{\frac{1}{2}},$$

assuming that $2P>M$.

7. A particle, of mass m, is attached to a fixed point on a smooth horizontal table by a fine elastic string, of natural length a and modulus of elasticity λ, and revolves uniformly on the table, the string being stretched to a length b; shew that the time of a small oscillation for a small additional extension of the string is $2\pi\sqrt{\dfrac{mab}{\lambda(4b-3a)}}$.

8. Two particles, of masses m_1 and m_2, are connected by a string, of length a_1+a_2, passing through a smooth ring on a horizontal table, and the particles are describing circles of radii a_1 and a_2 with angular velocities ω_1 and ω_2 respectively. Shew that $m_1a_1\omega_1^2=m_2a_2\omega_2^2$, and that the small oscillation about this state takes place in the time

$$2\pi\sqrt{\frac{m_1+m_2}{3(m_1\omega_1^2+m_2\omega_2^2)}}.$$

9. A particle, of mass m, on a smooth horizontal table is attached by a fine string through a hole in the table to a particle of mass m' which hangs freely. Find the condition that the particle m may describe a circle uniformly, and shew that, if m' be slightly disturbed in a vertical direction, the period of the resulting oscillation is $2\pi\sqrt{\dfrac{(m+m')a}{3m'g}}$ where a is the radius of the circle.

10. On a wire in the form of a parabola, whose latus-rectum is $4a$ and whose axis is vertical and vertex downwards, is a bead attached to the focus by an elastic string of natural length $\frac{a}{2}$, whose modulus is equal to the weight of the bead. Shew that the time of a small oscillation is

$$2\pi\sqrt{\frac{a}{g}}.$$

11. At the corners of a square whose diagonal is $2a$, are the centres of four equal attractive forces equal to any function $m.f(x)$ of the distance x of the attracted particle m; the particle is placed in one of the diagonals very near the centre; shew that the time of a small oscillation is

$$\pi\sqrt{2}\left\{\frac{1}{a}f(a)+f'(a)\right\}^{-\frac{1}{2}}.$$

12. Three particles, of equal mass m, are connected by equal elastic strings and repel one another with a force μ times the distance. In equilibrium each string is double its natural length; shew that if the particles are symmetrically displaced (so that the three strings always form an equilateral triangle) they will oscillate in period $2\pi\sqrt{\frac{m}{3\mu}}$.

13. Every point of a fine uniform circular ring repels a particle with a force which varies inversely as the square of the distance; shew that the time of a small oscillation of the particle about its position of equilibrium at the centre of the ring varies as the radius of the ring.

14. A uniform straight rod, of length $2a$, moves in a smooth fixed tube under the attraction of a fixed particle, of mass m, which is at a distance c from the tube. Shew that the time of a small oscillation is

$$2\pi\sqrt[4]{\frac{(a^2+c^2)^3}{\gamma^2m^2}}.$$

15. A uniform straight rod is perpendicular to the plane of a fixed uniform circular ring and passes through its centre; every particle of the ring attracts every particle of the rod with a force varying inversely as the square of the distance; find the time of a small oscillation about the position of equilibrium, the motion being perpendicular to the plane of the ring.

16. A particle, of mass M, hangs at the end of a vertical string, of length l, from a fixed point O, and attached to it is a second string which passes over a small pulley, in the same horizontal plane as O and distant l from O, and is attached at its other end to a mass m, which is small compared with M. When m is allowed to drop, shew that the system oscillates about a mean position with a period $2\pi\left[1+\frac{m}{8M}(2+\sqrt{2})\right]\sqrt{\frac{l}{g}}$ approximately, and find the mean position.

17. A heavy particle hangs in equilibrium suspended by an elastic string whose modulus of elasticity is three times the weight of the particle. It is then slightly displaced; shew that its path is a small arc of a parabola. If the displacement be in a direction making an angle $\cot^{-1} 4$ with the horizon, shew that the arc is the portion of a parabola cut ofl by the latus-rectum.

117. *A particle of mass m moves in a straight line under a force mn^2 (distance) towards a fixed point in the straight line and under a small resistance to its motion equal to $m.\mu$ (velocity); to find the motion.*

The equation of motion is

$$m\frac{d^2x}{dt^2}=-m.n^2x-m.\mu\frac{dx}{dt},$$

i.e.
$$\frac{d^2x}{dt^2}+\mu\frac{dx}{dt}+n^2x=0 \quad\text{.........................(1).}$$

[This is clearly the equation of motion if the particle is moving so that x is increasing.

If as in the second figure the particle is moving so that x decreases, *i.e.* towards the left, the frictional resistance is towards the right, and equals $m.\mu v$. But in this case $\frac{dx}{dt}$ is negative, so that the value of v is $-\frac{dx}{dt}$, the frictional resistance is thus $m\mu\left(-\frac{dx}{dt}\right)\rightarrow$. The equation of motion is then

$$m\frac{d^2x}{dt^2}=-mn^2x+m\mu\left(-\frac{dx}{dt}\right),$$

which again becomes (1).

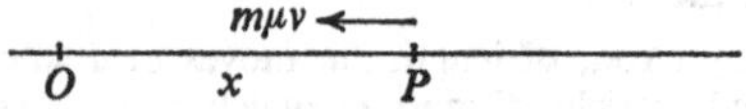

$m\mu v \leftarrow$

O x P

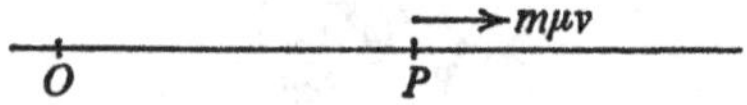

$\rightarrow m\mu v$

O P

Hence (1) gives the motion for all positions of P to the right of O, irrespective of the direction in which P is moving.

Similarly it can be shewn to be the equation of motion for positions of P to the left of O, whatever be the direction in which P is moving.]

To solve (1), put $x=Le^{pt}$, and we have

$$p^2+\mu p+n^2=0,$$

giving
$$p=-\frac{\mu}{2}\pm i\sqrt{n^2-\frac{\mu^2}{4}}.$$

$$\therefore\ x=e^{-\frac{\mu}{2}t}\left[Le^{\sqrt{n^2-\frac{\mu^2}{4}}it}+L'e^{-\sqrt{n^2-\frac{\mu^2}{4}}it}\right],$$

i.e.
$$x=Ae^{-\frac{\mu}{2}t}\cos\left[\sqrt{n^2-\frac{\mu^2}{4}}t+B\right] \quad\text{..............(2),}$$

where A and B are arbitrary constants.

If μ be small, then $Ae^{-\frac{\mu}{2}t}$ is a *slowly* varying quantity, so that (2) approximately represents a simple harmonic motion of period

$2\pi \div \sqrt{n^2 - \frac{\mu^2}{4}}$, whose amplitude, $Ae^{-\frac{\mu}{2}t}$, is a slowly decreasing quantity. Such a motion is called a *damped* oscillation and μ measures the damping.

This period depends on the square of μ, so that, to the first order of approximation, this small frictional resistance has no effect on the period of the motion. Its effect is chiefly seen in the decreasing amplitude of the motion, which $= A\left(1 - \frac{\mu}{2}t\right)$ when squares of μ are neglected, and therefore depends on the first power of μ.

Such a vibration as the above is called a free vibration. It is the vibration of a particle which moves under the action of no external periodic force.

If μ be not small compared with n, the motion cannot be so simply represented, but for all values of μ, $< 2n$, the equation (2) gives the motion.

From (2) we have, on differentiating, that $\dot{x} = 0$ when

$$\tan\left[\sqrt{n^2 - \frac{\mu^2}{4}}\,t + B\right] = -\frac{\mu}{\sqrt{4n^2 - \mu^2}} = \tan\alpha \text{ (say)} \ldots\ldots(3),$$

giving solutions of the form

$$\sqrt{n^2 - \frac{\mu^2}{4}}\,t + B = \alpha, \quad \pi + \alpha, \quad 2\pi + \alpha, \ldots .$$

Hence $\dot{x}$ is zero, that is the velocity vanishes, at the ends of periods of time differing by $\pi \div \sqrt{n^2 - \frac{\mu^2}{4}}$.

The times of oscillation thus still remain constant, though they are greater than when there is no frictional resistance.

If the successive values of t obtained from (3) are t_1, t_2, t_3, ... then the corresponding values of (2) are

$$Ae^{-\frac{\mu t_1}{2}} \cos\alpha, \quad -Ae^{-\frac{\mu t_2}{2}} \cos\alpha, \quad Ae^{-\frac{\mu t_3}{2}} \cos\alpha, \ldots$$

so that the amplitudes of the oscillations form a decreasing G.P. whose common ratio $= e^{-\frac{\mu}{2}(t_2 - t_1)} = e^{-\frac{\mu\pi}{2} \div \sqrt{n^2 - \frac{\mu^2}{4}}}$.

If $\mu > 2n$, the form of the solution changes; for now

$$p = -\frac{\mu}{2} \pm \sqrt{\frac{\mu^2}{4} - n^2},$$

and the general solution is

$$x=e^{-\frac{\mu t}{2}}\left[Le^{\sqrt{\left(\frac{\mu^2}{4}-n^2\right)}.t}+L'e^{-\sqrt{\left(\frac{\mu^2}{4}-n^2\right)}.t}\right]$$

$$=e^{-\frac{\mu t}{2}}A_1\cosh\left[\sqrt{\left(\frac{\mu^2}{4}-n^2\right)}.t+B_1\right].$$

In this case the motion is no longer oscillatory.

If $\mu=2n$, we have by the rules of Differential Equations

$$x=Le^{-nt}+\underset{\gamma=0}{\text{Lt}}\ Me^{-(n+\gamma)t}$$

$$=Le^{-nt}+\underset{\gamma=0}{\text{Lt}}\ Me^{-nt}(1-\gamma t+\text{squares})$$

$$=L_1e^{-nt}+M_1te^{-nt}=e^{-nt}(L_1+M_1t).$$

Ex. The time of oscillation of a particle when there is no frictional resistance is $1\frac{1}{2}$ secs.; if there be a frictional resistance equal to $\frac{1}{4}\times m\times$ velocity, find the consequent alteration in the period and the factor which gives the ratio of successive maximum amplitudes.

118. The motion of the last article may be represented graphically; let time t be represented by distances measured along the horizontal axis and the displacement x of the particle by the vertical ordinates. Then any displacement such as that of the last article will be represented as in the figure.

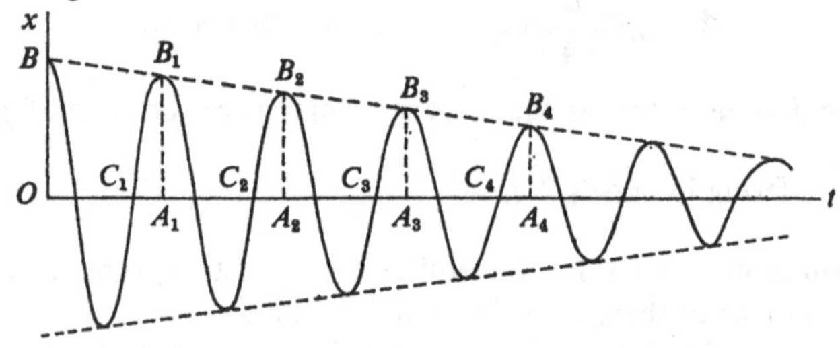

The dotted curve on which all the ends of the maximum ordinates lie is $x=\pm Ae^{-\frac{\mu}{2}t}\cos\alpha$. The times A_1A_2, A_2A_3, A_3A_4, ... of successive periods are equal, whilst the corresponding maximum ordinates A_1B_1, A_2B_2, ... form a decreasing geometrical progression whose ratio

$$=\frac{A\cos\alpha e^{-\frac{\mu}{2}.OA_2}}{A\cos\alpha e^{-\frac{\mu}{2}.OA_1}}=e^{-\frac{\mu}{2}.\tau},$$

where τ is the time of an oscillation.

If we have a particle moving with a damped vibration of this character, and we make it automatically draw its own displacement curve as in the above figure, we can from the curve determine the forces acting on it. For measuring the successive distances C_1C_2, C_2C_3, ..., etc., and taking their mean, we have the periodic time τ which we found in the last article to be $2\pi \div \sqrt{n^2-\frac{\mu^2}{4}}$, so that

$$\frac{4\pi^2}{\tau^2}=n^2-\frac{\mu^2}{4}.$$

Again, measuring the maximum ordinates A_1B_1, A_2B_2, A_3B_3, ..., finding the values of $\frac{A_2B_2}{A_1B_1}, \frac{A_3B_3}{A_2B_2}$ ..., and taking their mean, λ, we have the value of the quantity $e^{-\frac{\mu\tau}{2}}$, so that

$$-\frac{\mu}{2}\tau=\log_e\lambda.$$

We thus have the values of n^2 and μ, giving the restorative force and the frictional resistance of the motion.

119. *A point is moving in a straight line with an acceleration μx towards a fixed centre in the straight line and with an additional acceleration L cos pt; to find the motion.*

The equation of motion is

$$\frac{d^2x}{dt^2}=-\mu x+L\cos pt.$$

The solution of this is

$$x=A\cos(\sqrt{\mu}t+B)+L\frac{1}{D^2+\mu}\cos pt$$

$$=A\cos[\sqrt{\mu}t+B]+\frac{L}{\mu-p^2}\cos pt \quad ....(1).$$

If the particle starts from rest at a distance a at zero time, we have $B=0$ and $A=a-\frac{L}{\mu-p^2}$.

$$\therefore x=\left[a-\frac{L}{\mu-p^2}\right]\cos\sqrt{\mu}t+\frac{L}{\mu-p^2}\cos pt \quad ....(2).$$

The motion of the point is thus compounded of two simple harmonic motions whose periods are $\frac{2\pi}{\sqrt{\mu}}$ and $\frac{2\pi}{p}$.

From the right-hand side of (2) it follows that, if p be nearly equal to $\sqrt{\mu}$, the coefficient $\frac{L}{\mu-p^2}$ becomes very great; in other words, the effect of the disturbing acceleration $L \cos pt$ becomes very important. It follows that the ultimate effect of a periodic disturbing force depends not only on its magnitude L, but also on its period, and that, if the period be nearly that of the free motion, its effect may be very large even though its absolute magnitude L be comparatively small.

If $p=\sqrt{\mu}$, the terms in (2) become infinite. In this case the solution no longer holds, and the second term in (1)

$$=L\frac{1}{D^2+\mu}\cos[\sqrt{\mu}t]=L\,\underset{\gamma=0}{\text{Lt}}\frac{1}{D^2+\mu}\cos(\sqrt{\mu}+\gamma)t$$

$$=L\,\underset{\gamma=0}{\text{Lt}}\frac{1}{\mu-(\sqrt{\mu}+\gamma)^2}\cos[\sqrt{\mu}+\gamma]t$$

$$=-L\frac{1}{2\sqrt{\mu}}[\text{something infinite}-t\sin\sqrt{\mu}t].$$

Hence, by the ordinary theory of Differential Equations, the solution is

$$x=A_1\cos[\sqrt{\mu}t+B_1]+\frac{L}{2\sqrt{\mu}}t\sin\sqrt{\mu}t.$$

If, as before, $x=a$ and $\dot{x}=0$ when $t=0$, this gives

$$x=a\cos\sqrt{\mu}t+\frac{L}{2\sqrt{\mu}}t\sin\sqrt{\mu}t,$$

and hence $$\dot{x}=\left(\frac{L}{2\sqrt{\mu}}-a\sqrt{\mu}\right)\sin\sqrt{\mu}t+\frac{L}{2}t\cos\sqrt{\mu}t.$$

It follows that the amplitude of the motion, and also the velocity, become very great as t gets large.

120. If, instead of a linear motion of the character of the previous article, we have an angular motion, as in the case of a simple pendulum, the equation of motion is

$$\frac{d^2\theta}{dt^2}=-\frac{g}{l}\theta+L\cos pt,$$

and the solution is similar to that of the last article.

In this case, if L be large compared with $\frac{g}{l}$ or if p be very nearly equal to $\sqrt{\frac{g}{l}}$, the free time of vibration, θ is no longer small throughout the motion and the equation of motion must be replaced by the more accurate equation.

$$\frac{d^2\theta}{dt^2}=-\frac{g}{l}\sin\theta+L\cos pt.$$

121. As an example of the accumulative effect of a periodic force whose period coincides with the free period of the system, consider the case of a person in a swing to whom a small impulse is applied when he is at the highest point of his swing. This impulse is of the nature of a periodic force whose period is just equal to that of the swing and the effect of such an impulse is to make the swing to move through a continually increasing angle.

If however the period of the impulse is not the same as that of the swing, its effect is sometimes to help, and sometimes to oppose, the motion.

If its period is very nearly, but not quite, that of the swing its effect is for many successive applications to increase the motion, and then for many further applications to decrease the motion. In this case a great amplitude of motion is at first produced, which is then gradually destroyed, and then produced again, and so on.

122. *A particle, of mass m, is moving in a straight line under a force mn^2 (distance) towards a fixed point in the straight line, and under a frictional resistance equal to $m.\mu$ (velocity) and a periodic force $mL\cos pt$; to find the motion.*

The equation of motion is

$$\frac{d^2x}{dt^2} = -n^2x - \mu\frac{dx}{dt} + L\cos pt,$$

i.e.
$$\frac{d^2x}{dt^2} + \mu\frac{dx}{dt} + n^2x = L\cos pt.$$

The complementary function is

$$Ae^{-\frac{\mu}{2}t}\cos\left[\sqrt{n^2 - \frac{\mu^2}{4}}\,t + B\right] \quad \text{..................(1)},$$

assuming $\mu < 2n$, and the particular integral

$$= \frac{1}{n^2 - p^2 + \mu D} L\cos pt = L\frac{(n^2-p^2)\cos pt + \mu p \sin pt}{(n^2-p^2)^2 + \mu^2 p^2}$$

$$= \frac{L\sin\epsilon}{\mu p}\cos(pt - \epsilon) \quad \text{..(2)},$$

where
$$\tan\epsilon = \frac{\mu p}{n^2 - p^2}.$$

The motion is thus compounded of two oscillations; the first is called the free vibration and the second the forced vibration.

Particular case. Let the period $\frac{2\pi}{p}$ of the disturbing force be equal to $\frac{2\pi}{n}$, the free period.

The solution is then, for the forced vibration,

$$x = \frac{L}{\mu n} \sin nt.$$

If, as is usually the case, μ is also small, this gives a vibration whose maximum amplitude is very large. Hence we see that a small periodic force *may*, if its period is nearly equal to that of the free motion of the body, produce effects out of all proportion to its magnitude.

Hence we see why there may be danger to bridges from the accumulative effect of soldiers marching over them in step, why ships roll so heavily when the waves are of the proper period, and why a railway-carriage may oscillate considerably in a vertical direction when it is travelling at such a rate that the time it takes to go the length of a rail is equal to a period of vibration of the springs on which it rests.

Many other phenomena, of a more complicated character, are explainable on similar principles to those of the above simple case.

123. There is a very important difference between the free vibration given by (1) and the forced vibration given by (2).

Suppose for instance that the particle was initially at rest at a given finite distance from the origin. The arbitrary constants A and B are then easily determined and are found to be finite. The factor $e^{-\frac{\mu}{2}t}$ in (1), which gradually diminishes as time goes on, causes the expression (1) to continually decrease and ultimately to vanish. Hence the free vibration gradually dies out.

The forced vibration (2) has no such diminishing factor but is a continually repeating periodic function. Hence finally it is the only motion of any importance.

124. *Small oscillations of a simple pendulum under gravity, where the resistance* $= \mu$ *(velocity)*^{2} *and* μ *is small.*

The equation of motion is

$$l\ddot{\theta} = -g\theta + \mu l^2 \dot{\theta}^2 \quad \text{.........................(1).}$$

[If the pendulum start from rest at an inclination α to the vertical, the same equation is found to hold until it comes to rest on the other side of the vertical.]

For a first approximation, neglect the small term $\mu l^2 \dot{\theta}^2$, and we have

$$\theta = A \cos\left[\sqrt{\frac{g}{l}}\, t + B\right].$$

For a second approximation, put this value of θ in the small terms on the right-hand side of (1), and it becomes

$$\ddot{\theta}+\frac{g}{l}\theta=\mu l.\frac{g}{l}A^2\sin^2\left[\sqrt{\frac{g}{l}}\,t+B\right]$$

$$=\frac{A^2\mu g}{2}\left[1-\cos\left(2\sqrt{\frac{g}{l}}\,t+2B\right)\right].$$

$$\therefore\ \theta=A\cos\left[\sqrt{\frac{g}{l}}\,t+B\right]+\frac{A^2\mu l}{2}+\frac{A^2\mu l}{6}\cos\left[2\sqrt{\frac{g}{l}}\,t+2B\right]\quad\text{...(2)},$$

where $$\alpha=A\cos B+\frac{A^2\mu l}{2}+\frac{A^2\mu l}{6}\cos 2B,$$

and $$0=-A\sin B-\frac{A^2\mu l}{3}\sin 2B.$$

$\therefore$ $B=0$, and $A=\alpha-\frac{2}{3}\alpha^2\mu l$, squares of μ being neglected.

Hence (2) gives

$$\theta=\left(\alpha-\frac{2}{3}\alpha^2\mu l\right)\cos\left(\sqrt{\frac{g}{l}}\,t\right)+\frac{\alpha^2\mu l}{2}+\frac{\alpha^2\mu l}{6}\cos\left(2\sqrt{\frac{g}{l}}\,t\right)\quad\text{...(3)},$$

and hence

$$\dot{\theta}=-\sqrt{\frac{g}{l}}\left(\alpha-\frac{2}{3}\alpha^2\mu l\right)\sin\left(\sqrt{\frac{g}{l}}\,t\right)-\frac{\alpha^2\mu l}{3}\sqrt{\frac{g}{l}}\sin\left(2\sqrt{\frac{g}{l}}\,t\right)$$

$$\text{............(4)}.$$

$\therefore$ $\dot{\theta}$ is zero when $\sin\sqrt{\frac{g}{l}}\,t=0$, *i.e.* when $t=\pi\sqrt{\frac{l}{g}}$.

The time of a swing from rest to rest is therefore unaltered by the resistance, provided the square of μ be neglected. Again, when $t=\pi\sqrt{\frac{l}{g}}$,

$$\theta=-\left(\alpha-\frac{2}{3}\alpha^2\mu l\right)+\frac{\alpha^2\mu l}{2}+\frac{\alpha^2\mu l}{6}=-\left(\alpha-\frac{4}{3}\alpha^2\mu l\right).$$

Hence the amplitude of the swing is diminished by $\frac{4}{3}\alpha^2\mu l$.

Let the pendulum be passing through the lowest point of its path at time

$$\sqrt{\frac{l}{g}}\left(\frac{\pi}{2}+T\right),\ \text{where } T \text{ is small.}$$

Then (3) gives

$$0=\left(a-\frac{2}{3}a^2\mu l\right)(-\sin T)+\frac{a^2\mu l}{2}-\frac{a^2\mu l}{6}\cos 2T,$$

i.e.

$$T\left(a-\frac{2}{3}a^2\mu l\right)=\frac{a^2\mu l}{2}-\frac{a^2\mu l}{6}=\frac{a^2\mu l}{3},$$

and

$$\therefore\ T=\frac{a\mu l}{3}.$$

Hence the time of swinging to the lowest point

$$=\sqrt{\frac{l}{g}}\left(\frac{\pi}{2}+\frac{a\mu l}{3}\right)$$

and of swinging up to rest again

$$=\pi\sqrt{\frac{l}{g}}-\sqrt{\frac{l}{g}}\left(\frac{\pi}{2}+\frac{a\mu l}{3}\right)=\sqrt{\frac{l}{g}}\left(\frac{\pi}{2}-\frac{a\mu l}{3}\right).$$

EXAMPLES

1. Investigate the rectilinear motion given by the equation

$$A\frac{d^4x}{dt^4}+B\frac{d^2x}{dt^2}+Cx+D=0,$$

and shew that it is compounded of two harmonic oscillations if the equation $Ay^2+By+C=0$ has real negative roots.

2. A particle is executing simple harmonic oscillations of amplitude a, under an attraction $\frac{\mu x}{a}$. If a small disturbing force $\frac{\nu x^3}{a^3}$ be introduced (the amplitude being unchanged) shew that the period is, to a first approximation, decreased in the ratio $1-\frac{3_\nu}{8_\mu}$: 1.

3. Two heavy particles, of masses m and m', are fixed to two points, A and B, of an elastic string OAB. The end O is attached to a fixed point and the system hangs freely. A small vertical disturbance being given to it, find the times of the resultant oscillations.

4. A particle hangs at rest at the end of an elastic string whose unstretched length is a. In the position of equilibrium the length of the string is b, and $\frac{2\pi}{n}$ is the time of an oscillation about this position. At time zero, when the particle is in equilibrium, the point of suspension begins to move so that its downward displacement at time t is $c \sin pt$. Shew that the length of the string at time t is

$$b-\frac{cnp}{n^2-p^2}\sin nt+\frac{cp^2}{n^2-p^2}\sin pt.$$

If $p=n$, shew that the length of the string at time t is

$$b-\frac{c}{2}\sin nt-\frac{nct}{2}\cos nt.$$

5. A helical spring supports a weight of 20 lbs. attached to its lower end; the natural length of the spring is 12 inches and the load causes it to extend to a length of $13\frac{1}{2}$ inches. The upper end of the spring is then given a vertical simple harmonic motion, the full extent of the displacement being 2 inches and 100 complete vibrations occurring in one minute. Neglecting air resistance and the inertia of the spring, investigate the motion of the suspended mass after the motion has become steady, and shew that the amplitude of the motion set up is about $3\frac{1}{2}$ inches.

6. If a pendulum oscillates in a medium the resistance of which varies as the velocity, shew that the oscillations are isochronous.

7. The time of a complete oscillation of a pendulum making small oscillations *in vacuo* is 2 seconds; if the angular retardation due to the air is $\cdot 04 \times$(angular velocity of the pendulum) and the initial amplitude is 1°, find the inclination of the pendulum to the vertical at any subsequent time, and shew that the amplitude will in 10 complete oscillations be reduced to $40'$ approximately. [$\log_{10} e = \cdot 4343$.]

8. The point of suspension of a simple pendulum of length l has a horizontal motion given by $x = a \cos mt$. Find the effect on the motion of the particle.

Consider in particular the motion when m^2 is equal, or nearly equal, to $\frac{g}{l}$. In the latter case if the pendulum be passing through its vertical position with angular velocity ω at zero time, shew that, so long as it is small, the inclination to the vertical at time t

$$= \omega\sqrt{\frac{l}{g}}\left[1 + \frac{ag}{2\omega l^2}t\right] \sin\sqrt{\frac{g}{l}}t.$$

[If O' be the position of the point of suspension at time t its acceleration is $\ddot{x}$. Hence the accelerations of P, the bob of the pendulum, are $l\ddot{\theta}$ perpendicular to $O'P$, $l\dot{\theta}^2$ along PO, and $\ddot{x}$ parallel to OO'.

Hence resolving perpendicular to $O'P$,

$$l\ddot{\theta} + \ddot{x}\cos\theta = -g\sin\theta = -g\theta,$$

i.e.

$$\ddot{\theta} = -\frac{g}{l}\theta + \frac{am^2}{l}\cos mt, \text{ since } \theta \text{ is small.}$$

Now solve as in Art. 119.]

9. The point of support of a simple pendulum, of weight w and length l, is attached to a massless spring which moves backwards and forwards in a horizontal line; shew that the time of vibration $= 2\pi\sqrt{\frac{l}{g}}\left(1 + \frac{w}{W}\right)$, where W is the weight required to stretch the spring a distance l.

10. Two simple pendulums, each of length a, are hung from two points in the same horizontal plane at a distance b apart; the bob of each is of mass m and the mutual attraction is $\frac{\lambda m^2}{(\text{dist.})^2}$, where λ is small compared with g; shew that, if the pendulums be started so that they are always moving in opposite directions, the the time of oscillation of each is $2\pi\sqrt{\frac{a}{g}}\left(1 + \frac{2\lambda ma}{b^3 g}\right)$ nearly, about a mean position inclined at $\frac{\lambda m}{gb^2}$ radians nearly to the vertical.

11. A pendulum is suspended in a ship so that it can swing in a plane at right angles to the length of the ship, its excursions being read off on a scale fixed to the ship. The free period of oscillation of the pendulum is one second and its point of suspension is 10 feet above the centre of gravity of the ship. Shew that when the ship is rolling through a small angle on each side of the vertical with a period of 8 secs., the apparent angular movement of the pendulum will be approximately 20 per cent. greater than that of the ship.

12. The point of suspension of a simple pendulum of length l moves in a horizontal circle of radius a with constant angular velocity ω; when the motion has become steady, shew that the inclination α to the vertical of the thread of the pendulum is given by the equation

$$\omega^2(a+l\sin\alpha)-g\tan\alpha=0.$$

13. A pendulum consists of a light elastic string with a particle at one end and fastened at the other. In the position of equilibrium the string is stretched to $\frac{4}{3}$ of its natural length l. If the particle is slightly displaced from the position of equilibrium and is then let go, trace its subsequent path and find the times of its component oscillations.

CHAPTER IX

MOTION IN THREE DIMENSIONS

125. *To find the accelerations of a particle in terms of polar coordinates.*

Let the coordinates of any point P be r, θ, and ϕ, where r is the distance of P from a fixed origin O, θ is the angle that OP makes with a fixed axis Oz, and ϕ is the angle that the plane zOP makes with a fixed plane zOx.

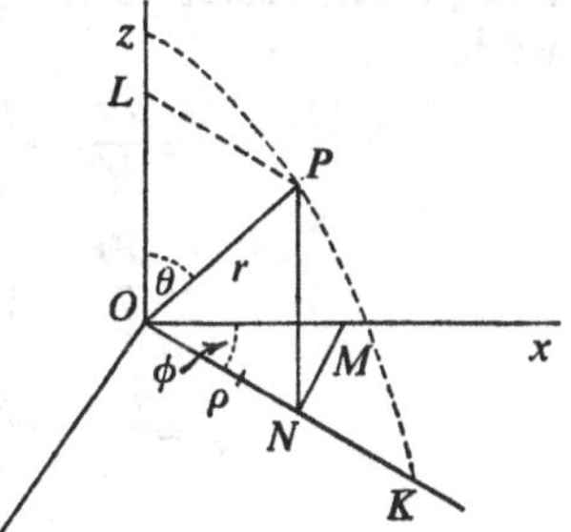

Draw PN perpendicular to the plane xOy and let $ON=\rho$.

Then the accelerations of P are $\frac{d^2x}{dt^2}$, $\frac{d^2y}{dt^2}$, and $\frac{d^2z}{dt^2}$, where x, y and z are the coordinates of P.

Since the polar coordinates of N, which is always in the plane xOy, are ρ and ϕ, its accelerations are, as in Art. 49,

$$\frac{d^2\rho}{dt^2}-\rho\left(\frac{d\phi}{dt}\right)^2 \text{ along } ON,$$

and
$$\frac{1}{\rho}\frac{d}{dt}\left(\rho^2\frac{d\phi}{dt}\right) \text{ perpendicular to } ON.$$

Also the acceleration of P relative to N is $\frac{d^2z}{dt^2}$ along NP.

Hence the accelerations of P are

$$\frac{d^2\rho}{dt^2}-\rho\left(\frac{d\phi}{dt}\right)^2 \text{ along } LP,$$

$$\frac{1}{\rho}\frac{d}{dt}\left(\rho^2\frac{d\phi}{dt}\right) \text{ perpendicular to the plane } zPK,$$

and
$$\frac{d^2z}{dt^2} \text{ parallel to } Oz.$$

Now, since $z=r\cos\theta$ and $\rho=r\sin\theta$, it follows, as in Art. 50, that accelerations $\frac{d^2z}{dt^2}$ and $\frac{d^2\rho}{dt^2}$, along and perpendicular to Oz in the plane

zPK, are equivalent to $\frac{d^2r}{dt^2}-r\left(\frac{d\theta}{dt}\right)^2$ along OP and $\frac{1}{r}\frac{d}{dt}\left(r^2\frac{d\theta}{dt}\right)$ perpendicular to OP in the plane zPK.

Also the acceleration $-\rho\left(\frac{d\phi}{dt}\right)^2$ along LP is equivalent to $-\rho \sin \theta\left(\frac{d\phi}{dt}\right)^2$ along OP and $-\rho \cos \theta \left(\frac{d\phi}{dt}\right)^2$ perpendicular to OP.

Hence if α, β, γ be the accelerations of P respectively along OP, perpendicular to OP in the plane zPK in the direction of θ increasing, and perpendicular to the plane zPK in the direction of ϕ increasing, we have

$$\alpha=\frac{d^2r}{dt^2}-r\left(\frac{d\theta}{dt}\right)^2-\rho \sin \theta\left(\frac{d\phi}{dt}\right)^2$$

$$=\frac{d^2r}{dt^2}-r\left(\frac{d\theta}{dt}\right)^2-r \sin^2 \theta\left(\frac{d\phi}{dt}\right)^2 \quad \text{.....................(1)},$$

$$\beta=\frac{1}{r}\frac{d}{dt}\left(r^2\frac{d\theta}{dt}\right)-\rho \cos \theta\left(\frac{d\phi}{dt}\right)^2$$

$$=\frac{1}{r}\frac{d}{dt}\left(r^2\frac{d\theta}{dt}\right)-r \sin \theta \cos \theta\left(\frac{d\phi}{dt}\right)^2 \quad \text{..................(2)},$$

and $$\gamma=\frac{1}{\rho}\frac{d}{dt}\left(\rho^2\frac{d\phi}{dt}\right)=\frac{1}{r \sin \theta}\frac{d}{dt}\left(r^2 \sin^2 \theta\frac{d\phi}{dt}\right) \quad \text{............(3)}.$$

126. Cylindrical coordinates.

It is sometimes convenient to refer the motion of P to the coordinates z, ρ, and ϕ, which are called cylindrical coordinates.

As in the previous article the accelerations are then

$$\frac{d^2\rho}{dt^2}-\rho\left(\frac{d\phi}{dt}\right)^2 \text{ along } LP,$$

$$\frac{1}{\rho}\frac{d}{dt}\left(\rho^2\frac{d\phi}{dt}\right) \text{ perpendicular to the plane } zPK,$$

and $$\frac{d^2z}{dt^2} \text{ parallel to } Oz.$$

127. *A particle is attached to one end of a string, of length l, the other end of which is tied to a fixed point O. When the string is inclined at an acute angle α to the downward-drawn vertical the particle is projected horizontally and perpendicular to the string with a velocity V; to find the resulting motion.*

In the expressions (1), (2), and (3) of Art. 125 for the accelerations we here have $r=l$.

The equations of motion are thus

$$-l\dot{\theta}^2 - l\sin^2\theta\dot{\phi}^2 = -\frac{T}{m}+g\cos\theta \quad \text{..............(1)},$$

$$l\ddot{\theta} - l\cos\theta\sin\theta\dot{\phi}^2 = -g\sin\theta \quad \text{................(2)},$$

and $$\frac{1}{\sin\theta}\frac{d}{dt}(\sin^2\theta\dot{\phi})=0 \quad \text{......................(3)}.$$

The last equation gives

$$\sin^2\theta\dot{\phi} = \text{constant} = \sin^2\alpha[\dot{\phi}]_0 = \frac{V\sin\alpha}{l} \quad \text{......(4)}.$$

On substituting for $\dot{\phi}$ in (2), we have

$$\ddot{\theta} - \frac{V^2\sin^2\alpha}{l^2}\frac{\cos\theta}{\sin^3\theta} = -\frac{g}{l}\sin\theta \quad \text{.................(5)}.$$

$$\therefore\ \dot{\theta}^2 + \frac{V^2\sin^2\alpha}{l^2}\cdot\frac{1}{\sin^2\theta} = \frac{2g}{l}\cos\theta + A,$$

where $$0 + \frac{V^2\sin^2\alpha}{l^2}\cdot\frac{1}{\sin^2\alpha} = \frac{2g}{l}\cos\alpha + A.$$

$$\therefore\ \dot{\theta}^2 = \frac{V^2\sin^2\alpha}{l^2}\left[\frac{1}{\sin^2\alpha} - \frac{1}{\sin^2\theta}\right] - \frac{2g}{l}(\cos\alpha - \cos\theta) \quad \text{......(6)}$$

$$= \frac{2g}{l}(\cos\alpha - \cos\theta)\left(2n^2\frac{\cos\alpha + \cos\theta}{\sin^2\theta} - 1\right),$$

where $V^2 = 4lgn^2$.

Hence $\dot{\theta}$ is again zero when

$$2n^2(\cos\alpha + \cos\theta) = \sin^2\theta,$$

i.e. when $$\cos\theta = -n^2 \pm \sqrt{1 - 2n^2\cos\alpha + n^4}.$$

The lower sign gives an inadmissible value for θ. The only inclination at which $\dot{\theta}$ again vanishes is when $\theta = \theta_1$, where

$$\cos\theta_1 = -n^2 + \sqrt{1 - 2n^2\cos\alpha + n^4}.$$

The motion is therefore confined between values α and θ_1, of θ.

The motion of the particle is always above or below the starting point, according as $\theta_1 \gtrless \alpha$,

i.e. according as $\cos\theta_1 \lessgtr \cos\alpha$,

i.e. „ „ $\sqrt{1 - 2n^2\cos\alpha + n^4} \lessgtr n^2 + \cos\alpha$,

i.e. „ „ $1 - 2n^2\cos\alpha \lessgtr \cos^2\alpha + 2n^2\cos\alpha$,

i.e. „ „ $n^2 \gtrless \dfrac{\sin^2\alpha}{4\cos\alpha}$,

i.e. „ „ $V^2 \gtrless lg\sin\alpha\tan\alpha$.

The tension of the string at any instant is now given by equation (1). In the foregoing it is assumed that T does not vanish during the motion.

The square of the velocity at any instant

$$=(l\dot{\theta})^2+(l\sin\theta\dot{\phi})^2=l^2(\dot{\theta}^2+\dot{\phi}^2\sin^2\theta).$$

Hence the Principle of Energy gives

$$\tfrac{1}{2}ml^2(\dot{\theta}^2+\dot{\phi}^2\sin^2\theta)=\tfrac{1}{2}mV^2-mgl\,(\cos\alpha-\cos\theta).$$

[On substituting for $\dot{\phi}$ from (4) we have equation (6).]

(1) then gives

$$\frac{T}{m}=g\cos\theta+\frac{(\text{vel.})^2}{l}=g\cos\theta+\frac{V^2-2gl\,(\cos\alpha-\cos\theta)}{l}$$

$$=\frac{V^2}{l}+g(3\cos\theta-2\cos\alpha).$$

128. In the previous example $\ddot{\theta}$ is zero when $\theta=\alpha$, *i.e.* the particle revolves at a constant depth below the centre O as in the ordinary conical pendulum, if $V^2=gl\dfrac{\sin^2\alpha}{\cos\alpha}$.

Suppose the particle to have been projected with this velocity, and when it is revolving steadily let it receive a small displacement in the plane NOP, so that the value of $\dot{\phi}$ was not instantaneously altered. Putting $\theta=\alpha+\psi$, where ψ is small, the equation (5) of the last article gives

$$\ddot{\psi}=\frac{g\sin^4\alpha}{l\cos\alpha}\,\frac{\cos(\alpha+\psi)}{\sin^3(\alpha+\psi)}-\frac{g}{l}\sin(\alpha+\psi)$$

$$=\frac{g\sin\alpha}{l}\left[\frac{1-\psi\tan\alpha}{(1+\psi\cot\alpha)^3}-(1+\psi\cot\alpha)\right],$$

neglecting squares of ψ,

$$=-\frac{g\sin\alpha}{l}\psi\,(\tan\alpha+4\cot\alpha)=-\frac{g}{l}\,\frac{1+3\cos^2\alpha}{\cos\alpha}\psi,$$

so that the time of a small oscillation about the position of relative equilibrium is $2\pi\sqrt{\dfrac{l}{g}\,\dfrac{\cos\alpha}{1+3\cos^2\alpha}}$.

Again, from (4), on putting $\theta=\alpha+\psi$, we have

$$\dot{\phi}=\sqrt{\frac{g}{l\cos\alpha}\,\frac{1}{(1+\psi\cot\alpha)^2}}=\sqrt{\frac{g}{l\cos\alpha}}[1-2\psi\cot\alpha],$$

so that during the oscillation there is a small change in the value of $\dot{\phi}$ whose period is the same as that of ψ.

129. *A particle moves on the inner surface of a smooth cone, of vertical angle 2α, being acted on by a force towards the vertex of the cone, and its direction of motion always cuts the generators at a constant angle β; find the motion and the law of force.*

Let $F.m$ be the force, where m is the mass of the particle, and R the reaction of the cone. Then in the accelerations of Art. 125 we have $\theta=\alpha$ and therefore $\dot{\theta}=0$.

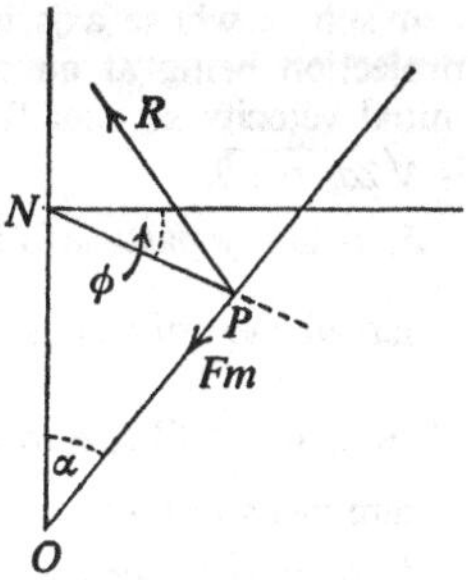

Hence the equations of motion are

$$\frac{d^2r}{dt^2}-r\sin^2\alpha\left(\frac{d\phi}{dt}\right)^2=-F \quad\text{.........(1)},$$

$$-r\sin\alpha\cos\alpha\left(\frac{d\phi}{dt}\right)^2=-\frac{R}{m} \quad\text{.........(2)},$$

and

$$\frac{\sin\alpha}{r}\frac{d}{dt}\left(r^2\frac{d\phi}{dt}\right)=0 \quad\text{.........(3)}.$$

Also, since the direction of motion always cuts OP at an angle β,

$$\therefore\ \frac{r\sin\alpha\dot{\phi}}{\dot{r}}=\tan\beta \quad\text{.........(4)}.$$

(3) gives

$$r^2\frac{d\phi}{dt}=\text{constant}=A \quad\text{.........(5)},$$

and therefore, from (4),

$$\frac{dr}{dt}=\sin\alpha\cot\beta.\frac{A}{r} \quad\text{.........(6)}.$$

Substituting in (1), we have

$$-F=-\sin^2\alpha\cot^2\beta.\frac{A^2}{r^3}-\sin^2\alpha.\frac{A^2}{r^3},$$

i.e.

$$F=\frac{A^2\sin^2\alpha}{\sin^2\beta}.\frac{1}{r^3}=\frac{\mu}{r^3} \quad\text{.........(7)}.$$

Also

$$v^2=\left(\frac{dr}{dt}\right)^2+r^2\sin^2\alpha\left(\frac{d\phi}{dt}\right)^2=\frac{A^2\sin^2\alpha}{r^2\sin^2\beta},$$

so that

$$v=\frac{\sqrt{\mu}}{r}.$$

Again, (2) gives $\dfrac{R}{m}=\dfrac{A^2\sin\alpha\cos\alpha}{r^3}=F\dfrac{\sin^2\beta\cos\alpha}{\sin\alpha}$.

From (4), the path is given by $r=r_0.e^{\sin\alpha\cot\beta.\phi}$.

EXAMPLES

1. A heavy particle moves in a smooth sphere; shew that, if the velocity be that due to the level of the centre, the reaction of the surface will vary as the depth below the centre.

2. A particle is projected horizontally along the interior surface of a smooth hemisphere whose axis is vertical and whose vertex is downwards; the point of projection being at an angular distance β from the lowest point, shew that the initial velocity so that the particle may just ascend to the rim of the hemisphere is $\sqrt{2ag \sec \beta}$.

3. A heavy particle is projected horizontally along the inner surface of a smooth spherical shell of radius $\frac{a}{\sqrt{2}}$ with velocity $\sqrt{\frac{7ag}{3}}$ at a depth $\frac{2a}{3}$ below the centre. Shew that it will rise to a height $\frac{a}{3}$ above the centre, and that the pressure on the sphere just vanishes at the highest point of the path.

4. A particle moves on a smooth sphere under no forces except the pressure of the surface; shew that its path is given by the equation $\cot \theta = \cot \beta \cos \phi$, where θ and ϕ are its angular coordinates.

5. A heavy particle is projected with velocity V from the end of a horizontal diameter of a sphere of radius a along the inner surface, the direction of projection making an angle β with the equator. If the particle never leaves the surface, prove that $3 \sin^2\beta < 2 + \left(\frac{V^2}{3ga}\right)^2$.

6. A particle constrained to move on a smooth spherical surface is projected horizontally from a point at the level of the centre so that its angular velocity relative to the centre is ω. If $\omega^2 a$ be very great compared with g, shew that its depth z below the level of the centre at time t is $\frac{2g}{\omega^2} \sin^2 \frac{\omega t}{2}$ approximately.

7. A thin straight hollow smooth tube is always inclined at an angle α to the upward drawn vertical, and revolves with uniform velocity ω about a vertical axis which intersects it. A heavy particle is projected from the stationary point of the tube with velocity $\frac{g}{\omega} \cot \alpha$; shew that in time t it has described a distance $\frac{g \cos \alpha}{\omega^2 \sin^2 \alpha}[1 - e^{-\omega \sin \alpha . t}]$. Find also the reaction of the tube.

8. A smooth hollow right circular cone is placed with its vertex downward and axis vertical, and at a point on its interior surface at a height h above the vertex a particle is projected horizontally along the surface with a velocity $\sqrt{\frac{2gh}{n^2+n}}$. Shew that the lowest point of its path will be at a height $\frac{h}{n}$ above the vertex of the cone.

9. A smooth circular cone, of angle $2a$, has its axis vertical and its vertex, which is pierced with a small hole, downwards. A mass M hangs at rest by a string which passes through the vertex, and a mass m attached to the upper end describes a horizontal circle on the inner surface of the cone. Find the time T of a complete revolution, and shew that small oscillations about the steady motion take place in the time

$$T \operatorname{cosec} \alpha \sqrt{\frac{M+m}{3m}}.$$

10. A smooth conical surface is fixed with its axis vertical and vertex downwards. A particle is in steady motion on its concave side in a horizontal circle and is slightly disturbed. Shew that the time of a small oscillation about this state of steady motion is $2\pi\sqrt{\dfrac{l}{3g \cos \alpha}}$, where α is the semi-vertical angle of the cone and l is the length of the generator to the circle of steady motion.

11. Three masses m_1, m_2 and m_3 are fastened to a string which passes through a ring, and m_1 describes a horizontal circle as a conical pendulum while m_2 and m_3 hang vertically. If m_3 drop off, shew that the instantaneous change of tension of the string is $\dfrac{gm_1m_3}{m_1+m_2}$.

12. A particle describes a rhumb-line on a sphere in such a way that its longitude increases uniformly; shew that the resultant acceleration varies as the cosine of the latitude and that its direction makes with the normal an angle equal to the latitude.

[A Rhumb-line is a curve on the sphere cutting all the meridians at a constant angle α; its equation is $\dfrac{\dot{\phi} \sin \theta}{\dot{\theta}} = \tan \alpha$.]

13. A particle moves on a smooth right circular cone under a force which is always in a direction perpendicular to the axis of the cone; if the particle describe on the cone a curve which cuts all the generators at a given constant angle, find the law of force and the initial velocity, and shew that at any instant the reaction of the cone is proportional to the acting force.

14. A point moves with constant velocity on a cone so that its direction of motion makes a constant angle with a plane perpendicular to the axis of the cone. Shew that the resultant acceleration is perpendicular to the axis of the cone and varies inversely as the distance of the point from the axis.

15. At the vertex of a smooth cone of vertical angle $2a$, fixed with its axis vertical and vertex downwards, is a centre of repulsive force $\dfrac{\mu}{(\text{distance})^4}$. A weightless particle is projected horizontally with velocity $\sqrt{\dfrac{2\mu \sin^3 \alpha}{c^3}}$ from a point, distant c from the axis, along the inside of the surface. Shew that it will describe a curve on the cone whose projection on a horizontal plane is $1-\dfrac{c}{r}=3 \tan \text{h}^2 \left(\dfrac{\theta}{2} \sin \alpha\right)$.

16. Investigate the motion of a conical pendulum when disturbed from its state of steady motion by a small vertical harmonic oscillation of the point of support. Can the steady motion be rendered unstable by such a disturbance?

17. A particle moves on the inside of a smooth sphere, of radius a, under a force perpendicular to and acting from a given diameter, which equals $\mu \dfrac{\sin \theta}{\cos^4 \theta}$ when the particle is at an angular distance θ from that diameter; if, when the angular distance of the particle is γ, it is projected with velocity $\sqrt{\mu a} \sec \gamma$ in a direction perpendicular to the plane through itself and the given diameter, shew that its path is a small circle of the sphere, and find the reaction of the sphere.

18. A particle moves on the surface of a smooth sphere along a rhumb-line, being acted on by a force parallel to the axis of the rhumb-line. Shew that the force varies inversely as the fourth power of the distance from the axis and directly as the distance from the medial plane perpendicular to the axis.

19. A particle moves on the surface of a smooth sphere and is acted on by a force in the direction of the perpendicular from the particle on a diameter and equal to $\frac{\mu}{(\text{distance})^3}$. Shew that it can be projected so that its path will cut the meridians at a constant angle.

20. A particle moves on the interior of a smooth sphere, of radius a, under a force producing an acceleration $\mu\omega^n$ along the perpendicular ω drawn to a fixed diameter. It is projected with velocity V along the great circle to which this diameter is perpendicular and is slightly disturbed from its path; shew that the new path will cut the old one m times in a revolution, where $m^2=4\left[1-\frac{\mu a^{n+1}}{V^2}\right]$.

21. A particle moves on a smooth cone under the action of a force to the vertex varying inversely as the square of the distance. If the cone be developed into a plane, shew that the path becomes a conic section.

22. A particle, of mass m, moves on the inner surface of a cone of revolution, whose semi-vertical angle is α, under the action of a repulsive force $\frac{m\mu}{(\text{distance})^3}$ from the axis; the moment of momentum of the particle about the axis being $m\sqrt{\mu}\tan\alpha$, shew that its path is an arc of a hyperbola whose eccentricity is $\sec\alpha$.

[With the notation of Art. 129 we obtain $\dot{\phi}^2=\frac{\mu}{\cos^2\alpha\sin^2\alpha}\cdot\frac{1}{r^4}$ and $\ddot{r}=\frac{\mu}{\cos^2\alpha\sin^2\alpha}\cdot\frac{1}{r^3}$, giving $\dot{r}^2=\frac{\mu}{\cos^2\alpha\sin^2\alpha}\left(\frac{1}{d^2}-\frac{1}{r^2}\right)$, where d is a constant. Hence $\left(\frac{dr}{d\phi}\right)^2=r^2.\frac{r^2-d^2}{d^2}$.

Hence $\phi=\gamma-\sin^{-1}\frac{d}{r}$. $\therefore \frac{d}{r}=\sin(\gamma-\phi)=\cos\phi$, if the initial plane for ϕ be properly chosen. This is the plane $x=d\sin\alpha$, which is a plane parallel to the axis of the cone. The locus is thus a hyperbolic section of the cone, the parallel section of which through the vertex consists of two straight lines inclined at 2α. Hence, etc.]

23. If a particle move on the inner surface of a right circular cone under the action of a force from the vertex, the law of repulsion being $m\mu\left[\frac{a\cos^2\alpha}{r^3}-\frac{1}{2r^2}\right]$, where $2a$ is the vertical angle of the cone, and if it be projected from an apse at distance a with velocity $\sqrt{\frac{\mu}{a}}\sin\alpha$, shew that the path will be a parabola.

[Show that the plane of the motion is parallel to a generator of the cone.]

24. A particle is constrained to move on a smooth conical surface of vertical angle 2α, and describes a plane curve under the action of an attraction to the vertex, the plane of the orbit cutting the axis of the cone at a distance a from the vertex. Shew that the attractive force must vary as $\frac{1}{r^2}-\frac{a\cos\alpha}{r^3}$.

25. A particle moves on a rough circular cylinder under the action of no external forces. Initially the particle has a velocity V in a direction making an angle α with the transverse plane of the cylinder; shew that the space described in time t is $\frac{a\sec^2\alpha}{\mu}\log\left[1+\frac{\mu V\cos^2\alpha}{a}t\right]$.

[Use the equations of Art. 126.]

130. *A point is moving along any curve in three dimensions; to find its accelerations along* (1) *the tangent to the curve,* (2) *the principal normal, and* (3) *the binormal.*

If (x, y, z) be the coordinates of the point at time t, the accelerations parallel to the axes of coordinates are $\ddot{x}$, $\ddot{y}$, and $\ddot{z}$.

Now $$\frac{dx}{dt}=\frac{dx}{ds}\frac{ds}{dt}.$$

$$\therefore \frac{d^2x}{dt^2}=\frac{dx}{ds}\frac{d^2s}{dt^2}+\frac{d^2x}{ds^2}\left(\frac{ds}{dt}\right)^2 \quad \text{.....................(1).}$$

So $$\frac{d^2y}{dt^2}=\frac{dy}{ds}\frac{d^2s}{dt^2}+\frac{d^2y}{ds^2}\left(\frac{ds}{dt}\right)^2 \quad \text{.....................(2),}$$

and $$\frac{d^2z}{dt^2}=\frac{dz}{ds}\frac{d^2s}{dt^2}+\frac{d^2z}{ds^2}\left(\frac{ds}{dt}\right)^2 \quad \text{.....................(3).}$$

The direction cosines of the tangent are $\frac{dx}{ds}$, $\frac{dy}{ds}$, and $\frac{dz}{ds}$.

Hence the acceleration along it

$$=\frac{dx}{ds}\frac{d^2x}{dt^2}+\frac{dy}{ds}\frac{d^2y}{dt^2}+\frac{dz}{ds}\frac{d^2z}{dt^2}$$

$$=\frac{d^2s}{dt^2}\left[\left(\frac{dx}{ds}\right)^2+\left(\frac{dy}{ds}\right)^2+\left(\frac{dz}{ds}\right)^2\right]+\left(\frac{ds}{dt}\right)^2\left[\frac{dx}{ds}\frac{d^2x}{ds^2}+\frac{dy}{ds}\frac{d^2y}{ds^2}+\frac{dz}{ds}\frac{d^2z}{ds^2}\right]$$

$$=\frac{d^2s}{dt^2} \quad \text{...(4),}$$

since $$\left(\frac{dx}{ds}\right)^2+\left(\frac{dy}{ds}\right)^2+\left(\frac{dz}{ds}\right)^2=1,$$

and therefore $$\frac{dx}{ds}\frac{d^2x}{ds^2}+\frac{dy}{ds}\frac{d^2y}{ds^2}+\frac{dz}{ds}\frac{d^2z}{ds^2}=0.$$

The direction cosines of the principal normal are $\rho\frac{d^2x}{ds^2}$, $\rho\frac{d^2y}{ds^2}$, and $\rho\frac{d^2z}{ds^2}$, where ρ is the radius of curvature.

Hence the acceleration along it

$$=\rho\frac{d^2x}{ds^2}\frac{d^2x}{dt^2}+\rho\frac{d^2y}{ds^2}\frac{d^2y}{dt^2}+\rho\frac{d^2z}{ds^2}\frac{d^2z}{dt^2}$$

$$=\rho\frac{d^2s}{dt^2}\left[\frac{dx}{ds}\frac{d^2x}{ds^2}+\frac{dy}{ds}\frac{d^2y}{ds^2}+\frac{dz}{ds}\frac{d^2z}{ds^2}\right]$$

$$=\rho\left(\frac{ds}{dt}\right)^2\left[\left(\frac{d^2x}{ds^2}\right)^2+\left(\frac{d^2y}{ds^2}\right)^2+\left(\frac{d^2z}{ds^2}\right)^2\right]$$

$$=\rho\left(\frac{ds}{dt}\right)^2\times\frac{1}{\rho^2}=\frac{1}{\rho}\left(\frac{ds}{dt}\right)^2 \quad\text{.............................(5).}$$

The direction cosines of the binormal are proportional to

$$\frac{dy}{ds}\frac{d^2z}{ds^2}-\frac{dz}{ds}\frac{d^2y}{ds^2},\quad \frac{dz}{ds}\frac{d^2x}{ds^2}-\frac{dx}{ds}\frac{d^2z}{ds^2},\text{ and }\frac{dx}{ds}\frac{d^2y}{ds^2}-\frac{dy}{ds}\frac{d^2x}{ds^2}.$$

On multiplying (1), (2), and (3) in succession by these and adding, the result is zero, *i.e.* the acceleration in the direction of the binormal vanishes.

The foregoing results might have been seen at once from equations (1), (2), (3). For if (l_1, m_1, n_1), (l_2, m_2, n_2), and (l_3, m_3, n_3) are the direction cosines of the tangent, the principal normal, and the binormal, these equations may be written

$$\frac{d^2x}{dt^2}=l_1\frac{d^2s}{dt^2}+l_2\left\{\frac{1}{\rho}\left(\frac{ds}{dt}\right)^2\right\},$$

$$\frac{d^2y}{dt^2}=m_1\frac{d^2s}{dt^2}+m_2\left\{\frac{1}{\rho}\left(\frac{ds}{dt}\right)^2\right\},$$

and

$$\frac{d^2z}{dt^2}=n_1\frac{d^2s}{dt^2}+n_2\left\{\frac{1}{\rho}\left(\frac{ds}{dt}\right)^2\right\}.$$

These equations shew that the accelerations along the axes are the components of

an acceleration $\frac{d^2s}{dt^2}$ along the tangent,

an acceleration $\frac{1}{\rho}\left(\frac{ds}{dt}\right)^2$ along the principal normal,

and nothing in the direction of the binormal.

We therefore see that, as in the case of a particle describing a plane curve, the accelerations are $\frac{d^2s}{dt^2}$, or $v\frac{dv}{ds}$, along the tangent and $\frac{v^2}{\rho}$ along the principal normal, which lies in the osculating plane of the curve.

131. *A particle moves in a curve, there being no friction, under forces such as occur in nature. Shew that the change in its kinetic energy as it passes from one position to the other is independent of the path pursued and depends only on its initial and final positions.*

Let X, Y, Z be the components of the forces. By the last article, resolving along the tangent to the path, we have

$$m\frac{d^2s}{dt^2}=X\frac{dx}{ds}+Y\frac{dy}{ds}+Z\frac{dz}{ds}.$$

$$\therefore \frac{1}{2}.m\left(\frac{ds}{dt}\right)^2=\int(Xdx+Ydy+Zdz).$$

Now, by Art. 95, since the forces are such as occur in nature, *i.e.* are one-valued functions of distances from fixed points, the quantity $Xdx+Ydy+Zdz$ is the differential of some function $\phi(x, y, z)$, so that

$$\frac{1}{2}mv^2=\frac{1}{2}m\left(\frac{ds}{dt}\right)^2=\phi(x, y, z)+C,$$

where $$\tfrac{1}{2}mv_0{}^2=\phi(x_0, y_0, z_0)+C,$$

(x_0, y_0, z_0) being the starting point and v_0 the initial velocity.

Hence $$\tfrac{1}{2}mv^2-\tfrac{1}{2}mv_0{}^2=\phi(x, y, z)-\phi(x_0, y_0, z_0).$$

The right-hand member of this equation depends only on the position of the initial point and on that of the point of the path under consideration, and is quite independent of the path pursued.

The reaction R of the curve in the direction of the principal normal is given by the equation

$$\frac{v^2}{\rho}=R,$$

where ρ is the radius of curvature of the curve.

132. *Motion on a smooth surface.* If the particle moves on a surface whose equation is $f(x, y, z)=0$, let the direction cosines of the normal at any point (x, y, z) of its path be (l_1, m_1, n_1), so that

$$\frac{l_1}{\frac{df}{dx}}=\frac{m_1}{\frac{df}{dy}}=\frac{n_1}{\frac{df}{dz}}=\frac{1}{\sqrt{\left(\frac{df}{dx}\right)^2+\left(\frac{df}{dy}\right)^2+\left(\frac{df}{dz}\right)^2}}.$$

Then, if R be the normal reaction, we have

$$m\frac{d^2x}{dt^2}=X+Rl_1,\quad m\frac{d^2y}{dt^2}=Y+Rm_1, \text{ and } m\frac{d^2z}{dt^2}=Z+Rn_1,$$

where X, Y, Z are the components of the impressed forces.

Multiplying these equations by $\frac{dx}{dt}, \frac{dy}{dt}, \frac{dz}{dt}$ and adding, we have

$$\frac{1}{2}m\frac{d}{dt}\left[\left(\frac{dx}{dt}\right)^2+\left(\frac{dy}{dt}\right)^2+\left(\frac{dz}{dt}\right)^2\right]=X\frac{dx}{dt}+Y\frac{dy}{dt}+Z\frac{dz}{dt};$$

for the coefficient of R

$$=l_1\frac{dx}{dt}+m_1\frac{dy}{dt}+n_1\frac{dz}{dt}$$

$$=\left(l_1\frac{dx}{ds}+m_1\frac{dy}{ds}+n_1\frac{dz}{ds}\right)\frac{ds}{dt}$$

$=\frac{ds}{dt}\times$ the cosine of the angle between a tangent line to the surface and the normal

$=0.$

Hence, on integration,

$$\tfrac{1}{2}mv^2=\int(Xdx+Ydy+Zdz),$$

as in the last article.

Also, on eliminating R, the path on the surface is given by

$$\frac{m\frac{d^2x}{dt^2}-X}{l_1}=\frac{m\frac{d^2y}{dt^2}-Y}{m_1}=\frac{m\frac{d^2z}{dt^2}-Z}{n_1},$$

giving two equations from which, by eliminating t, we should get a second surface cutting the first in the required path.

133. *Motion under gravity of a particle on a smooth surface of revolution whose axis is vertical.*

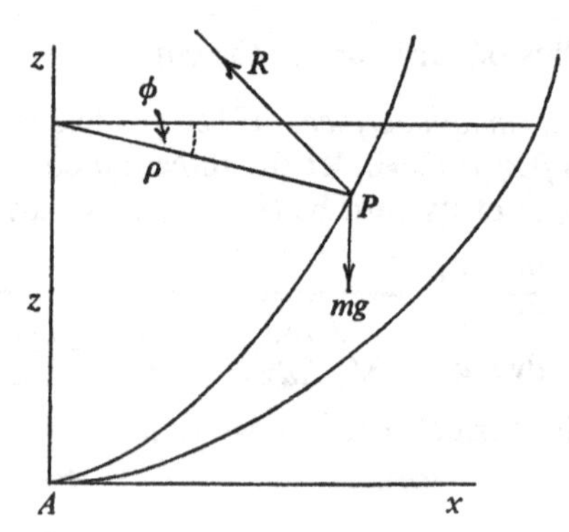

Use the coordinates z, ρ, and ϕ of Art. 126, the z-axis being the axis of revolution of the surface. The second equation of that article gives $\frac{1}{\rho}\frac{d}{dt}\left(\rho^2\frac{d\phi}{dt}\right)=0$,

i.e.
$$\rho^2\frac{d\phi}{dt}=\text{constant}=h \quad \text{.......................(1).}$$

Also, if s be the arc AP measured from any fixed point A, the velocities of P are $\frac{ds}{dt}$ along the tangent at P to the generating curve, and $\rho\frac{d\phi}{dt}$ perpendicular to the plane zAP. Hence the Principle of Energy gives

$$\frac{1}{2}\left\{\left(\frac{ds}{dt}\right)^2+\rho^2\left(\frac{d\phi}{dt}\right)^2\right\}=\text{const.}-gz \quad \text{.................(2).}$$

Equations (1) and (2) give the motion.

Equation (1) states that the moment of the momentum of the particle about the axis of z is constant.

By equating the forces parallel to Oz to $m\frac{d^2z}{dt^2}$, we easily have the value of the reaction R.

If the equation to the generating curve be $z=f(\rho)$ then, since

$$\dot{s}^2=\dot{z}^2+\dot{\rho}^2=[1+\{f'(\rho)\}^2]\left(\frac{d\rho}{d\phi}\right)^2\dot{\phi}^2,$$

equation (2) easily gives

$$\frac{1}{\rho^4}\left(\frac{d\rho}{d\phi}\right)^2[1+\{f'(\rho)\}^2]+\frac{1}{\rho^2}=\text{constant}-\frac{2g}{h^2}f(\rho),$$

which gives the differential equation of the projection of the motion on a horizontal plane.

EXAMPLES

1. A smooth helix is placed with its axis vertical and a small bead slides down it under gravity; shew that it makes its first revolution from rest in time $2\sqrt{\frac{\pi a}{g\sin\alpha\cos\alpha}}$, where α is the angle of the helix.

2. A particle, without weight, slides on a smooth helix of angle α and radius a under a force to a fixed point on the axis equal to $m\mu$ (distance). Shew that the reaction of the curve cannot vanish unless the greatest velocity of the particle is $a\sqrt{\mu}\sec\alpha$.

3. A smooth paraboloid is placed with its axis vertical and vertex downwards, the latus-rectum of the generating parabola being $4a$. A heavy particle is projected horizontally with velocity V at a height h above the lowest point; shew that the particle is again moving horizontally when its height is $\frac{V^2}{2g}$. Shew also that the reaction of the paraboloid at any point is inversely proportional to the corresponding radius of curvature of the generating parabola.

4. A particle is describing steadily a circle, of radius b, on the inner surface of a smooth paraboloid of revolution whose axis is vertical and vertex downwards, and is slightly disturbed by an impulse in a plane through the axis; shew that its period of oscillation about the steady motion is $\pi\sqrt{\dfrac{l^2+b^2}{gl}}$, where l is the semi-latus-rectum of the paraboloid.

5. A particle moving on a paraboloid of revolution under a force parallel to the axis crosses the meridians at a constant angle; shew that the force varies inversely as the fourth power of the distance from the axis.

6. A particle moves on a smooth paraboloid of revolution under the action of a force directed to the axis which varies inversely as the cube of the distance from the axis; shew that the equation of the projection of the path on the tangent plane at the vertex of the paraboloid may, under certain conditions of projection, be written

$$\sqrt{4a^2+r^2}+a\log\frac{\sqrt{4a^2+r^2}-2a}{\sqrt{4a^2+r^2}+2a}=k.\theta,$$

where $4a$ is the latus-rectum of the generating parabola.

7. A particle moves on a right circular cone under no forces; show that, whatever be the initial motion, the projection of the path on a plane perpendicular to the axis is one of the similar curves given by $r\sin n\theta=c$.

8. A smooth heavy particle moves on a surface of revolution formed by the revolution of the curve $x^2y=a^3$ about the axis of y, which is vertical with its positive direction downwards. Shew that, if projected with a suitable speed from any point, the particle will cross all the meridians at the same angle.

9. A heavy particle is projected horizontally along a smooth surface of revolution whose equation in cylindrical coordinates is $8z^3=27ar^2$, the axis of z being vertical and upwards.

Prove that, if the normal at the point of projection is inclined at 45° to the vertical and the particle leaves the surface where the normal is inclined at 60° to the vertical, the velocity of projection must be $\sqrt{\dfrac{ga}{525}}$.

CHAPTER X

MISCELLANEOUS

THE HODOGRAPH. MOTION ON REVOLVING CURVES. IMPULSIVE TENSIONS OF STRINGS

134. The Hodograph. If from any fixed point O we draw a straight line OQ which is parallel to, and proportional to, the velocity of any. moving point P, the locus of Q is called a hodograph of the motion of P.

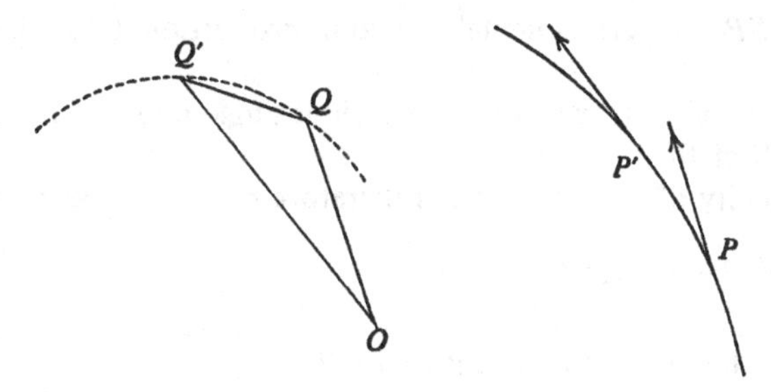

If P' be a consecutive point on the path and OQ' be parallel and proportional to the velocity at P', then the change of the velocity in passing from P to P' is, by Art. 3, represented by QQ'.

If τ be the time of describing the arc PP', then the acceleration of P

$$=\underset{\tau-0}{\text{Limit}}\,\frac{QQ'}{\tau}=\text{velocity of } Q \text{ in the hodograph.}$$

Hence the velocity of Q in the hodograph represents, both in magnitude and direction, the acceleration of P in its path.

It follows that the velocity, or coordinate, of Q in any direction is proportional to the acceleration, or velocity, of P in the same direction.

The same argument holds if the motion of P is not coplanar.

If at any moment x and y be the coordinates of the moving point P, and ξ and η those of the corresponding point Q of the hodograph, we have

$$\therefore\ \xi=\lambda\frac{dx}{dt}, \text{ and } \quad \eta=\lambda\frac{dy}{dt},$$

where λ is a constant.

The values of $\frac{dx}{dt}$ and $\frac{dy}{dt}$ being then known in terms of t, we eliminate t between these equations and have the locus of (ξ, η), *i.e.* the hodograph.

So for three-dimensional motion.

135. *The hodograph of a central orbit is a reciprocal of the orbit with respect to the centre of force S turned through a right angle about S.*

Let SY be the perpendicular to the tangent at any point P of the orbit. Produce SY to P' so that $SY.SP'=k^2$, a constant; the locus of P' is therefore the reciprocal of the path with respect to S.

By Art. 54, the velocity v of $P=\frac{h}{SY}=\frac{h}{k^2}.SP'$.

Hence SP' is perpendicular to, and proportional to, the velocity of P.

The locus of P' turned through a right angle about S is therefore a hodograph of the motion.

The velocity of P' in its path is therefore perpendicular to and equal to $\frac{k^2}{h}$ times the acceleration of P,

i.e. it$=\frac{k^2}{h}\times$ the central acceleration of P.

EXAMPLES

1. A particle describes a parabola under gravity; shew that the hodograph of its motion is a straight line parallel to the axis of the parabola and described with uniform velocity.

2. A particle describes a conic section under a force to its focus; shew that the hodograph is a circle which passes through the centre of force when the path is a parabola.

3. If the path be an ellipse described under a force to its centre, shew that the hodograph is a similar ellipse.

4. A bead moves on the arc of a smooth vertical circle starting from rest at the highest point. Shew that the equation to the hodograph is $r=\lambda \sin\frac{\theta}{2}$.

5. Shew that the hodograph of a circle described under a force to a point on the circumference is a parabola.

6. The hodograph of an orbit is a parabola whose ordinate increases uniformly. Shew that the orbit is a semi-cubical parabola.

7. A particle slides down in a thin cycloidal tube, whose axis is vertical and vertex the highest point; shew that the equation to the hodograph is of the form $r^2=2g[a+b\cos 2\theta)$, the particle starting from any point of the cycloid. If it start from the highest point, shew that the hodograph is a circle.

8. A particle describes an equiangular spiral about a centre of force at the pole; shew that its hodograph is also an equiangular spiral.

9. If a particle describe a lemniscate under a force to its pole, shew that the equation to the hodograph is $r^2=a^2 \sec^3 \frac{\pi-2\theta}{3}$.

10. If the hodograph be a circle described with constant angular velocity about a point on its circumference, shew that the path is a cycloid.

11. Shew that the only central orbits whose hodographs can also be described as central orbits are those where the central acceleration varies as the distance from the centre.

[In Art. 135 if SP meet the tangent at P' in Y', then SY' is perpendicular to $Y'P'$ and $=\frac{k^2}{SP}$. The hodograph is described with central acceleration to S if the velocity of $P' \times SY'$ is constant, *i.e.* if the central acceleration of $P \times \frac{k^2}{SP}$ is constant. Hence the result.]

12. If the path be a helix whose axis is vertical, described under gravity, shew that the hodograph is a curve described on a right circular cone whose semi-vertical angle is the complement of the angle of the helix.

136. Motion on a Revolving Curve. *A given curve turns in its own plane about a given fixed point O with constant angular velocity ω; a small bead P moves on the curve under the action of given forces whose components along and perpendicular to OP are X and Y; to find the motion.*

Let OA be a line fixed in the plane of the curve, and OB a line fixed with respect to the curve, which at zero time coincided with OA, so that, at time t, $\angle AOB=\omega t$.

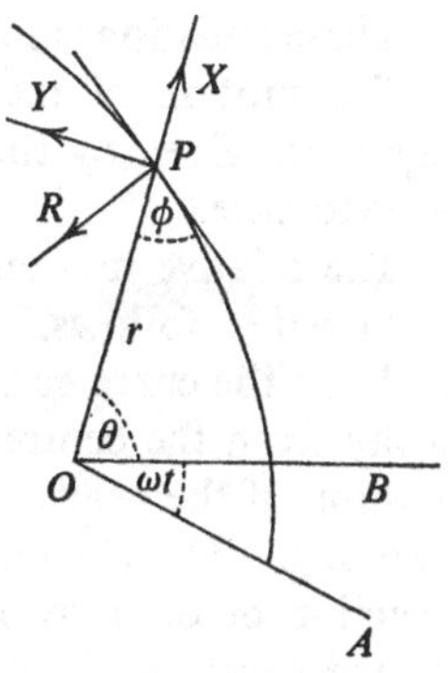

At time t let the bead be at P, where $OP=r$, and let ϕ be the angle between OP and the tangent to the curve.

Then, by Art. 49, the equations of motion are

$$\frac{d^2r}{dt^2}-r\left(\frac{d\theta}{dt}+\omega\right)^2=\frac{X}{m}-\frac{R}{m}\sin\phi,$$

and

$$\frac{1}{r}\frac{d}{dt}\left\{r^2\left(\frac{d\theta}{dt}+\omega\right)\right\}=\frac{Y}{m}+\frac{R}{m}\cos\phi.$$

These give

$$\frac{d^2r}{dt^2}-r\left(\frac{d\theta}{dt}\right)^2=r\omega^2+2r\omega\frac{d\theta}{dt}+\frac{X}{m}-\frac{R}{m}\sin\phi,$$

and

$$\frac{1}{r}\frac{d}{dt}\left(r^2\frac{d\theta}{dt}\right)=-2\omega\frac{dr}{dt}+\frac{Y}{m}+\frac{R}{m}\cos\phi.$$

Let v be the velocity of the bead relative to the wire, so that

$$v \cos \phi = \frac{dr}{dt} \quad \text{and} \quad v \sin \phi = r\frac{d\theta}{dt}.$$

The equations of motion are then

$$\frac{d^2r}{dt^2} - r\left(\frac{d\theta}{dt}\right)^2 = \omega^2 r + \frac{X}{m} - \frac{R'}{m}\sin\phi \quad \text{..............(1)},$$

and

$$\frac{1}{r}\frac{d}{dt}\left(r^2\frac{d\theta}{dt}\right) = \frac{Y}{m} + \frac{R'}{m}\cos\phi \quad \text{....................(2)},$$

where

$$\frac{R'}{m} = \frac{R}{m} - 2\omega v \quad \text{.............................(3)}.$$

These equations give the motion of the particle relative to the curve. Now suppose that the curve, instead of rotating, were at rest, and that the bead moved on it under the action of the same forces X and Y as in the first case together with an additional force $m\omega^2 r$ along OP, and let S be the new normal reaction. The equations of motion are now

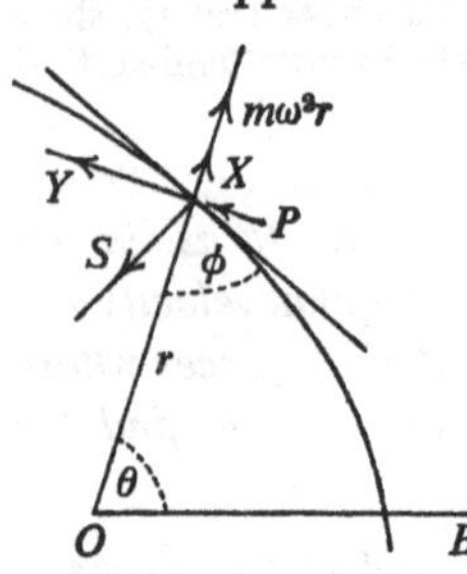

$$\frac{d^2r}{dt^2} - r\left(\frac{d\theta}{dt}\right)^2 = \omega^2 r + \frac{X}{m} - \frac{S}{m}\sin\phi \quad \text{.........(4)},$$

and

$$\frac{1}{r}\frac{d}{dt}\left(r^2\frac{d\theta}{dt}\right) = \frac{Y}{m} + \frac{S}{m}\cos\phi \quad \text{............(5)}.$$

These equations are the same as (1) and (2) with S substituted for R'.

The motion of the bead relative to the curve in the first case is therefore given by the same equations as the absolute motion in the second case.

The relative motions in the case of a revolving curve may thus be obtained as follows.

Treat the curve as fixed, and put on an additional force on the bead, away from the centre of rotation and equal to $m\omega^2.r$; then find the motion of the bead; this will be the relative motion when the curve is rotating. The reaction of the curve, so found, will not be the actual reaction of the moving curve. To get the latter we must by (3) add to the reaction, found by the foregoing process, the quantity $2m\omega v$, where v is the velocity of the bead relative to the curve.

The above process is known as that of "reducing the moving curve to rest." When the moving curve has been reduced to rest, the best accelerations to use are, in general, the tangential and normal ones of Art. 88.

137. If the angular velocity ω be not constant the work is the same, except that in equation (2) there is one extra term $-r\frac{d\omega}{dt}$ on the right-hand side. In this case, in addition to the force $m\omega^2 r$, we must put on another force $-mr\frac{d\omega}{dt}$ at right angles to OP, and the curve is then "reduced to rest."

138. *Ex.* *A smooth circular tube contains a particle, of mass m, and lies on a smooth table. The tube starts rotating with constant angular velocity ω about an axis perpendicular to the plane of the tube which passes through the other end, O, of the diameter through the initial position, A, of the particle. Shew that in time t the particle will have described an angle ϕ about the centre of the tube equal to* $4\tan^{-1}\left(\tanh\frac{\omega t}{2}\right)$. *Shew also that the reaction between the tube and the particle is then equal to* $2ma\omega^2\cos\frac{\phi}{2}\left(3\cos\frac{\phi}{2}-2\right)$.

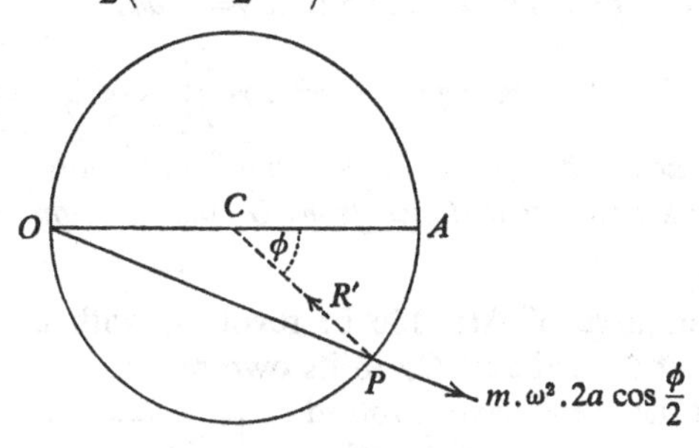

P being the position of the particle at time t, and $\phi = \angle ACP$, we may treat the tube as at rest if we assume an additional force along OP equal to $m\omega^2.OP$, *i.e.* $m\omega^2.2a\cos\frac{\phi}{2}$. R' being the normal reaction in this case, we have, on taking tangential and normal accelerations,

$$a\ddot{\phi} = -\omega^2.2a\cos\frac{\phi}{2}\sin\frac{\phi}{2} \quad \text{.....................(1)},$$

and

$$a\dot{\phi}^2 = \frac{R'}{m} - \omega^2.2a\cos^2\frac{\phi}{2} \quad \text{.......................(2)}.$$

(1) gives

$$\dot{\phi}^2 = 2\omega^2\cos\phi + A \quad \text{..............................(3)}.$$

Now, if the tube revolves ↻, then, since the particle was initially at rest, its initial *relative* velocity was $\omega.OA$, *i.e.* $\omega.2a$, in the opposite direction.

Hence $\qquad \dot{\phi}=2\omega$ initially.

Therefore (3) gives $\qquad \dot{\phi}^2=2\omega^2(1+\cos\phi)$.

$$\therefore\ 2\omega t=\int\frac{d\phi}{\cos\frac{\phi}{2}}=2\log\tan\left(\frac{\pi}{4}+\frac{\phi}{4}\right),\text{ the constant vanishing.}$$

$$\therefore\ e^{\omega t}=\frac{1+\tan\frac{\phi}{4}}{1-\tan\frac{\phi}{4}},\text{ so that }\tan\frac{\phi}{4}=\frac{e^{\omega t}-1}{e^{\omega t}+1}=\tanh\left(\frac{\omega t}{2}\right).$$

Also (2) gives $\qquad \dfrac{R'}{m}=6a\omega^2\cos^2\dfrac{\phi}{2}.$

Now, by equation (3) of Art. 136, the real reaction R is given by

$$\frac{R}{m}=\frac{R'}{m}+2\omega v,$$

where v is the velocity of the particle relative to the tube in the direction in which the tube revolves, *i.e.* ↻, so that $v=-a\dot{\phi}$.

$$\therefore\ R=R'-2m\omega a\dot{\phi}=2ma\omega^2\cos\frac{\phi}{2}\left[3\cos\frac{\phi}{2}-2\right].$$

Ex. If the particle be initially at rest indefinitely close to O, shew that, when it is at its greatest distance from O, the reaction of the curve is $10ma\omega^2$.

139. Let the curve of Art. 136 be revolving with uniform angular velocity ω about a fixed axis Oy in its own plane.

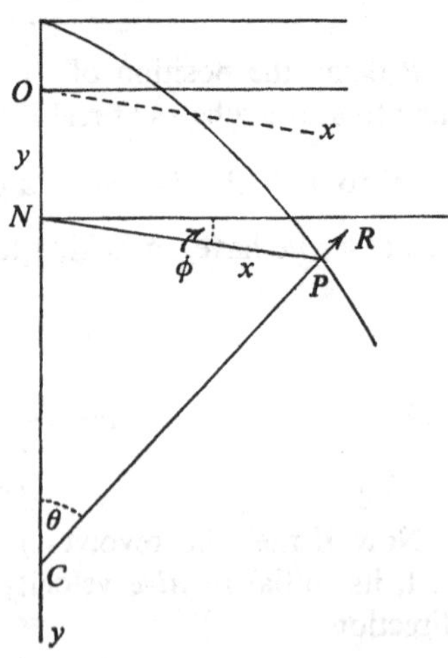

At time t let the curve have revolved through an angle ϕ, and let y and x be the coordinates of the bead measured along and perpendicular to the fixed axis. Let R be the normal reaction in the plane of the curve, and S the reaction perpendicular to the plane of the curve.

The equations of motion of Art. 126 then become, since

$$\dot{\phi}=\omega,$$

$$\frac{d^2x}{dt^2}=x\omega^2+\frac{R}{m}\sin\theta+\frac{X}{m}\ \ldots\ldots\ldots\ldots\ldots(1),$$

$$\frac{1}{x}\frac{d}{dt}(x^2\omega)=\frac{S}{m}+\frac{Z}{m}\ \ldots\ldots\ldots\ldots\ldots\ldots\ldots(2),$$

and $$\frac{d^2y}{dt^2}=-\frac{R}{m}\cos\theta+\frac{Y}{m} \quad \text{..........................(3)},$$

where X, Y, and Z are the components of the impressed forces parallel to x and y, and perpendicular to the plane of the curve, and θ is the inclination of the normal to the axis of y.

Equations (1) and (3), which give the motion of the bead relative to the wire, are the same equations as we should have *if we assumed the wire to be at rest and put on an additional force* $m\omega^2\times$ *distance perpendicular to and away from the axis of rotation.*

Hence, applying this additional force, we may treat the wire as at rest, and use whatever equations are most convenient.

140. As a numerical example, let the curve be a smooth circular wire revolving about its vertical diameter, C being its centre and a its radius. Let the bead start from rest at a point indefinitely close to the highest point of the wire. Treat the circle as at rest, and put on an additional force

$$m\omega^2 . NP (=m\omega^2 . a\sin\theta)$$

along NP.

Taking tangential and normal accelerations, we then have

$$a\ddot{\theta}=\omega^2 a\sin\theta\cos\theta+g\sin\theta \quad \text{..................(4)},$$

and $$a\dot{\theta}^2=-\frac{R}{m}+g\cos\theta-\omega^2 a\sin^2\theta \quad \text{..................(5)}.$$

(4) gives $$a\dot{\theta}^2=\omega^2 a\sin^2\theta+2g(1-\cos\theta),$$

and then (5) gives $$\frac{R}{m}=g(3\cos\theta-2)-2\omega^2 a\sin^2\theta.$$

Also the reaction S perpendicular to the plane of the wire is, by (2), given by

$$\frac{S}{m}=2\dot{x}\omega=2\omega\frac{d}{dt}(a\sin\theta)=2\omega a\cos\theta . \dot{\theta}$$

$$=2\omega\cos\theta\sqrt{\omega^2 a^2\sin^2\theta+2ga(1-\cos\theta)}.$$

EXAMPLES

1. A particle is placed in a smooth straight tube which is suddenly set rotating, with constant angular velocity ω, about a point O in its plane of motion, which is at a perpendicular distance a from the tube; shew that the distance described along the tube by the particle in time t is $a\sinh\omega t$, the particle and O being initially at a distance a apart; shew also that the reaction between the tube and the particle then is $ma\omega^2[2\cosh\omega t-1]$.

2. A circular tube, of radius a, revolves uniformly about a vertical diameter with angular velocity $\sqrt{\frac{ng}{a}}$, and a particle is projected from its lowest point with velocity just sufficient to carry it to the highest point; shew that the time of describing the first quadrant is

$$\sqrt{\frac{a}{(n+1)g}}\log\left[\sqrt{n+2}+\sqrt{n+1}\right].$$

3. A particle P moves in a smooth circular tube, of radius a, which turns with uniform angular velocity ω about a vertical diameter; if the angular distance of the particle at any time t from the lowest point is θ, and if it be at rest relative to the tube when $\theta=\alpha$, where $\cos\frac{\alpha}{2}=\frac{1}{\omega}\sqrt{\frac{g}{a}}$, then, at any subsequent time t, $\cot\frac{\theta}{2}=\cot\frac{\alpha}{2}\cosh\left(\omega t\sin\frac{\alpha}{2}\right)$.

4. A thin circular wire is made to revolve about a vertical diameter with constant angular velocity. A smooth ring slides on the wire, being attached to its highest point by an elastic string whose natural length is equal to the radius of the wire. If the ring be slightly displaced from the lowest point, find the motion, and shew that it will reach the highest point if the modulus of elasticity is four times the weight of the ring.

5. A smooth circular wire rotates with uniform angular velocity ω about its tangent line at a point A. A bead, without weight, slides on the wire from a position of rest at a point of the wire very near A. Shew that the angular distance on the wire traversed in a time t after passing the point opposite A is $2\tan^{-1}\omega t$.

6. A small bead slides on a circular arc, of radius a, which revolves with constant angular velocity ω about its vertical diameter. Find its position of stable equilibrium according as $\omega^2 \gtrless \frac{g}{a}$, and shew that the time of a small oscillation about its position of equilibrium is, for the two cases, respectively equal to $\frac{2\pi\omega a}{\sqrt{\omega^4a^2-g^2}}$ and $\frac{2\pi}{\sqrt{\frac{g}{a}-\omega^2}}$.

7. A parabolic wire, whose axis is vertical and vertex downwards, rotates about its axis with uniform angular velocity ω. A ring is placed at any point of it in relative rest; shew that it will move upwards or downwards according as $\omega^2 \gtrless \frac{g}{2a}$, and will remain at rest if $\omega^2=\frac{g}{2a}$ where $4a$ is the latus-rectum of the parabola.

8. A tube in the form of the cardioid $r=a(1+\cos\theta)$ is placed with its axis vertical and cusp uppermost, and revolves round the axis with angular velocity $\sqrt{\frac{g}{a}}$. A particle is projected from the lowest point of the tube along the tube with velocity $\sqrt{3ga}$; shew that the particle will ascend until it is on a level with the cusp.

9. A smooth plane tube, revolving with angular velocity ω about a point O in its plane, contains a particle of mass m, which is acted upon by a force $m\omega^2r$ towards O; shew that the reaction of the tube is $A+\frac{B}{\rho}$, where A and B are constants and and ρ is the radius of curvature of the tube at the point occupied by the particle.

10. An elastic string, whose unstretched length is a, of mass ma and whose modulus of elasticity is λ, has one end fastened to an extremity A of a straight smooth tube within which the string rests; if the tube revolve with uniform angular velocity in a horizontal plane about the end A, shew that when the string is in equilibrium its length is $a\dfrac{\tan\theta}{\theta}$, where

$$\theta = a\omega\sqrt{\frac{m}{\lambda}}.$$

11. A bead is at rest on an equiangular spiral of angle α at a distance a from the pole. The plane of the spiral is horizontal and the spiral is made to revolve about a vertical line through its pole with uniform angular velocity ω. Prove that the bead comes to a position of relative rest at a distance $a\cos\alpha$ from the pole and that the reaction of the curve is then $\frac{1}{2}m\omega^2 a\sin 2a$. Shew also that when the bead is again at its original distance from the pole, the reaction is $m\omega^2 a\sin\alpha\,(3+\sin^2\alpha)$.

12. A particle, of mass m, is placed on a horizontal table which is lubricated with oil so that the force on the particle due to viscous friction is mku, where u is the velocity of the particle relative to the table. The table is made to revolve with uniform angular velocity ω about a vertical axis. Shew that, by properly adjusting the circumstances of projection, the equations to the path of the particle on the table will be

$$x = e^{\left(\alpha-\frac{k}{2}\right)t}\cos(\omega-\beta)t,\quad y = -e^{\left(\alpha-\frac{k}{2}\right)t}\sin(\omega-\beta)t,$$

where
$$\alpha + i\beta = \sqrt{\frac{k^2}{4} + i\omega k}.$$

[With the notation of Art. 51, the equations of motion are

$$(D^2+kD-\omega^2)x - 2\omega Dy = 0 \quad\text{and}\quad (D^2+kD-\omega^2)y + 2\omega Dx = 0.$$

Hence $[D^2+kD-\omega^2+2i\omega D](x+iy)=0$. Now solve in the usual manner.]

13. A smooth horizontal plane revolves with angular velocity ω about a vertical axis to a point of which is attached the end of a light string, extensible according to Hooke's Law and of natural length d just sufficient to reach the plane. The string is stretched, and after passing through a small ring at the point where the axis meets the plane is attached to a particle of mass m which moves on the plane. If the mass is initially at rest on the plane, shew that its path relative to the plane is a hypo-cycloid generated by the rolling of a circle of radius $\frac{1}{2}a\{1-\omega(md\lambda^{-1})^{\frac{1}{2}}\}$ on a circle of radius a, where a is the initial extension and λ the coefficient of elasticity of the string.

14. A particle slides in a smooth straight tube which is made to rotate with uniform angular velocity ω about a vertical axis. If the particle start from relative rest from the point where the shortest distance between the axis and the tube meets the tube, shew that in time t the particle has moved through a distance

$$\frac{2g}{\omega^2}\cot\alpha\,\mathrm{cosec}\,\alpha\,\sinh^2(\tfrac{1}{2}\omega t\sin\alpha),$$

where α is the inclination of the tube to the vertical.

141. Impulsive tensions of chains.

A chain is lying in the form of a given curve on a smooth horizontal plane; to one end of it is applied a given impulsive tension in the direction of the tangent; to find the consequent impulsive tension at any other point of the curve, and the initial motion of the point.

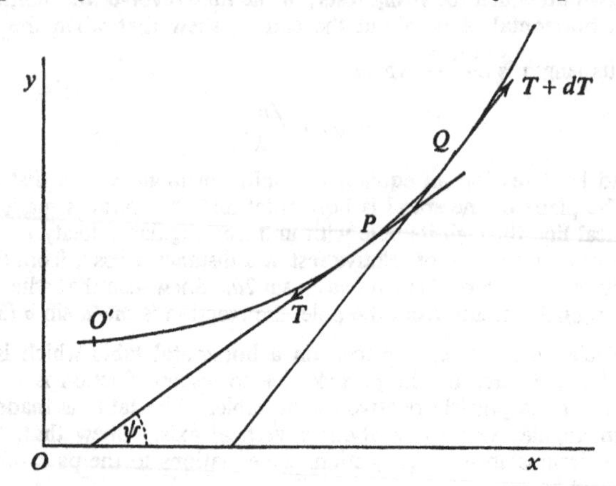

Let PQ be any element δs of the chain, s being the length of the arc $O'P$ measured from any fixed point O'.

Let T and $T+\delta T$ be the impulsive tensions at P and Q.

The resolved part of the tension at P parallel to Ox is $T\dfrac{dx}{ds}$, and this is clearly a function of the arc s. Let it be $f(s)$.

The resolved part of the tension at Q

$$=f(s+\delta s)=f(s)+\delta s f'(s)+\frac{\delta s^2}{1.2}f''(s)+\dots$$

$$=T\frac{dx}{ds}+\delta s\frac{d}{ds}\left(T\frac{dx}{ds}\right)+\dots,$$

by Taylor's Theorem.

Hence, if m be the mass of the chain at P per unit of length, and u, v the initial velocities of the element PQ parallel to the axes, we have

$$m\delta s.u=\left[T\frac{dx}{ds}+\delta s\frac{d}{ds}\left(T\frac{dx}{ds}\right)+\dots\right]-T\frac{dx}{ds},$$

i.e., in the limit when δs is indefinitely small,

$$mu=\frac{d}{ds}\left(T\frac{dx}{ds}\right) \quad \dots\dots\dots\dots\dots\dots\dots\dots(1).$$

So

$$mv=\frac{d}{ds}\left(T\frac{dy}{ds}\right) \quad \dots\dots\dots\dots\dots\dots\dots\dots(2).$$

Again, since the string is inextensible, the velocity of P along the direction PQ (*i.e.* along the tangent at P ultimately) must be equal to the velocity of Q in the same direction.

Hence

$$u \cos \psi + v \sin \psi = (u + \delta u) \cos \psi + (v + \delta v) \sin \psi.$$

$$\therefore \quad \delta u \cos \psi + \delta v \sin \psi = 0,$$

i.e. $$\frac{du}{ds}\frac{dx}{ds} + \frac{dv}{ds}\frac{dy}{ds} = 0 \quad \text{.........................(3).}$$

142. Tangential and Normal Resolutions.

Somewhat easier equations are obtained if we assume v_s and v_ρ as the initial tangential and normal velocities at P.

Resolving along the tangent, we have

$$m\delta s \,.\, v_s = (T + dT) \cos d\psi - T$$
$$= dT + \text{small quantities of the second order,}$$

i.e., in the limit, $$\frac{dT}{ds} = mv_s \quad \text{............................(1).}$$

So, resolving along the normal,

$$m\delta s \,.\, v_\rho = (T + dT) \sin \delta\psi = T\delta\psi + \ldots,$$

i.e., in the limit, $$\frac{T}{\rho} = mv_\rho \quad \text{................................(2)}$$

where ρ is the radius of curvature.

The condition of inextensibility gives

$$v_s = (v_s + \delta v_s) \cos \delta\psi - (v_\rho + \delta v_\rho) \sin \delta\psi,$$

i.e. $$0 = \delta v_s - v_\rho \delta\psi,$$

i.e. $$\frac{dv_s}{ds} = v_\rho \frac{d\psi}{ds} = \frac{v_\rho}{\rho}, \text{ in the limit} \quad \text{....................(3).}$$

By eliminating v_ρ and v_s from (1), (2), and (3), we have

$$\frac{d}{ds}\left[\frac{1}{m}\frac{dT}{ds}\right] = \frac{1}{m} \cdot \frac{T}{\rho^2} \quad \text{.........................(4).}$$

The initial form of the chain being given, ρ is known as a function of s; also m is either a constant when the chain is uniform or it is a known function of s. Equation (4) thus determines T with two arbitrary constants in the result; they are determined from the fact that T is at one end equal to the given terminal impulsive tension, and at the other end is zero.

Hence T is known and then (1) and (2) determine the initial velocities of each element.

Ex. A uniform chain is hanging in the form of a catenary, whose ends are at the same horizontal level; to each end is applied tangentially an impulse T_0; find the impulse at each point of the chain and its initial velocity.

In this curve $s=c\tan\psi$, so that $\rho=\frac{ds}{d\psi}=c\sec^2\psi$.

$$\therefore \frac{dT}{ds}=\frac{dT}{d\psi}\cdot\frac{\cos^2\psi}{c},$$

and
$$\frac{d^2T}{ds^2}=\frac{d^2T}{d\psi^2}\frac{\cos^4\psi}{c^2}-\frac{2\sin\psi\cos^3\psi}{c^2}\frac{dT}{d\psi}.$$

Hence equation (4) gives

$$\frac{d^2T}{d\psi^2}-2\tan\psi\frac{dT}{d\psi}=T,$$

i.e.
$$\frac{d^2T}{d\psi^2}\cos\psi-\sin\psi\frac{dT}{d\psi}=T\cos\psi+\sin\psi\frac{dT}{d\psi}.$$

$$\therefore \frac{dT}{d\psi}\cos\psi=T\sin\psi+A.$$

$$\therefore T\cos\psi=A\psi+B \quad\text{.......................(1)},$$

where A and B are constants.

Also, from equations (1) and (2),

$$mv_s=\frac{\cos^2\psi}{c}\cdot\frac{dT}{d\psi}=\frac{1}{c}[A\cos\psi+A\psi\sin\psi+B\sin\psi] \quad\text{......(2)},$$

and
$$mv_\rho=\frac{T}{c}\cos^2\psi=\frac{A\psi+B}{c}\cos\psi \quad\text{.................(3)}.$$

Now, by symmetry, the lowest point can have no tangential motion, so that v_s, must vanish when $\psi=0$. Therefore $A=0$.

Also if ψ_0 be the inclination of the tangent at either end, then, from (1),

$$B=T_0\cos\psi_0.$$

$$\therefore T=T_0\frac{\cos\psi_0}{\cos\psi}=\frac{T_0\cos\psi_0}{c}.y,$$

so that the impulsive tension at any point varies as its ordinate.

Also $\quad v_s=\frac{T_0}{mc}\cos\psi_0\sin\psi$, and $v_\rho=\frac{T_0}{mc}\cos\psi_0\cos\psi$.

The velocity of the point considered, parallel to the directrix of the catenary,

$$=v_s\cos\psi-v_\rho\sin\psi=0.$$

Hence all points of the catenary start moving perpendicularly to the directrix, *i.e.* in a vertical direction, and the velocity at any point

$$=\sqrt{v_s^2+v_\rho^2}=\frac{T_0}{mc}\cos\psi_0.$$

143. *Motion of a chain free to move in one plane.*

With the figure of Art. 141 let X, Y be the forces acting on an element PQ of a chain in directions parallel to the axes of coordinates. Let u and v be the component velocities of the point P parallel to the axes. Let T be the tension at P. Its component parallel to $Ox=T\frac{dx}{ds}=f(s)$, where s is the arc $O'P$.

If $PQ=\delta s$, the tension at Q parallel to Ox

$$=f(s+\delta s)=f(s)+\delta s f'(s)+\ldots.$$

Hence, if m be the mass of the chain per unit of length, the equation of motion of PQ is

$$m\delta s.\frac{du}{dt}=\text{the forces parallel to } Ox$$

$$=m\delta sX+\{f(s)+\delta sf'(s)+\ldots\}-f(s)$$

$$=m\delta s.X+\delta s\frac{d}{ds}\left(T\frac{dx}{ds}\right)+\ldots.$$

Dividing by δs, and neglecting powers of δs, we have

$$m\frac{du}{dt}=mX+\frac{d}{ds}\left(T\frac{dx}{ds}\right) \quad \ldots\ldots\ldots\ldots\ldots\ldots\ldots(1).$$

So

$$m\frac{dv}{dt}=mY+\frac{d}{ds}\left(T\frac{dy}{ds}\right) \quad \ldots\ldots\ldots\ldots\ldots\ldots\ldots(2).$$

Assuming the chain to be inextensible, it follows that the velocity of P in the direction PQ must be the same as that of Q in the same direction.

$$\therefore\ u\frac{dx}{ds}+v\frac{dy}{ds}=(u+\delta u)\frac{dx}{ds}+(v+\delta v)\frac{dy}{ds},$$

i.e.

$$0=\delta u.\frac{dx}{ds}+\delta v\frac{dy}{ds},$$

and hence

$$\frac{du}{ds}\frac{dx}{ds}+\frac{dv}{ds}\frac{dy}{ds}=0 \quad \ldots\ldots\ldots\ldots\ldots\ldots\ldots(3).$$

These equations give x, y, and T in terms of s and t, *i.e.* they give the position at time t of any element whose arcual distance is s.

EXAMPLES

1. A uniform chain, in the form of a semi-circle, is plucked at one end with an impulsive tension T_0. Shew that the impulsive tension at an angular distance θ from this end is $T_0\dfrac{\sinh(\pi-\theta)}{\sinh \pi}$.

2. A chain lies in the form of the curve $r=ae^{\frac{3\theta}{2}}$ from $\theta=0$ to $\theta=\beta$, and receives a tangential impulse T_0 at the point where $\theta=0$, the other end being free; shew that the impulsive tension at any point is

$$T_0\frac{e^{\frac{5\beta-\theta}{2}}-e^{2\theta}}{e^{\frac{5\beta}{2}}-1}.$$

3. A uniform chain is in the form of that portion of the plane curve cutting every radius vector at an angle $\tan^{-1}\frac{3}{8}$, which lies between 1 and 256 units distance from the pole. If tangential impulsive tensions of 2 and 1 units are simultaneously applied at the points nearest to and farthest from the pole, shew that the impulsive tension at a point distant 81 units from the pole is $\frac{2}{3}\sqrt{3}$ units.

4. A ring of inelastic wire, of radius r, is supported in a horizontal position and a sphere of radius R and mass m falls vertically into it with velocity V, centre above centre. Shew that the impulsive tension in the wire is $\dfrac{mV}{2\pi}\dfrac{r}{\sqrt{R^2-r^2}}$.

5. A smooth semi-circular tube, of radius a, contains a heavy inextensible chain, of length $a\pi$, which fits it exactly. The tube is fixed in a vertical plane with its diameter horizontal and its vertex upwards, the ends being open. A slight disturbance in a vertical plane takes place. If the length of the chain which at any moment has slipped out of the tube be $a\psi$ (where $\psi<\pi$), prove that

$$a\pi\frac{d^2\psi}{dt^2}=g(\psi+\sin\psi),$$

and find where at any moment the tension of the chain is greatest.

6. A fine uniform chain, of length l and mass ml, moves in a smooth tube in the form of an equiangular spiral of angle α, under a repulsion force from the pole equal to $\mu\times$distance per unit of mass. Prove that the motion of the chain is the same as that of a particle placed at the middle point of its length, and that the tension at an arcual distance x from either end of the chain is $\frac{1}{2}m\mu x(l-x)\cos^2\alpha$.

DYNAMICS OF A RIGID BODY

CHAPTER XI

MOMENTS AND PRODUCTS OF INERTIA. PRINCIPAL AXES

144. If r be the perpendicular distance from any given line of any element m of the mass of a body, then the quantity Σmr^2 is called the **moment of inertia** of the body about the given line.

In other words, the moment of inertia is thus obtained; take each element of the body, multiply it by the square of its perpendicular distance from the given line; and add together all the quantities thus obtained.

If this sum be equal to Mk^2, where M is the total mass of the body, then k is called the Radius of Gyration about the given line. It has sometimes been called the Swing-Radius.

If three mutually perpendicular axes Ox, Oy, Oz be taken, and if the coordinates of any element m of the system referred to these axes be x, y and z, then the quantities Σmyz, Σmzx, and Σmxy are called the **products of inertia** with respect to the axes y and z, z and x, and x and y respectively.

Since the distance of the element from the axis of x is $\sqrt{y^2+z^2}$, the moment of inertia about the axis of x

$$=\Sigma m(y^2+z^2).$$

145. *Simple cases of Moments of Inertia.*

I. *Thin uniform rod of mass M and length 2a.* Let AB be the rod, and PQ any element of it such that $AP=x$ and $PQ=\delta x$. The mass of PQ is $\frac{\delta x}{2a}.M$.

Hence the moment of inertia about an axis through A perpendicular to the rod

$$=\Sigma\frac{\delta x}{2a}.M.x^2=\frac{M}{2a}\int_0^{2a}x^2dx$$

$$=\frac{M}{2a}\cdot\frac{1}{3}[2a]^3=M.\frac{4a^2}{3}.$$

Similarly, if O be the centre of the rod, $OP=y$ and $PQ=\delta y$, the moment of inertia of the rod about an axis through O perpendicular to the rod

$$=\Sigma\frac{\delta y}{2a}.M.y^2=\frac{M}{2a}\int_{-a}^{+a}y^2dy$$

$$=\frac{M}{2a}.\frac{1}{3}\left[y^3\right]_{-a}^{+a}=M.\frac{a^2}{3}.$$

II. *Rectangular lamina.* Let $ABCD$ be the lamina, such that $AB=2a$ and $AD=2b$, whose centre is O. By drawing a large number of lines parallel to AD we obtain a large number of strips, each of which is ultimately a straight line.

The moment of inertia of each of these strips about an axis through O parallel to AB is (by I) equal to its mass multiplied by $\frac{b^2}{3}$. Hence the sum of the moments of the strips, *i.e.* the moments of inertia of the rectangle, about the same line is $M\frac{b^2}{3}$.

So its moment of inertia, about an axis through O parallel to the side $2b$, is

$$M\frac{a^2}{3}.$$

If x and y be the coordinates of any point P of the lamina referred to axes through O parallel to AB and AD respectively, these results give

$$\Sigma my^2=\text{moment of inertia about } Ox=M\frac{b^2}{3}, \text{ and } \Sigma mx^2=M\frac{a^2}{3}.$$

The moment of inertia of the lamina about an axis through O perpendicular to the lamina

$$=\Sigma m.OP^2=\Sigma m(x^2+y^2)=M\frac{a^2+b^2}{3}.$$

III. *Rectangular parallelopiped.* Let the lengths of its sides be $2a$, $2b$, and $2c$. Consider an axis through the centre parallel to the side $2a$, and conceive the solid as made up of a very large number of thin parallel rectangular slices all perpendicular to this axis; each of these slices has sides $2b$ and $2c$ and hence its moment of inertia about the axis is its mass multiplied by $\frac{b^2+c^2}{3}$. Hence the moment of inertia of the whole body is the whole mass multiplied by

$$\frac{b^2+c^2}{3}, \text{ i.e. } M\frac{b^2+c^2}{3}.$$

IV. *Circumference of a circle.* Let Ox be any axis through the centre O, P any point of the circumference such that $xOP=\theta$, PQ an element $a\delta\theta$; then the moment of inertia about Ox

$$=\Sigma\left[\frac{a\delta\theta}{2\pi a}M\right].a^2\sin^2\theta=\frac{Ma^2}{2\pi}\int_0^{2\pi}\sin^2\theta d\theta$$

$$=4\times\frac{Ma^2}{2\pi}\int^{\frac{\pi}{2}}\sin^2\theta d\theta=\frac{2Ma^2}{\pi}.\frac{1}{2}.\frac{\pi}{2}=M\frac{a^2}{2}.$$

V. *Circular disc of radius a.* The area contained between concentric circles of radii r and $r+\delta r$ is $2\pi r\delta r$ and its mass is thus $\frac{2\pi rdr}{\pi a^2}.M$; its moment of inertia about a diameter by the previous article $=\frac{2rdr}{a^2}M.\frac{r^2}{2}$.

Hence the required moment of inertia

$$=\frac{M}{a^2}\int_0^a r^3dr=\frac{M}{a^2}.\frac{a^4}{4}=M.\frac{a^2}{4}.$$

So for the moment about a perpendicular diameter.

The moment of inertia about an axis through the centre perpendicular to the disc $=$ (as in II) the sum of these $=M.\frac{a^2}{2}$.

Elliptic disc of axes 2a, 2b. Taking slices made by lines parallel to the axis of y, the moment of inertia about the axis of x clearly

$$=2\int_{\frac{\pi}{2}}^0\left[\frac{2b\sin\phi d(a\cos\phi)}{\pi ab}M\right].\frac{b^2\sin^2\phi}{3}$$

$$=\frac{4}{3}\frac{Mb^2}{\pi}\int_0^{\frac{\pi}{2}}\sin^4\phi d\phi=M.\frac{b^2}{4}.$$

So the moment of inertia about the axis of $y=M.\frac{a^2}{4}$.

VI. *Hollow sphere.* Let it be formed by the revolution of the circle of IV about the diameter. Then the moment of inertia about the diameter

$$=\Sigma\left[\frac{a\delta\theta.2\pi a\sin\theta}{4\pi a^2}M\right]a^2\sin^2\theta=\frac{Ma^2}{2}\int_0^{\pi}\sin^3\theta.d\theta$$

$$=2.\frac{Ma^2}{2}.\frac{2}{3.1}=\frac{2Ma^2}{3}.$$

VII. *Solid sphere.* The volume of the thin shell included between spheres of radii r and $r+\delta r$ is $4\pi r^2\delta r$, and hence its mass is

$$\frac{4\pi r^2 \delta r}{4\pi \frac{a^3}{3}} M, \text{ i.e. } \frac{3r^2\delta r}{a^3} M.$$

Hence, by VI, the required moment of inertia about a diameter

$$= \int_0^a \frac{3r^2\delta r}{a^3} M . \frac{2r^2}{3} = \frac{2M}{a^3} . \frac{a^5}{5} = M . \frac{2a^2}{5}.$$

VIII. *Solid ellipsoid about any principal axis.* Let the equation to the ellipsoid be $\frac{x^2}{a^2}+\frac{y^2}{b^2}+\frac{z^2}{c^2}=1$. Consider any slice included between planes at distances x and $x+\delta x$ from the centre and parallel to the plane through the axes of y and z.

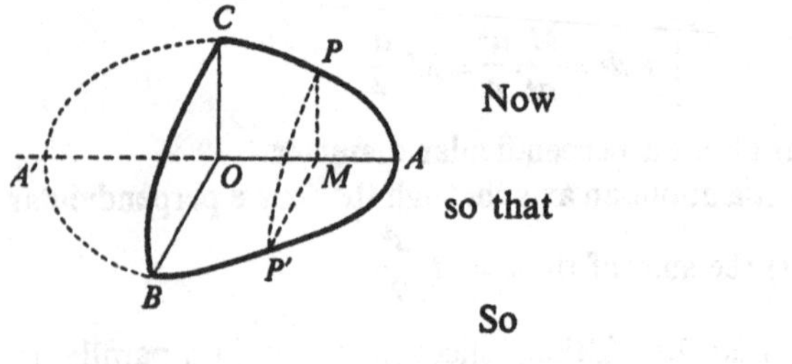

The area of the section through PMP' is

$$\pi . MP . MP'.$$

Now

$$\frac{PM^2}{OC^2}+\frac{OM^2}{OA^2}=1,$$

so that

$$MP = c\sqrt{1-\frac{x^2}{a^2}}.$$

So

$$MP' = b\sqrt{1-\frac{x^2}{a^2}}.$$

Hence the volume of the thin slice

$$= \pi bc\left(1-\frac{x^2}{a^2}\right) . \delta x.$$

Also its moment of inertia about the perpendicular to its plane

$$= \text{its mass} \times \frac{MP^2+MP'^2}{4}$$

$$= \pi bc\delta x . \left(1-\frac{x^2}{a^2}\right)^2 . \frac{b^2+c^2}{4} . \rho.$$

Hence the required moment of inertia

$$= \pi bc\frac{b^2+c^2}{4}\int_{-a}^{+a}\left(1-\frac{x^2}{a^2}\right)^2 \delta x . \rho$$

$$= \pi abc . \frac{b^2+c^2}{4} . \frac{16}{15}\rho = \left(\frac{4}{3}\pi abc\rho\right) \times \frac{b^2+c^2}{5}$$

$$= M \times \frac{b^2+c^2}{5}.$$

146. Dr Routh has pointed out a simple rule for remembering the moments of inertia of many of the simpler bodies, *viz.*

The moment of inertia about an axis of symmetry is

$$\text{Mass} \times \frac{\text{the sum of the squares of the perpendicular semi-axes}}{3,\ 4,\ \text{or}\ 5},$$

the denominator to be 3, 4, or 5 according as the body is rectangular, elliptical (including circular) or ellipsoidal (including spherical).

147. *If the moments and products of inertia about any line, or lines, through the centre of inertia G of a body are known, to obtain the corresponding quantities for any parallel line or lines.*

Let GX, GY, GZ be any three axes through the centre of gravity, and OX', OY', OZ' parallel axes through any point O. Let the coordinates of any element m of the body be x, y, z referred to the first three axes, and x', y', and z' referred to the second set. Then if f, g, and h be the coordinates of G referred to OX', OY', and OZ', we have

$$x' = x + f, \quad y' = y + g, \text{ and } z' = z + h.$$

Hence the moment of inertia of the body with regard to OX'

$$= \Sigma m(y'^2 + z'^2) = \Sigma m[y^2 + z^2 + 2yg + 2zh + g^2 + h^2] \quad \text{......(1).}$$

Now $$\Sigma m . 2yg = 2g . \Sigma my.$$

Also, by Statics, $\dfrac{\Sigma my}{\Sigma m}$ = the y-coordinate of the centre of inertia referred to G as origin $= 0$. Hence $\Sigma m . 2yg = 0$ and similarly $\Sigma m . 2zh = 0$.

Hence, from (1),

the moment of inertia with regard to OX'

$= \Sigma m(y^2 + z^2) + M(g^2 + h^2)$

= the moment of inertia with regard to GX + the moment of inertia of a mass M placed at G about the axis OX'.

Again, the product of inertia about the axes OX' and OY'

$= \Sigma mx'y' = \Sigma m(x + f)(y + g)$

$= \Sigma m[xy + g . x + fy + fg]$

$= \Sigma mxy + Mfg$

= the product of inertia about GX and GY + the product of inertia of a mass M placed at G about the axes OX' and OY'.

Cor. It follows from this article that of all axes drawn in a given direction the one through the centre of inertia is the one such that the moment of inertia about it is a minimum.

Exs. The moment of inertia of the arc of a complete circle about a tangent

$$=M\frac{a^2}{2}+Ma^2=\frac{3M}{2}a^2.$$

The moment of inertia of a solid sphere about a tangent

$$=M.\frac{2a^2}{5}+Ma^2=\frac{7}{5}Ma^2.$$

148. *If the moments and products of inertia of a body about three perpendicular and concurrent axes are known, to find the moment of inertia about any other axis through their meeting point.*

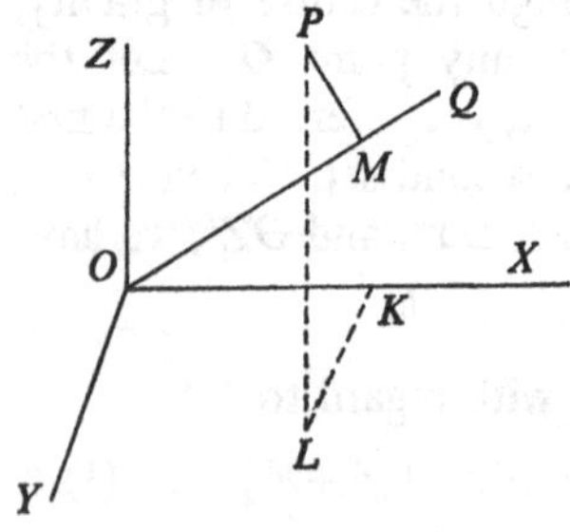

Let OX, OY, OZ be the three given axes, and let A, B and C be the moments of inertia with respect to them, and D, E and F the products of inertia with respect to the axes of y and z, of z and x, and x and y respectively.

Let the moment of inertia be required about OQ, whose direction-cosines with respect to OX, OY and OZ are l, m, and n.

Take any element m' of the body at P whose coordinates are x, y, and z, so that $OK=x$, $KL=y$, and $LP=z$.

Draw PM perpendicular to the axis OQ.

Then $$PM^2=OP^2-OM^2.$$

Now $$OP^2=x^2+y^2+z^2,$$

and OM = the projection on OQ of the straight line OP

= the projection on OQ of the broken line $OKLP$

$$=l.OK+m.KL+n.LP=lx+my+nz.$$

Hence the required moment of inertia about OM

$$=\Sigma m'.PM^2=\Sigma m'[x^2+y^2+z^2-(lx+my+nz)^2]$$

$$=\Sigma m'.\begin{bmatrix}x^2(m^2+n^2)+y^2(n^2+l^2)+z^2(l^2+m^2)\\ -2mnyz-2nlzx-2lmxy\end{bmatrix},$$

since $l^2+m^2+n^2=1$,

$$=l^2\Sigma m'(y^2+z^2)+m^2\Sigma m'(z^2+x^2)+n^2.\Sigma m'(x^2+y^2)$$
$$-2mn\Sigma m'yz-2nl\Sigma m'zx-2lm\Sigma m'xy$$

$$=Al^2+Bm^2+Cn^2-2Dmn-2Enl-2Flm.$$

149. As a particular case of the preceding article consider the case of a plane lamina.

Let A, B be its moments of inertia about two lines OX and OY at right angles, and F its product of inertia about the same two lines so that

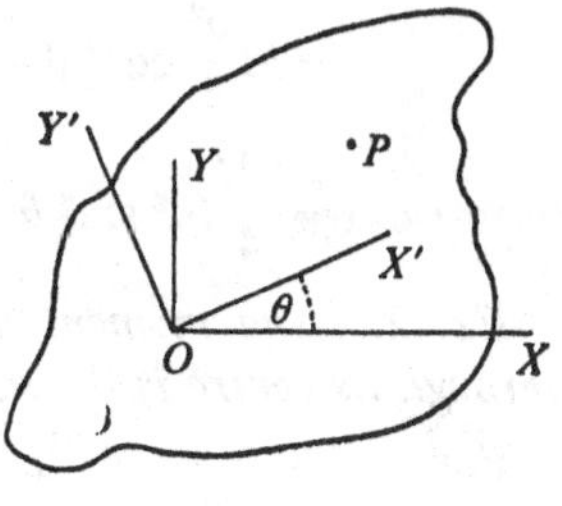

$$A=\Sigma my^2,\ B=\Sigma mx^2,\ F=\Sigma mxy.$$

If (x', y') are the coordinates of a point P referred to new axes OX' and OY', where $\angle XOX'=\theta$, then

$$x=x'\cos\theta-y'\sin\theta,$$

and
$$y=x'\sin\theta+y'\cos\theta.$$

$$\therefore\ x'=x\cos\theta+y\sin\theta \quad \text{and} \quad y'=y\cos\theta-x\sin\theta.$$

Hence the moment of inertia about OX'

$$\begin{aligned}&=\Sigma my'^2=\Sigma m(y\cos\theta-x\sin\theta)^2\\&=\cos^2\theta.\Sigma my^2+\sin^2\theta.\Sigma mx^2-2\sin\theta\cos\theta.\Sigma mxy\\&=A\cos^2\theta+B\sin^2\theta-2F\sin\theta\cos\theta.\end{aligned}$$

The product of inertia about OX' and OY'

$$\begin{aligned}&=\Sigma mx'y'=\Sigma m(x\cos\theta+y\sin\theta)(y\cos\theta-x\sin\theta)\\&=\Sigma m[y^2\sin\theta\cos\theta-x^2\sin\theta\cos\theta+xy(\cos^2\theta-\sin^2\theta)]\\&=(A-B)\sin\theta\cos\theta+F\cos 2\theta.\end{aligned}$$

In the case of a plane lamina, if A and B be the moments of inertia about any two perpendicular lines lying in it, the moment of inertia about a line through their intersection perpendicular to the plane

$$=\Sigma m(x^2+y^2)=\Sigma my^2+\Sigma mx^2=A+B.$$

150. *Ex.* 1. *Find the moment of inertia of an elliptic area about a line CP inclined at θ to the major axis, and about a tangent parallel to CP.*

The moments of inertia, A and B, about the major and minor axes are, as in Art. 145, $M\frac{b^2}{4}$ and $M\frac{a^2}{4}$. Hence the moment of inertia about CP

$$=M\frac{b^2}{4}\cos^2\theta+M\frac{a^2}{4}\sin^2\theta, \text{ since } F=0, \text{ by symmetry.}$$

The perpendicular CY upon a tangent parallel to $CP=\frac{ab}{CP}$.

Hence, by Art. 147, the moment of inertia about this tangent

$$=M\frac{b^2}{4}\cos^2\theta+M\frac{a^2}{4}\sin^2\theta+M\frac{a^2b^2}{CP^2}$$

$$=M\frac{b^2}{4}\cos^2\theta+M\frac{a^2}{4}\sin^2\theta+Ma^2b^2\left[\frac{\cos^2\theta}{a^2}+\frac{\sin^2\theta}{b^2}\right]$$

$$=\frac{5M}{4}(a^2\sin^2\theta+b^2\cos^2\theta).$$

Ex. 2. *The moment of inertia of a uniform cube about any axis through its centre is the same*

For $A=B=C$, and $D=E=F=0$,

Therefore, from Art. 148, $I=A(l^2+m^2+n^2)=A$.

EXAMPLES

Find the moments of inertia of the following:

1. A rectangle about a diagonal and any line through the centre.

2. A circular area about a line in its own plane whose perpendicular distance from its centre is c.

3. The arc of a circle about (1) the diameter bisecting the arc, (2) an axis through the centre perpendicular to its plane, (3) an axis through its middle point perpendicular to its plane.

4. An isosceles triangle about a perpendicular from the vertex upon the opposite side.

5. Any triangular area ABC about a perpendicular to its plane through A.

[*Result.* $\frac{M}{12}(3b^2+3c^2-a^2)$.]

6. The area bounded by $r^2=a^2\cos 2\theta$ about its axis.

[*Result.* $\frac{Ma^2}{16}\left(\pi-\frac{8}{3}\right)$.]

7. A right circular cylinder about (1) its axis, (2) a straight line through its centre of gravity perpendicular to its axis.

8. A rectangular parallelopiped about an edge.

9. A hollow sphere about a diameter, its external and internal radii being a and b.

[*Result.* $\frac{2M}{5}\frac{a^5-b^5}{a^3-b^3}$.]

10. A truncated cone about its axis, the radii of its ends being a and b.

[*Result.* $\frac{3M}{10}\frac{a^5-b^5}{a^3-b^3}$.]

11. Shew that the moment of inertia of a right solid cone, whose height is h and the radius of whose base is a, is $\frac{3Ma^2}{20}\cdot\frac{6h^2+a^2}{h^2+a^2}$ about a slant side, and $\frac{3M}{80}(h^2+4a^2)$ about a line through the centre of gravity of the cone perpendicular to its axis.

12. Shew that the moment of inertia of a parabolic area (of latus-rectum $4a$) cut off by an ordinate at distance h from the vertex is $\frac{3}{7}Mh^2$ about the tangent at the vertex, and $\frac{4}{5}Mah$ about the axis.

13. Shew that the moment of inertia of a paraboloid of revolution about its axis is $\frac{M}{3}\times$ the square of the radius of its base.

14. Find the moment of inertia of the homogeneous soid ellipsoid bounded by $\frac{x^2}{a^2}+\frac{y^2}{b^2}+\frac{z^2}{c^2}=1$ about the normal at the point (x', y', z').

15. Shew that the moment of inertia of a thin homogeneous ellipsoidal shell (bounded by similar, similarly situated and concentric ellipsoids) about an axis is $M\frac{b^2+c^2}{3}$, where M is the mass of the shell.

16. Shew that the moment of inertia of a regular polygon of n sides about any straight line through its centre is $\frac{Mc^2}{24}\frac{2+\cos\frac{2\pi}{n}}{1-\cos\frac{2\pi}{n}}$, where n is the number of sides and c is the length of each.

17. A solid body, of density ρ, is in the shape of the solid formed by the revolution of the cardioid $r=a(1+\cos\theta)$ about the initial line; shew that its moment of inertia about a straight line through the pole perpendicular to the initial line is $\frac{352}{105}\pi\rho a^5$.

18. A closed central curve revolves round any line Ox in its own plane which does not intersect it; shew that the moment of inertia of the solid of revolution so formed about Ox is equal to $M(a^2+3k^2)$, where M is the mass of the solid generated, a is the distance from Ox of the centre C of the curve, and k is the radius of gyration of the curve about a line through C parallel to Ox.

Prove a similar theorem for the moment of inertia of the surface generated by the arc of the curve.

19. The moment of inertia about its axis of a solid rubber tyre, of mass M and circular cross-section of radius a, is $\frac{M}{4}(4b^2+3a^2)$, where b is the radius of the core. If the tyre be hollow, and of small uniform thickness, shew that the corresponding result is $\frac{M}{2}(2b^2+3a^2)$.

151. Momental ellipsoid. Along a line OQ drawn through any point O take a distance OQ, such that the moment of inertia of the body about OQ may be inversely proportional to the square of OQ. The result of Art. 148 then gives

$$Al^2+Bm^2+Cn^2-2Dmn-2Enl-2Flm,$$

$$\propto \frac{1}{OQ^2}=\frac{M.K^4}{OQ^2},$$

where M is the mass of the body and K is some linear factor.

If (x, y, z) are the coordinates of Q referred to the axes OX, OY, OZ this gives

$$Ax^2+By^2+Cz^2-2Dyz-2Ezx-2Fxy=MK^4 \quad \text{.........(1).}$$

The locus of the point Q is thus an ellipsoid, which is called the momental ellipsoid of the body at the point O.

Since the position of Q is obtained by a physical definition, which is independent of any particular axes of coordinates, we arrive at the same ellipsoid whatever be the axes OX, OY, OZ with which we start.

It is proved in books on Solid Geometry that for every ellipsoid there can be found three perpendicular diameters such that, if they be taken as axes of coordinates, the resulting equation of the ellipsoid has no terms involving yz, zx, or xy, and the new axes of coordinates are then called the Principal Axes of the Ellipsoid.

Let the momental ellipsoid (1) when referred to its principal axes have as equation

$$A'x^2+B'y^2+C'z^2=MK^4 \quad \text{....................(2).}$$

The products of inertia with respect to these new axes must be zero; for if any one of them, say D', existed, then there would, as in equation (1), be a term $-2D'yz$ in (2).

Hence we have the following important proposition: *For any body whatever there exists at each point O a set of three perpendicular axes (which are the three principal diameters of the momental ellipsoid at O) such that the products of inertia of the body about them, taken two at a time, all vanish.*

These three axes are called the **principal axes** of the body at the point O; also a plane through any two of these axes is called a principal plane of the body.

152. It is also shewn in Solid Geometry that of the three principal axes of an ellipsoid, one is the maximum radius vector of the ellipsoid, and another is the minimum. Since the square of the radius vector of the momental ellipsoid is inversely proportional to the corresponding moment of inertia of the body, it follows that of the three principal axes, one has the minimum moment of inertia and another has the maximum.

If the three principal moments of inertia at O are equal, the ellipsoid of inertia becomes a sphere, all radii of which are equal; in this case all moments of inertia about lines through O are equal.

Thus, in the case of a cube of side $2a$, the principal moments of inertia at its centre are equal, and hence the moment of inertia about any line through its centre is the same and equal to $M.\frac{2a^2}{3}$.

If the body be a lamina, the section of the momental ellipsoid at any point of the lamina, which is made by the plane of the lamina, is called the momental ellipse at the point.

If the two principal moments in this case are the same, the momental ellipse becomes a circle, and the moments of inertia of the lamina about all lines through O are the same.

153. *To shew that the moments and products of inertia of a uniform triangle about any lines are the same as the moments and products of inertia, about the same lines, of three particles placed at the middle points of the sides, each equal to one-third the mass of the triangle.*

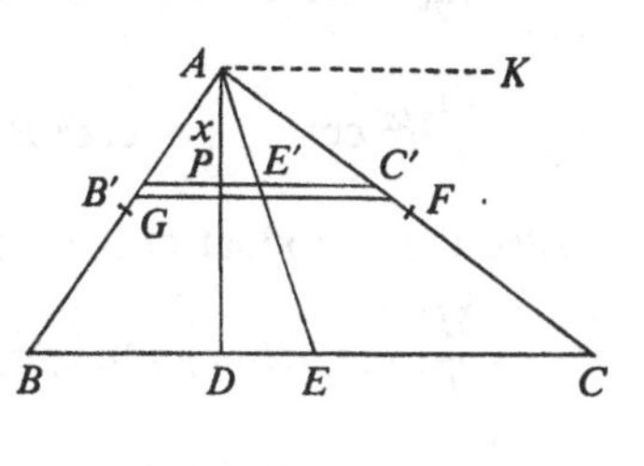

Divide up the triangle ABC into narrow slips by a large number of straight lines parallel to its base.

Let $x, = AP$, be the distance of one of them from A. Then $B'C' = \frac{x}{h}a$, where $AD = h$, and the mass M of the triangle $= \frac{1}{2}ah.\rho$, where ρ is the density.

The moment of inertia about a line AK parallel to BC

$$= \int_0^h \left[\frac{xa\rho}{h}.dx\right]x^2 = \frac{a\rho}{h}.\frac{1}{4}h^4 = M\frac{h^2}{2} \quad \text{..............(1).}$$

The moment of inertia about AD, by Art. 147,

$$= \int_0^h \left[\frac{xa\rho}{h}dx\right]\left[\frac{1}{3}\left(\frac{xa}{2h}\right)^2 + \frac{x^2}{h^2}DE^2\right],$$

where E is the middle point of BC,

$$= \frac{1}{4}\rho ah\left[\frac{a^2}{12} + DE^2\right] = \frac{M}{24}[(b \cos C + c \cos B)^2$$

$$+ 3(b \cos C - c \cos B)^2]$$

$$= \frac{M}{6}[b^2 \cos^2 C + c^2 \cos^2 B - bc \cos B \cos C] \quad \text{....................(2).}$$

The product of inertia about AK, AD

$$= \int_0^h \frac{xa\rho}{h}dx.x.PE', \text{ by Art. 147,}$$

$$= \frac{a\rho}{h^2}\int_0^h x^3.DE.dx = \frac{a\rho}{h^2}.\frac{1}{4}h^4.DE = \frac{Mh}{4}[b \cos C - c \cos B] \text{...(3).}$$

If there be put three particles, each of mass $\frac{M}{3}$, at E, F, and G, the middle points of the sides, their moment of inertia about AK

$$=\frac{M}{3}\left[h^2+\left(\frac{h}{2}\right)^2+\left(\frac{h}{2}\right)^2\right]=M\frac{h^2}{2}.$$

Their moment of inertia about AD

$$=\frac{M}{3}\left[\left(\frac{a}{2}-c\cos B\right)^2+\left(\frac{b}{2}\cos C\right)^2+\left(\frac{c}{2}\cos B\right)^2\right]$$

$$=\frac{M}{12}[(b\cos C-c\cos B)^2+b^2\cos^2 C+c^2\cos^2 B]$$

$$=\frac{M}{6}[b^2\cos^2 C+c^2\cos^2 B-bc\cos B\cos C].$$

Also their product of inertia about AK, AD

$$=\frac{M}{3}\left[AD.DE+\frac{1}{2}AD.\frac{1}{2}DC-\frac{1}{2}AD.\frac{1}{2}BD\right]$$

$$=\frac{Mh}{3}\left[DE+\frac{b\cos C-c\cos B}{4}\right]=\frac{Mh}{4}[b\cos C-c\cos B].$$

The moments and products of inertia of the three particles about AK, AD are thus the same as those of the triangle.

Hence, by Art. 149, the moments of inertia about any line through A are the same; and also, by the same article, the products of inertia about any two perpendicular lines through A are the same.

Also it is easily seen that the centre of inertia of the three particles coincides with the centre of inertia of the triangle.

Hence, by Art. 147, it follows that the moments and products of inertia about any lines through the common centre of gravity are the same for the two systems, and therefore also, by the same article, the moments and products about any two other perpendicular lines in the plane of the triangle are the same.

Finally, the moment of inertia about a perpendicular to the plane of the triangle through any point P is equal to the sum of the moments about any two perpendicular lines through P in the plane of the triangle, and is thus the same for the two systems.

154. Two mechanical systems such as the triangle and the three particles of the preceding article, which are such that their moments of inertia about all lines are the same, are said to be **equi-momental,** or kinetically equivalent.

If two systems have the same centre of inertia, the same mass, and the same principal axes and the same principal moments at their centre of inertia, it follows, by Arts. 147 and 148, that their moments of inertia about any straight line are the same, and hence that the systems are equi-momental.

EXAMPLES

1. The momental ellipsoid at the centre of an elliptic plate is

$$\frac{x^2}{a^2}+\frac{y^2}{b^2}+z^2\left[\frac{1}{a^2}+\frac{1}{b^2}\right]=\text{const.}$$

2. The momental ellipsoid at the centre of a solid ellipsoid is

$$(b^2+c^2)x^2+(c^2+a^2)y^2+(a^2+b^2)z^2=\text{const.}$$

3. The equation of the momental ellipsoid at the corner of a cube, of side $2a$, referred to its principal axes is $2x^2+11(y^2+z^2)=\text{const.}$

4. The momental ellipsoid at a point on the rim of a hemisphere is

$$2x^2+7(y^2+z^2)-\tfrac{15}{4}xz=\text{const.}$$

5. The momental ellipsoid at a point on the circular edge of a solid cone is $(3a^2+2h^2)x^2+(23a^2+2h^2)y^2+26a^2z^2-10ahxz=\text{const.}$, where h is the height and a the radius of the base.

6. Find the principal axes of a right circular cone at a point on the circumference of the base; and shew that one of them will pass through its centre of gravity if the vertical angle of the cone is $2\tan^{-1}\frac{1}{2}$.

7. Shew that a uniform rod, of mass m, is kinetically equivalent to three particles, rigidly connected and situated one at each end of the rod and one at its middle point, the masses of the particles being $\frac{1}{6}m$, $\frac{1}{6}m$ and $\frac{2}{3}m$.

8. $ABCD$ is a uniform parallelogram, of mass M; at the middle points of the four sides are placed particles each equal to $\frac{M}{6}$, and at the intersection of the diagonals a particle, of mass $\frac{M}{3}$; shew that these five particles and the parallelogram are equi-momental systems.

9. Shew that any lamina is dynamically equivalent to three particles, each one-third of the mass of the lamina, placed at the corners of a maximum triangle inscribed in the ellipse, whose equation referred to the principal axes at the centre of inertia is $\frac{x^2}{B}+\frac{y^2}{A}=2$, where mA and mB are the principal moments of inertia about Ox and Oy and m is the mass.

10. Shew that there is a momental ellipse at an angular point of a triangular area which touches the opposite side at its middle point and bisects the adjacent sides. [Use Art. 153.]

11. Shew that there is a momental ellipse at the centre of inertia of a uniform triangle which touches the sides of the triangle at the middle points.

12. *Shew that a uniform tetrahedron is kinetically equivalent to four particles, each of mass* $\frac{M}{20}$, *at the vertices of the tetrahedron, and a fifth particle, of mass* $\frac{4M}{5}$, *placed at its centre of inertia.*

Let $OABC$ be the tetrahedron and through the vertex O draw any three rectangular axes OX, OY, OZ. Let the coordinates of A, B and C referred to these axes be (x_1, y_1, z_1), (x_2, y_2, z_2), and (x_3, y_3, z_3), so that the middle point of BC is $\left(\frac{x_2+x_3}{2}, \frac{y_2+y_3}{2}, \frac{z_2+z_3}{2}\right)$. Take any section PQR parallel to ABC at a perpendicular distance ξ from O; its area is $A_0\frac{\xi^2}{p^2}$, where A_0 is the area of ABC and p is the perpendicular from O on ABC. By Art. 153 the moment of inertia of a thin slice, of thickness $d\xi$, about Ox

$=$ that of three particles, each $\frac{1}{3}.A_0\rho.\frac{\xi^2}{p^2}.d\xi$ placed at the middle points of QR, RP, and PQ

$$=\frac{1}{3}.A_0\rho.\frac{\xi^2}{p^2}d\xi\left[\left(\frac{\xi}{p}.\frac{y_2+y_3}{2}\right)^2+\text{two similar terms}\right.$$
$$\left.+\left(\frac{\xi}{p}.\frac{z_2+z_3}{2}\right)^2+\text{two similar terms}\right].$$

Hence, on integrating with respect to ξ from O to p, the moment of inertia about OX of the whole tetrahedoh

$$=\frac{1}{15}.A_0\rho.\frac{p}{4}[\{(y_2+y_3)^2+(z_2+z_3)^2\}+\text{two similar expressions}]$$
$$=\frac{M}{10}\begin{bmatrix} y_1^2+y_2^2+y_3^2+y_2y_3+y_3y_1+y_1y_2 \\ +z_1^2+z_2^2+z_3^2+z_2z_3+z_3z_1+z_1z_2\end{bmatrix} \quad \ldots\ldots(1).$$

Now the moment of inertia about OX of four particles, each of mass $\frac{M}{20}$, at the vertices of the tetrahedron, and of $\frac{4M}{5}$ at its centre of inertia

$=\frac{M}{20}[(y_1^2+z_1^2)+\text{two similar terms}]+\frac{4M}{5}\left[\left(\frac{y_1+y_2+y_3}{4}\right)^2+\left(\frac{z_1+z_2+z_3}{4}\right)^2\right]$, and this, on reduction, equals (1). Similarly for the moments about the axes OY and OZ.

In a similar manner the product of inertia of the tetrahedron about OY, OZ

$$=\frac{M}{20}[(y_2+y_3)(z_2+z_3)+\text{two similar expressions}] \quad \ldots\ldots(2),$$

and that of the five particles

$$=\frac{M}{20}[y_2z_2+y_3z_3+y_1z_1]+\frac{4M}{5}\frac{(y_1+y_2+y_3)}{4}\frac{(z_1+z_2+z_3)}{4},$$

and this is easily seen to be equal to (2).

Also it follows at once that the centre of inertia of the tetrahedron coincides with the centre of inertia of the five particles.

The two systems are therefore equal momental, for, by Arts. 147 and 148, their moments of inertia about any straight line is the same.

13. Shew that a tetrahedron is kinetically equivalent to six particles at the middle points of its edges, each $\frac{1}{10}$th of the mass of the tetrahedron and one at the centroid $\frac{2}{5}$th of the mass of the tetrahedron.

155. *To find whether a given straight line is, at any point of its length, a principal axis of a given material system, and, if so, to find the other two principal axes.*

Take the given straight line as the axis of z, with any point O on it as origin, and any two perpendicular lines OX, OY as the other two axes.

Assume that OZ is a principal axis at a point C of its length and let CX', CY' be the other two principal axes where CX' is inclined at an angle θ to a line parallel to OX. Let OC be h.

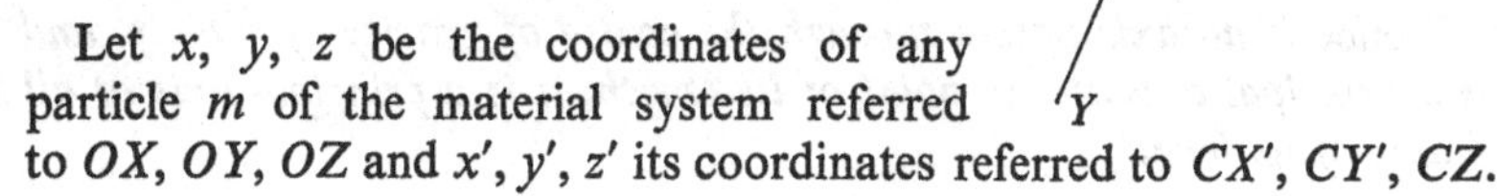

Let x, y, z be the coordinates of any particle m of the material system referred to OX, OY, OZ and x', y', z' its coordinates referred to CX', CY', CZ.

Then

$$z=z'+h, \quad x=x'\cos\theta-y'\sin\theta, \quad \text{and} \quad y=x'\sin\theta+y'\cos\theta,$$

so that

$$x'=x\cos\theta+y\sin\theta, \quad y'=-x\sin\theta+y\cos\theta, \quad \text{and} \quad z'=z-h.$$

$$\therefore \Sigma my'z'=\Sigma m(-xz\sin\theta+yz\cos\theta+hx\sin\theta-hy\cos\theta)$$
$$=D\cos\theta-E\sin\theta+Mh(\bar{x}\sin\theta-\bar{y}\cos\theta) \quad \text{......(1)},$$

with the notation of Art. 148,

$$\Sigma mz'x'=\Sigma m[xz\cos\theta+yz\sin\theta-hx\cos\theta-hy\sin\theta]$$
$$=D\sin\theta+E\cos\theta-Mh(\bar{x}\cos\theta+\bar{y}\sin\theta)\text{.........(2)},$$

and

$$\Sigma mx'y'=\Sigma m[-x^2\sin\theta\cos\theta+xy(\cos^2\theta-\sin^2\theta)+y^2\sin\theta\cos\theta]$$
$$=\tfrac{1}{2}\sin 2\theta(A-B)+F\cos 2\theta \quad \text{..............................(3)}.$$

If CX', CY', CZ are principal axes, the quantities (1), (2), (3) must vanish.

The latter gives $$\tan 2\theta=\frac{2F}{B-A} \quad \text{..........................(4)}.$$

From (1) and (2) we have

$$\frac{E\sin\theta-D\cos\theta}{\bar{x}\sin\theta-\bar{y}\cos\theta}=\frac{D\sin\theta+E\cos\theta}{\bar{x}\cos\theta+\bar{y}\sin\theta}=Mh.$$

These give $$\frac{E}{\bar{x}}=\frac{D}{\bar{y}} \quad \text{..............................(5)},$$

and $$h=\frac{D}{M\bar{y}}=\frac{E}{M\bar{x}} \quad \text{...........................(6)}.$$

(5) is the condition that must hold so that the line OZ may be a principal axis at some point of its length, and then, if it be satisfied, (6) and (4) give the position of the point and the directions of the other principal axes at it.

156. If an axis be a principal axis at a point O of its length it is not, in general, a principal axis at any other point. For if it be a principal axis at O then D, E, and F are all zero; equation (6) of the previous article then gives $h=0$, *i.e.* there is no other such point as C, except when $\bar{x}=0$ and $\bar{y}=0$. In this latter case the axis of z passes through the centre of gravity and the value of h is indeterminate, *i.e.* the axis of z is a principal axis at any point of its length.

Hence *if an axis passes through the centre of gravity of a body, and is a principal axis at any point of its length, it is a principal axis at all points of its length.*

157. If the body be a lamina, as in the figure of Art. 149, the principal axes at a point O are a normal OZ to its plane, and two lines OX', OY' inclined at an angle θ to OX and OY.

In this case, since z is zero for every point of the lamina, both D and E vanish. Hence equation (6) of Art. 155 gives $h=0$, and θ is given by

$$\tan 2\theta=\frac{2F}{B-A}.$$

As a numerical illustration, take the case of the triangle of Art. 153.

Here $$A=M\frac{h^2}{2};\quad B=\frac{M}{6}\begin{bmatrix} b^2\cos^2 C+c^2\cos^2 B \\ -bc\cos B\cos C\end{bmatrix};$$

and $$F=\frac{Mh}{4}(b\cos C-c\cos B).$$

The inclination θ of one of the principal axes to AK is then given by the above formula.

158. The principal axes at any point P of a lamina may be constructed as follows.

The plane of the lamina being the plane of the paper, let G be its centre of inertia and GX, GY the principal axes at G, the moments of inertia about which are A and B, A being greater than B.

On GX take points S and H, such that

$$GS=GH=\sqrt{\frac{A-B}{M}}.$$

Then, by Art. 147, the moment of inertia about SY', parallel to GY, $=B+M.GS^2=A$, so that the moments of inertia about SX and SY' are both equal to A.

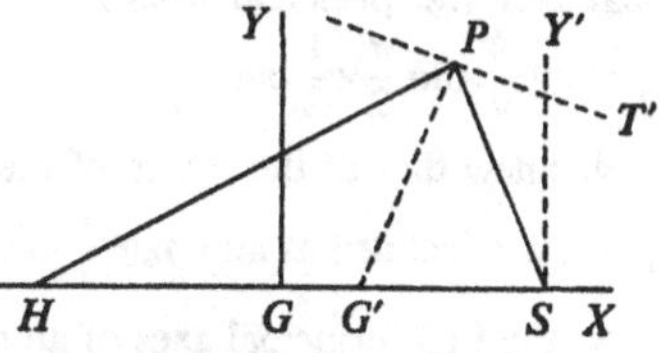

Also the product of inertia about SX, SY'

$$=\Sigma m(x-GS).y=\Sigma mxy-GS.\Sigma my=0,$$

since GX and GY are the principal axes at G, and since G is the centre of inertia.

Hence S is a point such that SX and SY' are the principal axes, and the moments about each are equal to A.

Hence, by Art. 149 or 152, any line through S in the plane of the paper is a principal axis at S, and the moment of inertia about it is A.

Similarly for any line through H.

Hence the moments of inertia about SP and HP are each equal to A. Also the normal at P to the lamina is clearly one of the principal axes at P, so that the other two lie in the plane of the lamina. If then we construct the momental ellipse at P, its radii-vectores in the directions PS and PH must be equal, since we have shewn that the moments of inertia about PS and PH are the same. Also in any ellipse equal radii-vectores are equally inclined to its principal axes, so that the latter bisect the angles between equal radii-vectores.

Hence the principal axes of the momental ellipse at P, *i.e.* the principal axes of the lamina at P in its plane, bisect the angles between PS and PH.

If then, with S and H as foci, we describe an ellipse to pass through any point P of the lamina, the principal axes of the lamina are the tangent and normal to this ellipse at P. The points S and H are hence known as the Foci of Inertia.

159. The proposition of the preceding article may be extended to any body, if G be the centre of inertia, GX, GY, and GZ its principal axes at G and P be any point in the plane of XY.

EXAMPLES

1. If A and B be the moments of inertia of a uniform lamina about two perpendicular axes, OX and OY, lying in its plane, and F be the product of inertia of the lamina about these lines, shew that the principal moments at O are equal to $\frac{1}{2}[A+B\pm\sqrt{(A-B)^2+4F^2}]$.

2. The lengths AB and AD of the sides of a rectangle $ABCD$ are $2a$ and $2b$; shew that the inclination to AB of one of the principal axes at A is $\frac{1}{2}\tan^{-1}\frac{3ab}{2(a^2-b^2)}$.

3. A wire is in the form of a semi-circle of radius a; shew that at an end of its diameter the principal axes in its plane are inclined to the diameter at angles $\frac{1}{2}\tan^{-1}\frac{4}{\pi}$ and $\frac{\pi}{2}\times\frac{1}{2}\tan^{-1}\frac{4}{\pi}$.

4. Shew that at the centre of the quadrant of an ellipse the principal axes in its plane are inclined at an angle $\frac{1}{2}\tan^{-1}\left(\frac{4}{\pi}\frac{ab}{a^2-b^2}\right)$ to the axes.

5. Find the principal axes of an elliptic area at any point of its bounding arc.

6. At the vertex C of a triangle ABC, which is right-angled at C, the principal axes are a perpendicular to the plane and two others inclined to the sides at an angle $\frac{1}{2}\tan^{-1}\frac{ab}{a^2-b^2}$.

7. ABC is a triangular area and AD is perpendicular to BC; E is the middle point of BC and O the middle point of DE; shew that BC is a principal axis of the triangle at O. [Use the property of Art. 153.]

8. A uniform square lamina is bounded by the axes of x and y and the lines $x=2c$, $y=2c$, and a corner is cut off by the line $\frac{x}{a}+\frac{y}{b}=2$. Shew that the principal axes at the centre of the square are inclined to the axis of x at angles given by $\tan 2\theta=\frac{ab-2(a+b)c+3c^2}{(a-b)(a+b-2c)}$.

9. A uniform lamina is bounded by a parabolic arc, of latus-rectum $4a$, and a double ordinate at a distance b from the vertex. If $b=\frac{a}{3}(7+4\sqrt{7})$, shew that two of the principal axes at the end of a latus-rectum are the tangent and normal there.

10. Shew that the principal axes at the node of a half-loop of the lemniscate $r^2=a^2\cos 2\theta$ are inclined to the initial line at angles

$$\frac{1}{2}\tan^{-1}\frac{1}{2} \quad\text{and}\quad \frac{\pi}{2}+\frac{1}{2}\tan^{-1}\frac{1}{2}.$$

11. The principal axes at a corner O of a cube are the line joining O to the centre of the cube and any two perpendicular lines.

12. If the vertical angle of a cone is 90°, the point at which a generator is a principal axis divides the generator in the ratio 3 : 7.
[Use Art. 159.]

13. Three rods AB, BC, and CD, each of mass m and length $2a$, are such that each is perpendicular to the other two. Shew that the principal moments of inertia at the centre of mass are ma^2, $\frac{11}{3}ma^2$ and $4ma^2$.

14. The length of the axis of a solid parabola of revolution is equal to the latus-rectum of the generating parabola. Prove that one principal axis at a point of the circular rim meets the axis of revolution at an angle $\frac{1}{2}\tan^{-1}\frac{2}{3}$.

CHAPTER XII

D'ALEMBERT'S PRINCIPLE

THE GENERAL EQUATIONS OF MOTION

160. We have already found that, if x, y, z be the coordinates of a particle m at time t, its motion is found by equating $m\frac{d^2x}{dt^2}$ to the force parallel to the axis of x, and similarly for the motion parallel to the axes of y and z.

If m be a portion of a rigid body its motion is similarly given, but in this case we must include under the forces parallel to the axes not only the external forces acting on the particle (such as its weight), but also the forces acting on the particle which are due to the actions of the rest of the body on it.

The quantity $m\frac{d^2x}{dt^2}$ is called the effective force acting on the particle parallel to the axis of x. [It is also sometimes called the kinetic reaction of the particle.]

Thus we may say that the x-component of the effective force is equivalent to the x-component of the external forces together with the x-component of the internal forces,

or again that the x-component of the reversed effective forces together with the x-components of the external and internal forces form a system in equilibrium.

So for the components parallel to the axes of y and z.

Hence the reversed effective force, the external force, and the internal force acting on any particle m of a body are in equilibrium.

So for all the other particles on the body.

Hence the reversed effective forces acting on each particle of the body, the external forces, and the internal forces of the body are in equilibrium. Now the internal forces of the body are in equilibrium amongst themselves; for by Newton's third law there is to every action an equal and opposite reaction.

Hence *the reversed effective forces acting on each particle of the body and the external forces of the system are in equilibrium.*

This is D'Alembert's principle. It was enunciated by him in his *Traité de Dynamique* published in the year 1743. It will be noted however that it is only a deduction from Newton's Third Law of Motion.

161. Let X, Y, Z be the components parallel to the axes of the external forces acting on a particle m whose coordinates are x, y, z at the time t.

Then the principle of the preceding article says that forces whose components are

$$X-m\frac{d^2x}{dt^2},\quad Y-m\frac{d^2y}{dt^2},\quad Z-m\frac{d^2z}{dt^2}$$

acting at the point (x, y, z), together with similar forces acting at each other such point of the body, form a system in equilibrium.

Hence, from the ordinary conditions of equilibrium (*Statics*, Art. 165), we have

$$\Sigma\left(X-m\frac{d^2x}{dt^2}\right)=0,$$

$$\Sigma\left(Y-m\frac{d^2y}{dt^2}\right)=0,$$

$$\Sigma\left(Z-m\frac{d^2z}{dt^2}\right)=0,$$

$$\Sigma\left[y\left(Z-m\frac{d^2z}{dt^2}\right)-z\left(Y-m\frac{d^2y}{dt^2}\right)\right]=0,$$

$$\Sigma\left[z\left(X-m\frac{d^2x}{dt^2}\right)-x\left(Z-m\frac{d^2z}{dt^2}\right)\right]=0,$$

and $$\Sigma\left[x\left(Y-m\frac{d^2y}{dt^2}\right)-y\left(X-m\frac{d^2x}{dt^2}\right)\right]=0.$$

These give $$\Sigma m\frac{d^2x}{dt^2}=\Sigma X \quad\text{.............................(1)},$$

$$\Sigma m\frac{d^2y}{dt^2}=\Sigma Y \quad\text{.............................(2)},$$

$$\Sigma m\frac{d^2z}{dt^2}=\Sigma Z \quad\text{.............................(3)},$$

$$\Sigma m\left(y\frac{d^2z}{dt^2}-z\frac{d^2y}{dt^2}\right)=\Sigma(yZ-zY) \quad\text{.................(4)},$$

$$\Sigma m\left(z\frac{d^2x}{dt^2}-x\frac{d^2z}{dt^2}\right)=\Sigma(zX-xZ) \quad\text{.................(5)},$$

and $$\Sigma m\left(x\frac{d^2y}{dt^2}-y\frac{d^2x}{dt^2}\right)=\Sigma(xY-yX) \quad\text{.................(6)}.$$

These are the equations of motion of any rigid body.

Equations (1), (2), and (3) state that the sums of the components, parallel to the axes of coordinates, of the effective forces are respectively equal to the sums of the components parallel to the same axes of the external impressed forces.

Equations (4), (5), (6) state that the sum of the moments about the axes of coordinates of the effective forces are respectively equal to the sums of the moments about the same axes of the external impressed forces.

162. *Motion of the centre of inertia, and motion relative to the centre of inertia.*

Let $(\bar{x}, \bar{y}, \bar{z})$ be the coordinates of the centre of inertia, and M the mass of the body.

Then $M\bar{x}=\Sigma mx$ throughout the motion, and therefore

$$M\frac{d^2\bar{x}}{dt^2}=\Sigma m\frac{d^2x}{dt^2}.$$

Hence equation (1) of the last article gives

$$M\frac{d^2\bar{x}}{dt^2}=\Sigma X \quad\text{.............................(1).}$$

So

$$M\frac{d^2\bar{y}}{dt^2}=\Sigma Y \quad\text{.............................(2),}$$

and

$$M\frac{d^2\bar{z}}{dt^2}=\Sigma Z \quad\text{.............................(3).}$$

But these are the equations of motion of a particle, of mass M, placed at the centre of inertia of the body, and acted on by forces parallel to, and equal to, the external forces acting on the different particles of the body.

Hence *the centre of inertia of a body moves as if all the mass of the body were collected at it, and as if all the external forces were acting at it in directions parallel to those in which they act.*

Next, let (x', y', z') be the coordinates, relative to the centre of inertia, G, of a particle of the body whose coordinates referred to the original axes were (x, y, z).

Then $\qquad x=\bar{x}+x', \quad y=\bar{y}+y' \quad\text{and}\quad z=\bar{z}+z'$

throughout the motion.

$$\therefore\ \frac{d^2x}{dt^2}=\frac{d^2\bar{x}}{dt^2}+\frac{d^2x'}{dt^2},\quad \frac{d^2y}{dt^2}=\frac{d^2\bar{y}}{dt^2}+\frac{d^2y'}{dt^2},\ \text{and}\ \frac{d^2z}{dt^2}=\frac{d^2\bar{z}}{dt^2}+\frac{d^2z'}{dt^2}.$$

$$\therefore\ y\frac{d^2z}{dt^2}-z\frac{d^2y}{dt^2}=(\bar{y}+y')\left(\frac{d^2\bar{z}}{dt^2}+\frac{d^2z'}{dt^2}\right)-(\bar{z}+z')\left(\frac{d^2\bar{y}}{dt^2}+\frac{d^2y'}{dt^2}\right).$$

Hence the equation (4) of the last article gives

$$\Sigma m\left(\bar{y}\frac{d^2\bar{z}}{dt^2}-\bar{z}\frac{d^2\bar{y}}{dt^2}\right)+\Sigma m\left(y'\frac{d^2z'}{dt^2}-z'\frac{d^2y'}{dt^2}\right)$$

$$+\Sigma m\left[\bar{y}\frac{d^2z'}{dt^2}+y'\frac{d^2\bar{z}}{dt^2}-\bar{z}\frac{d^2y'}{dt^2}-z'\frac{d^2\bar{y}}{dt^2}\right]$$

$$=\Sigma[(\bar{y}+y')Z-(\bar{z}+z')Y]\text{...(4).}$$

Now $\dfrac{\Sigma my'}{\Sigma m}$ = the y-coordinate of the centre of inertia referred to G as origin $=0$,

and therefore $\Sigma my'=0$ and $\Sigma m\dfrac{d^2y'}{dt^2}=0$;

so $\Sigma mz'=0$ and $\Sigma m\dfrac{d^2z'}{dt^2}=0$.

Hence (4) gives

$$M\left[\bar{y}\frac{d^2\bar{z}}{dt^2}-\bar{z}\frac{d^2\bar{y}}{dt^2}\right]+\Sigma m\left(y'\frac{d^2z'}{dt^2}-z'\frac{d^2y'}{dt^2}\right)$$

$$=\Sigma[\bar{y}Z-\bar{z}Y+y'Z-z'Y]\quad\text{...(5).}$$

But equations (2) and (3) give

$$M\left[\bar{y}\frac{d^2\bar{z}}{dt^2}-\bar{z}\frac{d^2\bar{y}}{dt^2}\right]=\Sigma[\bar{y}Z-\bar{z}Y].$$

∴ (5) gives

$$\Sigma m\left[y'\frac{d^2z'}{dt^2}-z'\frac{d^2y'}{dt^2}\right]=\Sigma[y'Z-z'Y]\quad\text{..............(6).}$$

But this equation is of the same form as equation (4) of the last article, and is thus the same equation as we should have obtained if we had regarded the centre of inertia as a fixed point.

Hence *the motion of a body about its centre of inertia is the same as it would be if the centre of inertia were fixed and the same forces acted on the body.*

163. The two results proved in the previous article shew us that the motion of translation of the body can be considered independently of the motion of rotation.

By the first result we see that the motion of the centre of inertia is to be found by the methods of Dynamics of a Particle.

By the second result we see that the motion of rotation is reduced to finding that of a body about a fixed point.

As a simple example, consider the case of a uniform stick thrown into the air in such a way that at the start its centre is moving in a given direction and at the same time it is rotating with given angular velocity about its centre. [Neglect the resistance of the air and suppose gracity to be constant.] By the first result the motion of the centre of inertia is the same as if there were applied at it all the external forces acting on the body in directions parallel to that in which they act. In this case these external forces are the weights of the various elements of the body; when applied at the centre of inertia they are equivalent to the total weight of the body. Hence the centre of the stick moves as if it were a particle of mass M acted on by a vertical force Mg, *i.e.* it moves just as a particle would under gravity if it were projected with the same velocity as the centre of the stick. Hence the path of the centre of the stick would be a parabola.

In a subsequent chapter it will be seen that the angular velocity of the stick will remain unaltered. Hence the centre of the stick will describe a parabola and the stick revolve uniformly about it.

As another example consider a shell which is in motion in the air and suppose that it bursts into fragments. The internal forces exerted by the explosion balance one another, and do not exert any influence on the motion of the centre of inertia of the shell. The centre of inertia therefore continues to describe the same parabola in which it was moving before the explosion. [The motion is supposed to be *in vacuo* and gravity to be constant.]

164. Equation (1) of Art. 161 may be written in the form

$$\frac{d}{dt}\left[\Sigma m\frac{dx}{dt}\right]=\Sigma(X),$$

i.e. $\frac{d}{dt}$ [Total momentum parallel to the axis of x]

= Sum of the impressed forces parallel to OX.

So for the other two axes.

Also (4) can be written

$$\frac{d}{dt}\left[\Sigma m\left(y\frac{dz}{dt}-z\frac{dy}{dt}\right)\right]=\Sigma(yZ-zY),$$

i.e. $\frac{d}{dt}$ [Total moment of momentum about the axis of x]

= Sum of the moments of the impressed forces about OX.

165. As an example of the application of D'Alembert's principle let us consider the following question.

A uniform rod OA, of length $2a$, free to turn about its end O, revolves with uniform angular velocity ω about the vertical OZ through O, and is inclined at a constant angle α to OZ; find the value of α.

Consider an element PQ of the rod, such that $OP=\xi$ and $PQ=d\xi$.

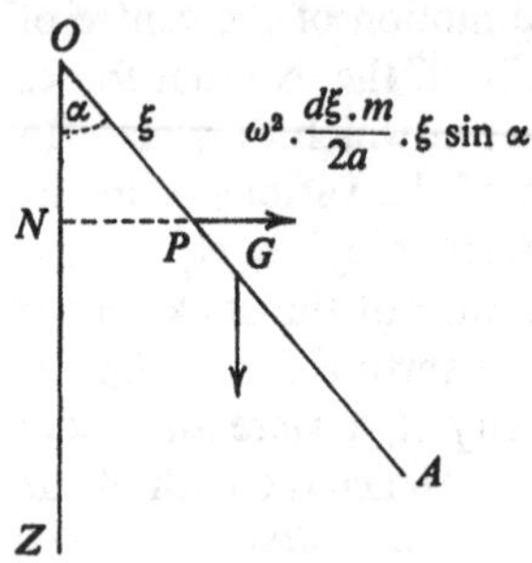

Draw PN perpendicular to OZ. By Elementary Dynamics, the acceleration of P is $\omega^2.PN$ along PN.

Hence the reversed effective force is

$$\left[\frac{d\xi}{2a}.m\right].\omega^2.\xi \sin\alpha \text{ as marked.}$$

All the reversed effective forces acting at different points of the rod, together with the external force, *i.e.* the weight mg, and the reactions at O, form a system of forces in statical equilibrium.

Taking moments about O to avoid the reactions, we therefore have

$$mg.a\sin\alpha = \text{moment about } O \text{ of all the several effective forces}$$

$$=\Sigma\left[\frac{d\xi}{2a}.m\right].\omega^2\xi\sin\alpha\times\xi\cos\alpha$$

$$=\frac{m\omega^2\sin\alpha\cos\alpha}{2a}\int_0^{2a}\xi^2.d\xi=m\omega^2\sin\alpha\cos\alpha.\frac{4a^2}{3}.$$

Hence either $\alpha=0$, or $\cos\alpha=\frac{3g}{4\omega^2 a}$. If $3g>4\omega^2 a$, *i.e.* if $\omega^2<\frac{3g}{4a}$, the second equation gives an impossible value for α, and the only solution in this case is $\alpha=0$, *i.e.* the rod hangs vertically. If $3g<4\omega^2 a$, then $\alpha=\cos^{-1}\frac{3g}{4\omega^2 a}$.

EXAMPLES

1. A plank, of mass M, is initially at rest along a line of greatest slope of a smooth plane inclined at an angle α to the horizon, and a man, of mass M', starting from the upper end walks down the plank so that it does not move; shew that he gets to the other end in time

$$\sqrt{\frac{2M'a}{(M+M')g\sin\alpha}},$$

where a is the length of the plank.

2. A rough uniform board, of mass m and length $2a$, rests on a smooth horizontal plane, and a man, of mass M, walks on it from one end to the other. Find the distance through which the board moves in this time.

[The centre of inertia of the system remains at rest.]

3. A rod revolving on a smooth horizontal plane about one end, which is fixed, breaks into two parts; what is the subsequent motion of the two parts?

4. A circular board is placed on a smooth horizontal plane, and a boy runs round the edge of it at a uniform rate; what is the motion of the centre of the board?

5. A rod, of length $2a$, is suspended by a string, of length l, attached to one end; if the string and rod revolve about the vertical with uniform angular velcoity, and their inclinations to the vertical be θ and ϕ respectively, shew that $\frac{3l}{a}=\frac{(4\tan\theta-3\tan\phi)\sin\phi}{(\tan\phi-\tan\theta)\sin\theta}$.

6. A thin circular disc, of mass M and radius a, can turn freely about a thin axis, OA, which is perpendicular to its plane and passes through a point O of its circumference. The axis OA is compelled to move in a horizontal plane with angular velocity ω about its end A. Shew that the inclination θ to the vertical of the radius of the disc through O is $\cos^{-1}\left(\frac{g}{a\omega^2}\right)$, unless $\omega^2<\frac{g}{a}$, and then θ is zero.

7. A thin heavy disc can turn freely about an axis in its own plane, and this axis revolves horizontally with a uniform angular velocity ω about a fixed point on itself. Shew that the inclination θ of the plane of the disc to the vertical is $\cos^{-1}\frac{gh}{k^2\omega^2}$, where h is the distance of the centre of inertia of the disc from the axis and k is the radius of gyration of the disc about the axis.

If $\omega^2<\frac{gh}{k^2}$, then the plane of the disc is vertical.

8. Two uniform spheres, each of mass M and radius a, are firmly fixed to the ends of two uniform thin rods, each of mass m and length l, and the other ends of the rods are freely hinged to a point O. The whole system revolves, as in the Governor of a Steam-Engine, about a vertical line through O with angular velocity ω. Shew that, when the motion is steady, the rods are inclined to the vertical at an angle θ given by the equation

$$\cos\theta=\frac{g}{\omega^2}\,\frac{M(l+a)+m\frac{l}{2}}{M(l+a)^2+m\frac{l^2}{3}}.$$

Impulsive Forces

166. When the forces acting on a body are very great and act for a very short time, we measure their effects by their impulses. If the short time during which an impulsive force X acts be T, its impulse is $\int_0^T X dt$.

In the case of impulsive forces the equations (1) to (6) of Art. 161 take a different form.

Integrating equations (1), we have

$$\left[\Sigma m\frac{dx}{dt}\right]_0^T=\int_0^T \Sigma X.dt=\Sigma\int_0^T Xdt.$$

If u and u' be the velocities of the particle m before and after the action of the impulsive forces, this gives

$$\Sigma m(u'-u)=\Sigma X',$$

where X' is the impulse of the force on m parallel to the axis of x.

This can be written

$$\Sigma mu'-\Sigma mu=\Sigma X' \qquad \text{.........................(1)},$$

i.e. the total change in the momentum parallel to the axis of x is equal to the total impulse of the external forces parallel to this direction.

Hence the change in the momentum parallel to Ox of the whole mass M, supposed collected at the centre of inertia and moving with it, is equal to the impulse of the external forces parallel to Ox.

So for the change in the motion parallel to the axes of y and z, the equations being

$$\Sigma mv'-\Sigma mv=\Sigma Y' \qquad \text{.......................(2)},$$

and

$$\Sigma mw'-\Sigma mw=\Sigma Z' \qquad \text{.......................(3)}.$$

Again, on integrating equation (4), we have

$$\left[\Sigma m\left(y\frac{dz}{dt}-z\frac{dy}{dt}\right)\right]_0^T=\Sigma\left[y\int_0^T Zdt-z\int_0^T Ydt\right],$$

i.e.

$$\Sigma m[y(w'-w)-z(v'-v)]=\Sigma[yZ'-zY'].$$

Hence

$$\Sigma m[yw'-zv']-\Sigma m[yw-zv]=\Sigma(yZ'-zY') \qquad \text{.........(4)}.$$

Hence the change in the moment of momentum about the axis of x is equal to the moment about the axis of x of the impulses of the external forces.

So for the other two axes, the equations being

$$\Sigma m(zu'-xw')-\Sigma m(zu-xw)=\Sigma(zX'-xZ') \qquad \text{.........(5)},$$

and

$$\Sigma m(xv'-yu')-\Sigma m(xv-yu)=\Sigma(xY'-yX') \qquad \text{.........(6)}.$$

167. The equations of Arts. 161 and 166 are the general equations of motion of a rigid body under finite and impulsive forces respectively, and always give the motion. They are not, however, in a form which can be easily applied to any given problem.

Different forms are found to be desirable, and will be obtained in the following chapters, for different classes of Problems.

CHAPTER XIII

MOTION ABOUT A FIXED AXIS

168. Let the fixed axis of rotation be a perpendicular OZ at O to the plane of the paper, and let a fixed plane through OZ cut the paper in OA. Let a plane ZOG, through OZ and fixed in the body, make an angle θ with the fixed plane, so that $\angle AOG = \theta$.

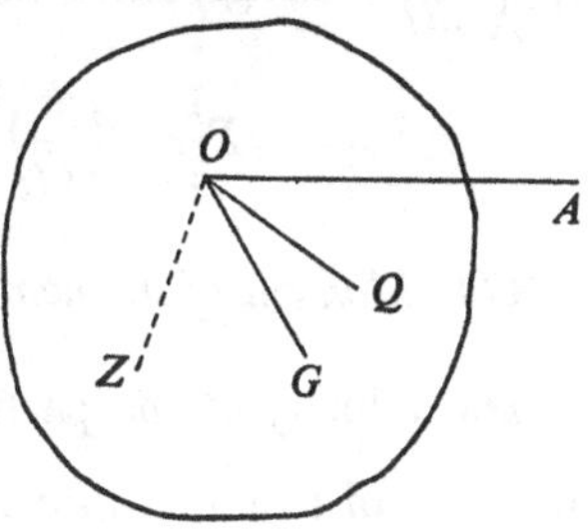

Let a plane through OZ and any point P of the body make an angle ϕ with ZOA and cut the plane of the paper in OQ, so that $\angle AOQ = \phi$.

As the body rotates about OZ the angle QOG remains the same always, so that the rate of change of θ is the same as that of ϕ.

$$\therefore \frac{d\phi}{dt} = \frac{d\theta}{dt}, \text{ and so } \frac{d^2\phi}{dt^2} = \frac{d^2\theta}{dt^2}.$$

If r be the distance, PM, of the particle P from the axis OZ, then, since P describes a circle about M as centre, its accelerations are $r\left(\frac{d\phi}{dt}\right)^2$ along PM and $r\frac{d^2\phi}{dt^2}$ perpendicular to PM.

Hence its effective forces in these directions are

$$mr\left(\frac{d\phi}{dt}\right)^2 \text{ and } mr\frac{d^2\phi}{dt^2}, \text{ i.e. } mr\left(\frac{d\theta}{dt}\right)^2 \text{ and } mr\frac{d^2\theta}{dt^2}.$$

Hence the moment of its effective forces about the axis Oz is

$$r \times mr\frac{d^2\theta}{dt^2}, \text{ i.e. } mr^2 \cdot \frac{d^2\theta}{dt^2}.$$

Hence the moment of the effective forces of the whole body about OZ is

$$\Sigma mr^2 \cdot \frac{d^2\theta}{dt^2}, \text{ i.e. } \frac{d^2\theta}{dt^2} \times \Sigma mr^2,$$

since $\frac{d^2\theta}{dt^2}$ is the same for all particles of the body.

Now Σmr^2 is the moment of inertia, Mk^2, of the body about the axis. Hence the required moment of the effective forces is $Mk^2 . \frac{d^2\theta}{dt^2}$, where θ

is the angle any plane through the axis which is fixed in the body makes with any plane through the axis which is fixed in space.

169. *Kinetic energy of the body.*

The velocity of the particle m is $r\frac{d\phi}{dt}$, *i.e.* $r\frac{d\theta}{dt}$. Its energy is therefore $\frac{1}{2}m\left(r\frac{d\theta}{dt}\right)^2$. Hence the total kinetic energy of the body

$$=\Sigma\frac{1}{2}mr^2\left(\frac{d\theta}{dt}\right)^2=\frac{1}{2}\left(\frac{d\theta}{dt}\right)^2\times\Sigma mr^2=\frac{1}{2}Mk^2\left(\frac{d\theta}{dt}\right)^2.$$

170. *Moment of momentum of the body about the fixed axis.*

The velocity of the particle m is $r\frac{d\theta}{dt}$ in a direction perpendicular to the line, of length r, drawn from m perpendicular to the axis. Hence the moment about the axis of the momentum of m is $mr\times r\frac{d\theta}{dt}$, *i.e.* $mr^2\frac{d\theta}{dt}$. Hence the moment of momentum of the body

$$=\Sigma mr^2\cdot\frac{d\theta}{dt}=\frac{d\theta}{dt}\cdot\Sigma mr^2=Mk^2\cdot\frac{d\theta}{dt}.$$

171. *To find the motion about the axis of rotation.*

Art. 161 tells us that in any motion the moment of the effective forces about the axis is equal to the moment of the impressed forces. Hence, if L be the moment of the impressed forces about the axis of rotation, in the sense which would cause θ to increase, we have

$$Mk^2\frac{d^2\theta}{dt^2}=L.$$

This equation on being integrated twice will give θ and $\frac{d\theta}{dt}$ in terms of the time t. The arbitrary constants which appear in the integration will be known if we are given the position of the plane ZOG, which is fixed in the body, and its angular velocity at any time.

172. *Ex.* 1. *A uniform rod, of mass m and length 2a, can turn freely about one end which is fixed; it is started with angular velocity ω from the position in which it hangs vertically; find the motion.*

The only external force is the weight Mg whose moment L about the fixed axis is $Mg.a \sin\theta$, when the rod has revolved through an angle θ, and this moment tends to lessen θ. Hence the equation of motion is

$$Mk^2\frac{d^2\theta}{dt^2} = -Mga\sin\theta,$$

or, since $k^2 = \frac{4a^2}{3}$, $\frac{d^2\theta}{dt^2} = -\frac{3g}{4a}\sin\theta$.

Integrating, we have

$$\frac{1}{2}\left(\frac{d\theta}{dt}\right)^2 = \frac{3g}{4a}\cos\theta + C, \text{ where } \frac{1}{2}\omega^2 = \frac{3g}{4a} + C.$$

$$\therefore \left(\frac{d\theta}{dt}\right)^2 = \omega^2 - \frac{3g}{2a}(1-\cos\theta) \quad \text{............(1)},$$

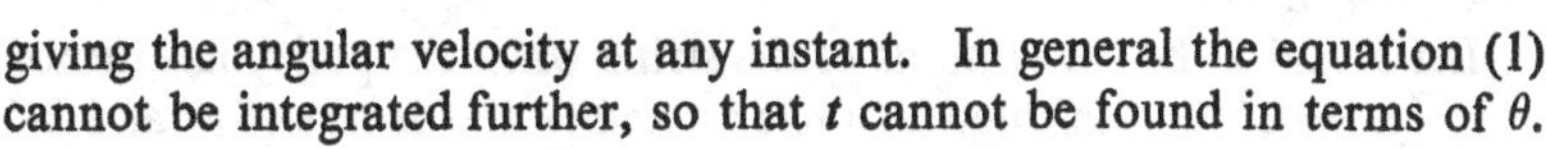

giving the angular velocity at any instant. In general the equation (1) cannot be integrated further, so that t cannot be found in terms of θ.

The angular velocity $\frac{d\theta}{dt}$ gets less and less as θ gets bigger, and just vanishes when $\theta = \pi$, *i.e.* when the rod is in its highest position, if $\omega = \sqrt{\frac{3g}{a}}$.

This is the least value of the angular velocity of the rod, when in its lowest position, so that the rod may just make complete revolutions. With this particular angular velocity the equation (1) gives

$$\left(\frac{d\theta}{dt}\right)^2 = \frac{3g}{2a}(1+\cos\theta) = \frac{3g}{a}\cdot\cos^2\frac{\theta}{2}.$$

$$\therefore t\sqrt{\frac{3g}{a}} = \int_0^\theta \frac{d\theta}{\cos\frac{\theta}{2}} = 2\left[\log\tan\left(\frac{\pi}{4}+\frac{\theta}{4}\right)\right]_0^\theta$$

$$= 2\log\tan\left(\frac{\pi}{4}+\frac{\theta}{4}\right),$$

giving the time of describing any angle θ in this particular case.

Energy and Work.

The equation (1) may be written in the form

$$\frac{1}{2}M\cdot\frac{4a^2}{3}\cdot\left(\frac{d\theta}{dt}\right)^2 - \frac{1}{2}M\cdot\frac{4a^2}{3}\cdot\omega^2 = -Mga(1-\cos\theta),$$

i.e. by Art. 169, the change in the kinetic energy of the body is equal to the work done against the weight of the body.

Ex. 2. *A fine string has two masses, M and M', tied to its ends and passes over a rough pulley, of mass m, whose centre is fixed; if the string does not slip over the pulley, shew that M will descend with acceleration* $\dfrac{M-M'}{M+M'+m\dfrac{k^2}{a^2}}\cdot g$, *where a is the radius and k the radius of gyration of the pulley.*

If the pulley be not sufficiently rough to prevent sliding, and M be the descending mass, shew that its acceleration is $\dfrac{M-M'e^{\mu\pi}}{M+M'e^{\mu\pi}}g$, *and that the pulley will now spin with an angular acceleration equal to* $\dfrac{2MM'ga(e^{\mu\pi}-1)}{mk^2(M+M'e^{\mu\pi})}$.

Let T and T' be the tensions of the string when the pulley has turned through an angle θ; and let the depths of M and M' below the centre of the pulley be x and y. Then, by Art. 171, the equation of motion of the pulley is

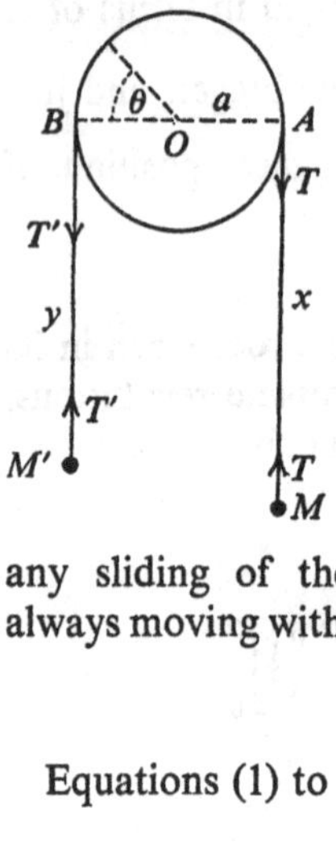

$$mk^2\ddot{\theta}=(T-T')a \quad \text{.........................(1).}$$

Also the equations of motion of the weights are

$$M\ddot{x}=Mg-T \quad \text{and} \quad M'\ddot{y}=M'g-T' \quad \text{......(2).}$$

Again $x+y$ is constant throughout the motion, so that

$$\ddot{y}=-\ddot{x} \quad \text{...............................(3).}$$

First, let the pulley be rough enough to prevent any sliding of the string, so that the string and pulley at A are always moving with the same velocity. Then $\dot{x}=a\dot{\theta}$ always, and therefore

$$\ddot{x}=a\ddot{\theta} \quad \text{...............................(4).}$$

Equations (1) to (4) give $\ddot{x}=a\ddot{\theta}=\dfrac{M-M'}{M+M'+m\dfrac{k^2}{a^2}}g$, giving the constant acceleration with which M descends.

If the pulley be a uniform disc, $k^2=\dfrac{a^2}{2}$, and this acceleration is

$$\frac{M-M'}{M+M'+\dfrac{m}{2}}g.$$

If it be a thin ring, $k^2=a^2$, and the acceleration is $\dfrac{M-M'}{M+M'+m}g$.

Secondly, let the pulley be not rough enough to prevent all sliding of the string. In this case equation (4) does not hold; instead, if μ be the coefficient of friction, we have (*Statics*, Art. 266),

$$T = T' . e^{\mu\pi} \text{ ....(5).}$$

Solving (2), (3), and (5), we have

$$T' e^{\mu\pi} = T = \frac{2MM' g e^{\mu\pi}}{M + M' e^{\mu\pi}}, \text{ and } \ddot{x} = \frac{M - M' e^{\mu\pi}}{M + M' e^{\mu\pi}} g,$$

and then (1) gives $\ddot{\theta} = \frac{2ga(e^{\mu\pi} - 1)}{mk^2} \cdot \frac{MM'}{M + M' e^{\mu\pi}}.$

The result of the first case might have been easily obtained by assuming the Principle of Work and Energy; in the second case it does not apply.

EXAMPLES

1. A cord, 10 feet long, is wrapped round the axle, whose diameter is 4 inches, of a wheel, and is pulled with a constant force equal to 50 lbs. weight, until all the cord is unwound. If the wheel is then rotating 100 times per minute, shew that its moment of inertia is $\frac{90g}{\pi^2}$ ft.-lb. units.

2. A uniform wheel, of weight 100 lbs. and whose radius of gyration about its centre is one foot, is acted upon by a couple equal to 10 ft.-lb. units for one minute; find the angular velocity produced.

Find also the constant couple which would in half-a-minute stop the wheel if it be rotating at the rate of 15 revolutions per second. Find also how many revolutions the wheel would make before stopping.

3. A wheel consists of a disc, of 3 ft. diameter and of mass 50 lbs., loaded with a mass of 10 lbs. attached to it at a point distant one foot from its centre; it is turning freely about its axis which is horizontal. If in the course of a single revolution its least angular velocity is at the rate of 200 revolutions per minute, shew that its maximum angular velocity is at the rate of about 204·4 revolutions per minute.

4. Two unequal masses, M and M', rest on two rough planes inclined at angles α and β to the horizon; they are connected by a fine string passing over a small pulley, of mass m and radius a, which is placed at the common vertex of the two planes; shew that the acceleration of either mass is

$$g[M(\sin \alpha - \mu \cos \alpha) - M'(\sin \beta + \mu' \cos \beta)] \div \left[M + M' + m\frac{k^2}{a^2}\right],$$

where μ, μ' are the coefficients of friction, k is the radius of gyration of the pulley about its axis, and M is the mass which moves downwards.

5. A uniform rod AB is freely movable on a rough inclined plane, whose inclination to the horizon is i and whose coefficient of friction is μ, about a smooth pin fixed through the end A; the bar is held in the horizontal position in the plane and allowed to fall from this position. If θ be the angle through which it falls from rest, shew that

$$\frac{\sin \theta}{\theta} = \mu \cot i.$$

6. A uniform vertical circular plate, of radius a, is capable of revolving about a smooth horizontal axis through its centre; a rough perfectly flexible chain, whose mass is equal to that of the plate and whose length is equal to its circumference, hangs over its rim in equilibrium; if one end be slightly displaced, shew that the velocity of the chain when the other end reaches the plate is $\sqrt{\frac{\pi a g}{6}}$.

[Use the Principle of Energy and Work.]

7. A uniform chain, of length 20 feet and mass 40 lbs., hangs in equal lengths over a solid circular pulley, of mass 10 lbs. and small radius, the axis of the pulley being horizontal. Masses of 40 and 35 lbs. are attached to the ends of the chain and motion takes place. Shew that the time taken by the smaller mass to reach the pulley is

$$\frac{\sqrt{15}}{4}\log_e(9+4\sqrt{5}) \text{ secs.}$$

8. *A heavy fly-wheel, rotating about a symmetrical axis, is slowing down under the friction of its bearings. During a certain minute its angular velocity drops to 90% of its value at the beginning of the minute. What will be the angular velocity at the end of the next minute on the assumption that the frictional moment is* (1) *constant,* (2) *proportional to the angular velocity,* (3) *proportional to the square of the angular velocity?*

Let I be the moment of inertia of the body about its axis, ω its angular velocity at any time t, and Ω its initial angular velocity. Let $x\Omega$ be the angular velocity at the end of the second minute.

(1) If F be the constant frictional moment, the equation of Art. 171 is

$$I\frac{d\omega}{dt}=-F.$$

$$\therefore\ I\omega=-Ft+C=-Ft+I\Omega,$$

where $$I.\frac{9}{10}.\Omega=-F.60+I\Omega,\quad\text{and}\quad I.x\Omega=-F.120+I\Omega.$$

$$\therefore\ x=\frac{80}{100}.$$

(2) If the frictional moment is $\lambda\omega$, the equation of motion is

$$I\frac{d\omega}{dt}=-\lambda\omega.$$

$$\therefore\ I\log\omega=-\lambda t+\text{const.}$$

$$\therefore\ \omega=Ce^{-\frac{\lambda}{I}.t}=\Omega e^{-\frac{\lambda}{I}.t},$$

where $$\frac{9\Omega}{10}=\Omega e^{-\frac{\lambda}{I}.60},\quad\text{and}\quad x\Omega=\Omega e^{-\frac{\lambda}{I}.120}.$$

$$\therefore\ x=\left(\frac{9}{10}\right)^2=\frac{81}{100}.$$

(3) Let the frictional moment be $\mu\omega^2$, so that the equation of motion is

$$I\frac{d\omega}{dt}=-\mu\omega^2.$$

$$\therefore\ I.\frac{1}{\omega}=\mu t+C=\mu t+\frac{I}{\Omega},$$

where $$I.\frac{10}{9\Omega}=\mu.60+\frac{I}{\Omega}, \text{ and } I.\frac{1}{x\Omega}=\mu.120+\frac{I}{\Omega}.$$

$$\therefore x=\frac{9}{11}=\frac{81\frac{9}{11}}{100}.$$

With the three suppositions the angular velocity at the end of two minutes is therefore 80, 81, and $81\frac{9}{11}$% of the initial angular velocity.

9. A fly-wheel, weighing 100 lbs. and having a radius of gyration of 3 ft., has a fan attached to its spindle. It is rotating at 120 revolutions per minute when the fan is suddenly immersed in water. If the resistance of the water be proportional to the square of the speed, and if the angular velocity of the fly-wheel be halved in three minutes, shew that the initial retarding couple is 20π ft.-poundals.

10. A fly-wheel, whose moment of inertia is I, is acted on by a variable couple $G \cos pt$; find the amplitude of the fluctuations in the angular velocity.

THE COMPOUND PENDULUM

173. *If a rigid body swing, under gravity, from a fixed horizontal axis, to shew that the time of a complete small oscillation is* $2\pi\sqrt{\frac{k^2}{hg}}$, *where k is its radius of gyration about the fixed axis, and h is the distance between the fixed axis and the centre of inertia of the body.*

Let the plane of the paper be the plane through the centre of inertia G perpendicular to the fixed axis; let it meet the axis in O and let θ be the angle between the vertical OA and the line OG, so that θ is the angle a plane fixed in the body makes with a plane fixed in space.

The moment L about the horizontal axis of rotation OZ of the impressed forces

=the sum of the moments of the weights of the component particles of the body

=the moment of the weight Mg acting at G

$=Mgh \sin\theta$, where $OG=h$,

and it acts so as to diminish θ.

Hence the equation of Art. 171 becomes

$$Mk^2\frac{d^2\theta}{dt^2}=-Mgh\sin\theta, \quad i.e. \ \frac{d^2\theta}{dt^2}=-\frac{gh}{k^2}\sin\theta \quad \text{.........(1).}$$

If θ be so small that its cubes and higher powers may be neglected, this equation becomes

$$\frac{d^2\theta}{dt^2}=-\frac{gh}{k^2}\theta \quad \text{..............................(2).}$$

The motion is now simple harmonic and the time of a complete oscillation is

$$\frac{2\pi}{\sqrt{\frac{gh}{k^2}}}, \quad i.e. \; 2\pi\sqrt{\frac{k^2}{hg}}.$$

By Art. 97 the time of oscillation is therefore the same as that of a simple pendulum of length $\frac{k^2}{h}$. This length is that of the *simple equivalent pendulum.*

Even if the oscillation of the compound pendulum be not small, it will oscillate in the same time as a simple pendulum of length $\frac{k^2}{h}$.

For the equation of motion of the latter is, by Art. 97,

$$\frac{d^2\theta}{dt^2}=\frac{-g\sin\theta}{\frac{k^2}{h}}=-\frac{gh}{k^2}\sin\theta \;\ldots\ldots\ldots\ldots\ldots\ldots\ldots(3),$$

which is the same equation as (1). Hence the motion given by (1) and (3) will always be the same if the initial conditions of the two motions are the same, *e.g.* if the two pendulums are instantaneously at rest when the value of θ is equal to the same value α in each case, or again if the angular velocities of the two pendulums are the same when each is passing through its position of stable equilibrium.

174. If from O we measure off, along OG, a distance OO_1, equal to the length of the simple equivalent pendulum $\frac{k^2}{h}$, the point O_1 is called the centre of oscillation.

We can easily shew that the centres of suspension and oscillation, O and O_1, are convertible, *i.e.* that if we suspend the body from O_1 instead of from O, then the body will swing in the same time as a simple pendulum of length O_1O.

For we have

$$OO_1=\frac{k^2}{OG}=\frac{K^2+OG^2}{OG},$$

where K is the radius of gyration about an axis through G parallel to the axis of rotation.

Hence $$K^2=OG.OO_1-OG^2=OG.GO_1 \;\ldots\ldots\ldots\ldots(1).$$

When the body swings about a parallel axis through O_1, let O_2 be the centre of oscillation. We then have, similarly,

$$K^2 = O_1G.GO_2 \quad \text{..........................(2).}$$

Comparing (1) and (2) we see that O_2 and O are the same point. Hence when O_1 is the centre of suspension, O is the centre of oscillation, so that the two points are convertible.

This property was used by Captain Kater in determining the value of g. His pendulum has two knife-edges, about either of which the pendulum can swing. It also has a movable mass, or masses, which can be adjusted so that the times of oscillation about the two knife-edges are the same. We then know that the distance, l, between the knife-edges is the length of the simple equivalent pendulum which would swing in the observed time of oscillation, T, of the compound pendulum. Hence g is obtained from the formula $T = 2\pi\sqrt{\frac{l}{g}}$.

For details of the experiment the Student is referred to practical books on Physics.

175. *Minimum time of oscillation of a compound pendulum.*

If K be the radius of gyration of the body about a line through the centre of inertia parallel to the axis of rotation, then

$$k^2 = K^2 + h^2.$$

Hence the length of the simple equivalent pendulum

$$= \frac{K^2 + h^2}{h} = h + \frac{K^2}{h}.$$

The simple equivalent pendulum is of minimum length, and therefore its time of oscillation least, when $\frac{d}{dh}\left(h + \frac{K^2}{h}\right) = 0$,

$$\textit{i.e.} \text{ when } 1 - \frac{K^2}{h^2} = 0, \quad \textit{i.e.} \text{ when } h = K,$$

and then the length of the simple equivalent pendulum is $2K$.

If $h = 0$ or infinity, *i.e.* if the axis of suspension either passes through the centre of inertia or be at infinity, the corresponding simple equivalent pendulum is of infinite length and the time of oscillation infinite.

The above gives only the minimum time of oscillation for axes of suspension which are drawn in a given direction. But we know, from Art. 152, that of all axes drawn through the centre of inertia G there is one such that the moment of inertia about it is a maximum, and another such that the moment of inertia about it is a minimum. If the latter axis be found and if the radius of gyration about it be K_1, then the axis about which the time is an absolute minimum will be parallel to it and at a distance K_1.

176. *Ex. Find the time of oscillation of a compound pendulum, consisting of a rod, of mass m and length a, carrying at one end a sphere, of mass* m_1 *and diameter 2b, the other end of the rod being fixed.*

Here $$(m+m_1)k^2=m.\frac{a^2}{3}+m_1\left[(a+b)^2+\frac{2b^2}{5}\right],$$

and $$(m+m_1)h=m.\frac{a}{2}+m_1(a+b).$$

Hence the length of the required simple pendulum

$$=\frac{k^2}{h}=\frac{m\frac{a^2}{3}+m_1\left[(a+b)^2+\frac{2b^2}{5}\right]}{m\frac{a}{2}+m_1(a+b)}.$$

177. Isochronism of Torsional Vibrations.

Suppose that a heavy uniform circular disc (or cylinder) is suspended by a fairly long thin wire, attached at one end to the centre C of the disc, and with its other end firmly fixed to a point O.

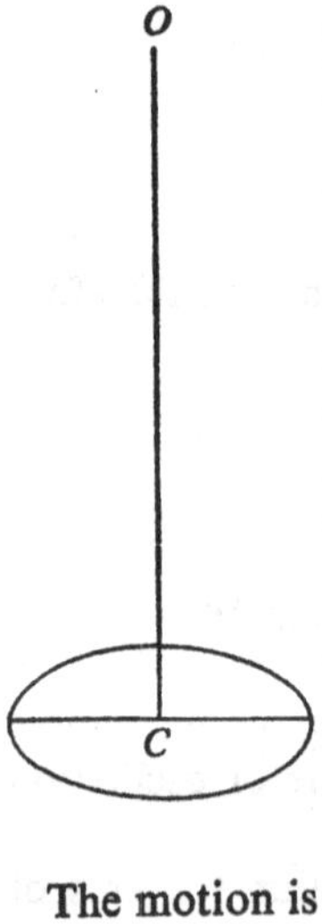

Let the disc be twisted through an angle α about OC, so that its plane is still horizontal, and let it be then left to oscillate.

We shall assume that the torsion-couple of the wire, *i.e.* the couple tending to twist the disc back towards its position of equilibrium, is proportional to the angle through which the disc has been twisted, so that the couple is $\lambda\theta$ when the disc is twisted through an angle θ.

Let M be the mass of the disc, and k its radius of gyration about the axis of rotation OC.

By Art. 171 the equation of motion is

$$Mk^2\frac{d^2\theta}{dt^2}=-\lambda\theta, \quad i.e. \ \ddot{\theta}=-\frac{\lambda}{Mk^2}\theta.$$

The motion is therefore simple harmonic, and the time of oscillation

$$=2\pi\div\sqrt{\frac{\lambda}{Mk^2}}=2\pi\sqrt{\frac{Mk^2}{\lambda}} \quad \text{.................(1).}$$

This time is independent of α, the amplitude of the oscillation.

We can hence test practically the truth of the assumption that the torsion-couple is $\lambda\theta$. Twist the disc through any angle α and, by taking the mean of a number of oscillations, find the corresponding time of oscillation. Repeat the experiment for different values of α, considerably differing from one another, and find the corresponding times of oscillation. These times are found in any given case to be approximately the same. Hence, from (1), the quantity λ is a constant quantity.

178. ***Experimental determination of moments of inertia.***

The moment of inertia of a body about an axis of symmetry may be determined experimentally by the use of the preceding article.

If the disc be weighed, and its diameter determined, then its Mk^2 is known. Let it be I.

Its time of oscillation is then T, where

$$T=2\pi\sqrt{\frac{I}{\lambda}} \quad \text{.............................(1).}$$

Let the body, whose moment of inertia I' about an axis of symmetry is to be found, be placed on the disc with this axis of symmetry coinciding with CO, and the time of oscillation T' determined for the compound body as in the previous article.

Then
$$T'=2\pi\sqrt{\frac{I+I'}{\lambda}} \quad \text{.........................(2).}$$

(1) and (2) give

$$\frac{I+I'}{I}=\frac{T'^2}{T^2}, \text{ i.e. } I'=I.\frac{T'^2-T^2}{T^2},$$

giving I' in terms of known quantities.

EXAMPLES

Find the lengths of the simple equivalent pendulums in the following cases, the axis being horizontal:

1. Circular wire; axis (1) a tangent, (2) a perpendicular to the plane of the wire at any point of its arc.

2. Circular disc; axis a tangent to it.

3. Elliptic lamina; axis a latus-rectum.

4. Hemisphere; axis a diameter of the base. $\left[\textit{Result. } \frac{16}{15}a.\right]$

5. Cube of side $2a$; axis (1) an edge, (2) a diagonal of one of its faces.

$$\left[\textit{Results. } (1)\ \frac{4}{3}\sqrt{2}a;\ (2)\ \frac{5a}{3}.\right]$$

6. Triangular lamina ABC; axis (1) the side BC, (2) a perpendicular to the lamina through the point A.

$$\left[\textit{Results.}\ (1)\ \tfrac{1}{2}b\sin C;\ (2)\ \frac{1}{4}\cdot\frac{3b^2+3c^2-a^2}{\sqrt{2b^2+2c^2-a^2}}.\right]$$

7. Cone; axis a diameter of the base. $\left[\textit{Result.}\ \frac{2+3\tan^2\alpha}{5}.h.\right]$

8. Three equal particles are attached to a weightless rod at equal distances a apart. The system is suspended from, and is free to turn about, a point of the rod distant x from the middle particle. Find the time of a small oscillation, and shew that it is least when $x = \cdot 82a$ nearly.

9. A bent lever, whose arms are of lengths a and b, the angle between them being α, makes small oscillations in its own plane about the fulcrum; shew that the length of the corresponding simple pendulum is

$$\frac{2}{3}\frac{a^3+b^3}{\sqrt{a^4+2a^2b^2\cos\alpha+b^4}}.$$

10. A solid homogeneous cone, of height h and vertical angle 2α, oscillates about a horizontal axis through its vertex; shew that the length of the simple equivalent pendulum is $\frac{h}{5}(4+\tan^2\alpha)$.

11. A sphere, of radius a, is suspended by a fine wire from a fixed point at a distance l from its centre; shew that the time of a small oscillation is given by $\pi\sqrt{\frac{5l^2+2a^2}{5lg}}\left[1+\frac{1}{4}\sin^2\frac{\alpha}{2}\right]$, where α represents the amplitude of the vibration.

12. A weightless straight rod ABC, of length $2a$, is movable about the end A which is fixed and carries two particles of the same mass, one fastened to the middle point B and the other to the end C of the rod. If the rod be held in a horizontal position and be then let go, shew that its angular velocity when vertical is $\sqrt{\frac{6g}{5a}}$, and that $\frac{5a}{3}$ is the length of the simple equivalent pendulum.

13. For a compound pendulum shew that there are three other axes of support, parallel to the original axis and intersecting the line from the centre of inertia perpendicular to the original axis, for which the time of oscillation is the same as about the original axis. What is the practical application of this result?

14. Find the law of graduation of the stem of the common metronome.

15. A simple circular pendulum is formed of a mass M suspended from a fixed point by a weightless wire of length l; if a mass m, very small compared with M, be knotted on to the wire at a point distant a from the point of suspension, shew that the time of a small vibration of the pendulum is approximately diminished by $\frac{m}{2M}\frac{a}{l}\left(1-\frac{a}{l}\right)$ of itself.

16. A given compound pendulum has attached to it a particle of small mass; shew that the greatest alteration in the time of the pendulum is made when it is placed at the middle point of the line bisecting the distance joining the centres of oscillation and of suspension; shew also that a small error in this point of attachment will not, to a first approximation, alter the weight of the particle to be added to make a given difference in the time of oscillation.

17. A uniform heavy sphere, whose mass is 1 lb. and whose radius is 3 inches, is suspended by a wire from a fixed point, and the torsion-couple of the wire is proportional to the angle through which the sphere is turned from the position of equilibrium. If the period of an oscillation be 2 secs., find the couple that will hold the sphere in equilibrium in the position in which it is turned through four right angles from the equilibrium-position.

18. A fly-wheel is hung up with its axis vertical by two long ropes parallel to and equidistant from the axis so that it can perform torsional vibrations. It is found that a static-couple of 50 ft.-lbs. will hold it when it is turned through $\frac{1}{10}$th of a radian, and that if it be turned through any small angle and let go it will make a complete oscillation in 5 secs. Shew that when this fly-wheel is revolving at the rate of 200 revolutions per minute the energy stored up in it will be about 31 ft.-tons.

179. Reactions of the axis of rotation. Let us first consider the simple case in which both the forces and the body are symmetrical with respect to the plane through the centre of gravity perpendicular to the fixed axis, *i.e.* with respect to the plane of the paper, and let gravity be the only external force.

By symmetry, the actions of the axis on the body must reduce to a single force acting at O in the plane of the paper; let the components of this single force be P and Q, along and perpendicular to GO.

By Art. 162 the motion of the centre of gravity G is the same as it would be if it were a particle of mass M acted on by all the external forces applied to it parallel to their original directions.

Now G describes a circle round O as centre, so that its accelerations along and perpendicular to GO are

$$h\left(\frac{d\theta}{dt}\right)^2 \quad \text{and} \quad h\frac{d^2\theta}{dt^2}.$$

Hence its equations of motion are

$$M.h\left(\frac{d\theta}{dt}\right)^2 = P - Mg\cos\theta \quad \ldots\ldots\ldots\ldots\ldots\ldots(1),$$

and

$$M.h\frac{d^2\theta}{dt^2} = Q - Mg\sin\theta \quad \ldots\ldots\ldots\ldots\ldots\ldots(2).$$

Also, as in Art. 171, we have

$$Mk^2\frac{d^2\theta}{dt^2} = -Mgh\sin\theta \ldots\ldots\ldots\ldots\ldots\ldots\ldots(3).$$

Q is given by eliminating $\frac{d^2\theta}{dt^2}$ between (2) and (3).

If (3) be integrated and the resulting constant determined from the initial conditions, we then, by (1), obtain P.

As a particular case let the body be a uniform rod, of length $2a$, turning about its end O, and let it start from the position in which it was vertically above O. In this case $h=a$, $k^2=a^2+\frac{a^2}{3}=\frac{4a^2}{3}$.

Hence equation (3) becomes

$$\ddot{\theta}=-\frac{3g}{4a}\sin\theta \quad \text{.........................(4).}$$

$$\therefore \frac{1}{2}\dot{\theta}^2=\frac{3g}{4a}\cos\theta+\text{const.}=\frac{3g}{4a}(1+\cos\theta) \quad \text{...........(5),}$$

since $\dot{\theta}$ is zero when $\theta=\pi$.

(1) and (5) give $$P=Mg\cdot\frac{3+5\cos\theta}{2}.$$

(2) and (4) give $$Q=\tfrac{1}{4}Mg\sin\theta.$$

Hence the resulting reaction of the fixed axis. When θ is zero, *i.e.* when the rod is in its lowest position, this reaction is four times the weight.

The vertical reaction for any position of the rod

$$=P\cos\theta+Q\sin\theta=Mg\left(\frac{1+3\cos\theta}{2}\right)^2,$$

and therefore vanishes when $\theta=\cos^{-1}(-\tfrac{1}{3})$.

The horizontal reaction $=P\sin\theta-Q\cos\theta=\tfrac{3}{4}Mg\sin\theta(2+3\cos\theta)$.

180. In the general case when either the external forces acting on the body, or the body itself, is not symmetrical about the axis of rotation we may proceed as follows.

Let the axis of rotation be taken as the axis of y, and let the body be attached to it at two points distant b_1 and b_2 from the origin. Let the component actions of the axis at these points parallel to the axes be X_1, Y_1, Z_1 and X_2, Y_2, Z_2 respectively.

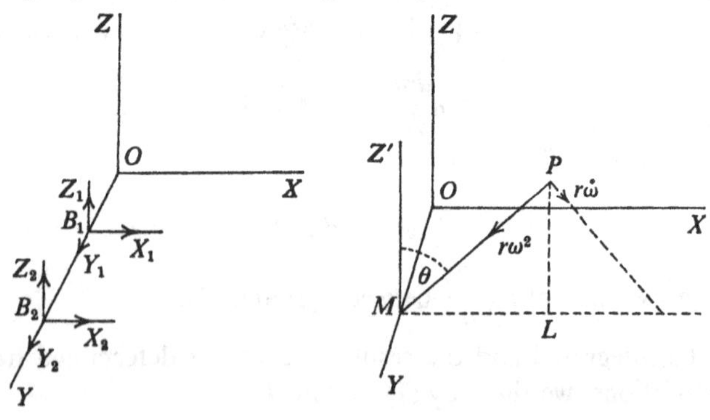

Let P be any point (x, y, z) of the body, whose perpendicular distance PM from OY is of length r and makes an angle θ with a line parallel to OZ.

Then during the motion P describes a circle about M as centre, so that r is constant throughout the motion and hence $\dot{r}$ is zero.

Now $\qquad x = r \sin\theta; \quad z = r\cos\theta.$

$$\therefore \dot{x} = r\cos\theta\dot{\theta}, \quad \text{and} \quad \dot{z} = -r\sin\theta\dot{\theta}.$$

$$\therefore \ddot{x} = -r\sin\theta\dot{\theta}^2 + r\cos\theta\ddot{\theta}; \quad \ddot{z} = -r\cos\theta\dot{\theta}^2 - r\sin\theta\ddot{\theta}.$$

Hence, if $\dot{\theta}$ be denoted by ω,

$$\ddot{x} = -x\omega^2 + z\dot{\omega}; \quad \ddot{y} = 0; \quad \ddot{z} = -z\omega^2 - x\dot{\omega}.$$

[These results may also be obtained by resolving parallel to the axes the accelerations of P, *viz.* $r\omega^2$ along PM and $r\dot{\omega}$ perpendicular to MP.]

The equations of motion of Art. 161 now become, if X. Y, Z are the components parallel to the axes of the external force acting at any point (x, y, z) of the body,

$$\Sigma X + X_1 + X_2 = \Sigma m\ddot{x} = \Sigma m[-x\omega^2 + z\dot{\omega}]$$
$$= -M\bar{x}.\omega^2 + M\bar{z}.\dot{\omega} \quad \ldots\ldots(1);$$

$$\Sigma Y + Y_1 + Y_2 = \Sigma m\ddot{y} = 0 \quad \ldots\ldots(2);$$

$$\Sigma Z + Z_1 + Z_2 = \Sigma m\ddot{z} = \Sigma m(-z\omega^2 - x\dot{\omega})$$
$$= -M\bar{z}.\omega^2 - M\bar{x}.\dot{\omega} \quad \ldots\ldots(3);$$

$$\Sigma(yZ - zY) + Z_1 b_1 + Z_2 b_2$$
$$= \Sigma m(y\ddot{z} - z\ddot{y}) = \Sigma my(-z\omega^2 - x\dot{\omega})$$
$$= -\omega^2\Sigma myz - \dot{\omega}\Sigma mxy \quad \ldots\ldots(4);$$

$$\Sigma(zX - xZ) = \Sigma m(z\ddot{x} - x\ddot{z})$$
$$= \Sigma m(-zx\omega^2 + z^2\dot{\omega} + xz\omega^2 + x^2\dot{\omega}) = \dot{\omega}.Mk^2 \quad \ldots\ldots(5),$$

where k is the radius of gyration about OY; and

$$\Sigma(xY - yX) - X_1 b_1 - X_2 b_2$$
$$= \Sigma m(x\ddot{y} - y\ddot{x}) = -\Sigma my(-x\omega^2 + z\dot{\omega})$$
$$= \omega^2\Sigma mxy - \dot{\omega}\Sigma myz \quad \ldots\ldots(6).$$

On integrating (5) we have the values of ω and $\dot{\omega}$, and then, by substitution, the right-hand members of equations (1) to (4) and (6) are given.

(1) and (6) determine X_1 and X_2.

(3) and (4) determine Z_1 and Z_2.

Y_1 and Y_2 are indeterminate but (2) gives their sum.

It is clear that the right-hand members of (4) and (6) would be both zero if the axis of rotation were a principal axis at the origin O; for then the quantities Σmxy and Σmyz would be zero.

In a problem of this kind the origin O should therefore be always taken at the point, if there be one, where the axis of rotation is a principal axis.

EXAMPLES

1. A thin uniform rod has one end attached to a smooth hinge and is allowed to fall from a horizontal position; shew that the horizontal strain on the hinge is greatest when the rod is inclined at an angle of 45° to the vertical, and that the vertical strain is then $\frac{11}{8}$ times the weight of the rod.

2. A heavy homogeneous cube, of weight W, can swing about an edge which is horizontal; it starts from rest being displaced from its unstable position of equilibrium; when the perpendicular from the centre of gravity upon the edge has turned through an angle θ, shew that the components of the action at the hinge along, and at right angles to, this perpendicular are $\frac{W}{2}(3-5\cos\theta)$ and $\frac{W}{4}\sin\theta$.

3. A circular area can turn freely about a horizontal axis which passes through a point O of its circumference and is perpendicular to its plane. If motion commences when the diameter through O is vertically above O, shew that, when the diameter has turned through an angle θ, the components of the strain at O along, and perpendicular to, this diameter are respectively $\frac{W}{3}(7\cos\theta-4)$ and $\frac{W}{3}\sin\theta$.

4. A uniform semi-circular arc, of mass m and radius a, is fixed at its ends to two points in the same vertical line, and is rotating with constant angular velocity ω. Shew that the horizontal thrust on the upper end is $m.\frac{g+\omega^2 a}{\pi}$.

5. A right cone, of angle 2α, can turn freely about an axis passing through the centre of its base and perpendicular to its axis; if the cone starts from rest with its axis horizontal, shew that, when the axis is vertical, the thrust on the fixed axis is to the weight of the cone as

$$1+\tfrac{1}{2}\cos^2\alpha \text{ to } 1-\tfrac{1}{3}\cos^2\alpha.$$

6. A regular tetrahedron, of mass M, swings about one edge which is horizontal. In the initial position the perpendicular from the centre of mass upon this edge is horizontal. Shew that, when this line makes an angle θ with the vertical, the vertical component of thrust is

$$\frac{Mg}{7}(2\sin^2\theta+17\cos^2\theta).$$

181. Motion about a fixed axis. Impulsive Forces.

By Art. 166 we have that the change in the moment of momentum about the fixed axis is equal to the moment L of the impulsive forces about this axis.

But, as in Art. 170, the moment of momentum of the body about the axis is $Mk^2.\Omega$, where Ω is the angular velocity and Mk^2 the moment of inertia about the axis.

Hence, if ω and ω' be the angular velocities about the axis just before and just after the action of the impulsive forces, this change is

$$Mk^2(\omega'-\omega),$$

and we have

$$Mk^2(\omega'-\omega)=L.$$

Ex. A uniform rod OA, of mass M and length 2a, rests on a smooth table and is free to turn about a smooth pivot at its end O; in contact with it at a distance b from O is an inelastic particle of mass m; a horizontal blow, of impulse P, is given to the rod at a distance x from O in a direction perpendicular to the rod; find the resulting instantaneous angular velocity of the rod and the impulsive actions at O and on the particle.

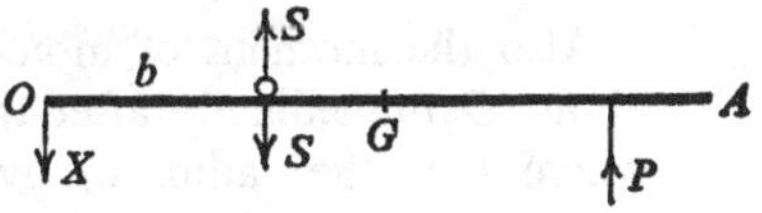

If ω be the angular velocity required and S the impulse of the action between the rod and particle, then, by the last article, we have

$$M\frac{4a^2}{3}\omega = P.x - S.b \quad \text{.......................(1).}$$

Also the impulse S communicates a velocity $b\omega$ to the mass m, so that

$$m.b\omega = S \quad \text{..............................(2).}$$

(1) and (2) give $$\omega = Px \bigg/ \left(M\frac{4a^2}{3} + mb^2\right).$$

Again, let X be the action at O on the rod. Then, since the change in the motion of the centre of gravity of the rod is the same as if all the impulsive forces were applied there,

$$\therefore\ M.a\omega = P - S - X.$$

$$\therefore\ X = P - (Ma + mb)\omega = P\left[1 - \frac{(Ma+mb)x}{M\frac{4a^2}{3} + mb^2}\right].$$

Also (2) gives $$S = \frac{mPbx}{M\frac{4a^2}{3} + mb^2}.$$

182. Centre of percussion. When the fixed axis of rotation is given and the body can be so struck that there is no impulsive action on the axis, any point on the line of action of the blow is called a centre of percussion.

As a simple case consider a thin uniform rod OA $(=2a)$ suspended freely from one end and struck by a horizontal blow at a point C, where OC is x and P is the impulse of the blow.

Let ω' be the instantaneous angular velocity communicated to the rod, and X the impulsive action upon the rod of the axis about which it rotates.

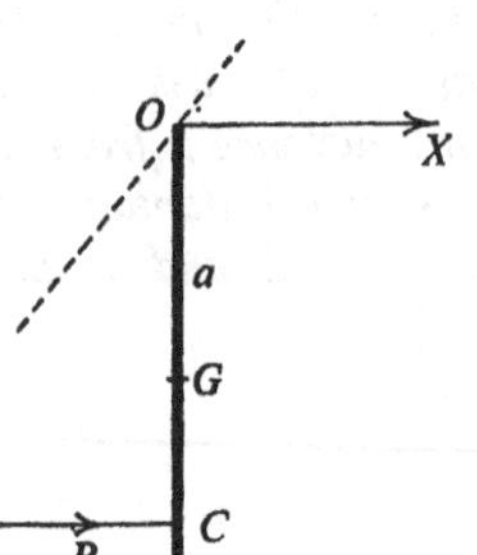

The velocity of the centre of gravity G immediately after the blow is $a\omega'$. Hence the result (1) of Art. 166 gives

$$Ma\omega' = P + X \text{..................(1).}$$

Also the moment of momentum of the rod about O immediately after the blow is $Mk^2\omega'$, where k is the radius of gyration of the rod about O, *i.e.* $k^2 = \frac{4a^2}{3}$.

Hence the result (4) of Art. 166 gives

$$Mk^2\omega' = P.x \text{..........................(2).}$$

Hence $$X = Ma\omega' - M\frac{k^2}{x}\omega' = Ma\omega' . \frac{x - \frac{k^2}{a}}{x} \text{..............(3).}$$

Hence X is zero, *i.e.* there is no impulsive action at O, when $x = \frac{k^2}{a}$, and then OC = the length of the simple equivalent pendulum (Art. 173). In this case C, the required point, coincides with the centre of oscillation, *i.e.* the centre of percussion with regard to the fixed axis coincides with the centre of oscillation with regard to the same axis.

If x be not equal to $\frac{k^2}{a}$, then, by (3), X is positive or negative according as x is greater or less then $\frac{k^2}{a}$, *i.e.* the impulsive stress at O on the body is in the same, or opposite, direction as the blow, according as the blow is applied at a point below or above the centre of percussion.

183. For the general case of the motion of a body free to move about an axis, and acted on by impulsive forces, we must use the fundamental equations of Art. 166.

With the notation and figure of Art. 180, let (X, Y, Z) be the components of the impulsive forces at any point (x, y, z) and (X_1, Y_1, Z_1) and (X_2, Y_2, Z_2) the components of the corresponding impulsive actions at B_1 and B_2.

Then, as in Art. 180,

$$u=\dot{x}=z\omega;\quad v=\dot{y}=0;\quad w=\dot{z}=-x\omega;$$
$$u'=z\omega';\quad v'=0;\ \text{and}\ w'=-x\omega',$$

where ω' is the angular velocity about OY after the blows.

The equations (1) to (6) of Art. 166 then become

$$\Sigma X+X_1+X_2=\Sigma mz\omega'-\Sigma mz\omega=M\bar{z}.(\omega'-\omega)\ \ldots\ldots(1);$$
$$\Sigma Y+Y_1+Y_2=0\ \ldots\ldots(2);$$
$$\Sigma Z+Z_1+Z_2=\Sigma m(-x\omega')-\Sigma m(-x\omega)$$
$$=-M\bar{x}.(\omega'-\omega)\ \ldots\ldots(3);$$
$$\Sigma(yZ-zY)+Z_1b_1+Z_2b_2=\Sigma m[-xy\omega']-\Sigma m[-xy\omega]$$
$$=-(\omega'-\omega).\Sigma mxy\ \ldots\ldots(4);$$
$$\Sigma(zX-xZ)=\Sigma m(z^2\omega'+x^2\omega')-\Sigma m(z^2\omega+x^2\omega)$$
$$=(\omega'-\omega).Mk^2\ \ldots\ldots(5);$$
and $$\Sigma(xY-yX)-X_1b_1-X_2b_2=\Sigma m(-yz\omega')-\Sigma m(-yz\omega)$$
$$=-(\omega'-\omega).\Sigma myz\ \ldots\ldots(6).$$

The rest of the solution is as in Art. 180.

184. Centre of Percussion. Take the fixed axis as the axis of y; let the plane of xy pass through the instantaneous position of the centre of inertia G; and let the plane through the point of application, Q, of the blow perpendicular to the fixed axis be the plane of xz, so that G is the point $(\bar{x}, \bar{y}, 0)$ and Q is the point $(\xi, 0, \zeta)$.

Let the components of the blow parallel to the axes be X, Y, and Z, and let us assume that there is no action on the axis of rotation.

The equations of the previous article then become

$$X=0\ \ldots\ldots(1),$$
$$Y=0\ \ldots\ldots(2),$$
$$Z=-M\bar{x}(\omega'-\omega)\ \ldots\ldots(3),$$
$$\zeta Y=(\omega'-\omega)\Sigma mxy\ \ldots\ldots(4),$$
$$\zeta X-\xi Z=(\omega'-\omega)Mk^2\ \ldots\ldots(5),$$
and $$\xi Y=-(\omega'-\omega)\Sigma myz\ \ldots\ldots(6).$$

Equations (1) and (2) shew that the blow must have no components parallel to the axes of x and y, *i.e.* it must be perpendicular to the plane through the fixed axis and the instantaneous position of the centre of inertia.

(4) and (6) then give $\Sigma mxy=0$, and $\Sigma myz=0$, so that the fixed axis must be a principal axis of the body at the origin, *i.e.* at the point where the plane through the line of action of the blow perpendicular to the fixed axis cuts it.

This is the essential condition for the existence of the centre of percussion. Hence, if the fixed axis is not a principal axis at some point of its length, there is no centre of percussion. If it be a principal axis at only one point of its length, then the blow must act in the plane through this point perpendicular to the axis of rotation.

Finally, (3) and (5) give $\xi=\dfrac{k^2}{\bar{x}}$.

It follows, therefore, from Art. 173, that when a centre of percussion does exist, its distance from the fixed axis is the same as that of the centre of oscillation for the case when the body oscillates freely about the fixed axis taken as a horizontal axis of suspension.

COROLLARY. In the particular case when $\bar{y}=0$ and the centre of inertia G lies on Ox, the line of percussion passes through the centre of oscillation. This is the case when the plane through the centre of inertia perpendicular to the axis of rotation cuts the latter at the point at which it is a principal axis, and therefore, by Art. 147, the axis of rotation is parallel to a principal axis at the centre of inertia.

The investigation of the three preceding Articles refers to impulsive stresses, *i.e.* stresses due to the blow, only; after the rotation has commenced there will be on the axis the ordinary finite stresses due to the motion.

185. A rough example of the foregoing article is found in a cricket-bat. This is not strictly movable about a single axis, but the hands of the batsman occupy only a small portion of the handle of the bat, so that we have an approximation to a single axis. If the bat hits the ball at the proper place, there is very little jar on the batsman's hands.

Another example is the ordinary hammer with a wooden handle; the principal part of the mass is collected in the iron hammer-head; the centre of percussion is situated in, or close to, the hammer-head, so that the blow acts at a point very near the centre of percussion, and the action on the axis of rotation, *i.e.* on the hand of the workman, is very slight accordingly. If the handle of the hammer were made of the same material as its head, the effect would be different.

186. *Ex. A triangle ABC is free to move about its side BC; find the centre of percussion.*

Draw AD perpendicular to BC, and let E be the middle point of BC and F the middle point of DE. Then, as in Ex. 7, page 194, F is the point at which BC is a principal axis.

If $AD=p$, then, by Art. 153, the moment of inertia about BC

$$=\frac{M}{3}\left[\left(\frac{p}{2}\right)^2+\left(\frac{p}{2}\right)^2\right], \text{ so that } k^2=\frac{p^2}{6}.$$

Also $$h=\frac{p}{3}, \text{ so that } \frac{k^2}{h}=\frac{p}{2}.$$

In the triangle draw FF' perpendicular to BC to meet AE in F', so that $FF'=\frac{1}{2}.AD=\frac{p}{2}$. Hence F' is the centre of percussion required. If we draw EE' perpendicular to BC and equal to $\frac{p}{2}$, then E' is the centre of oscillation for a rotation about BC as a horizontal axis of suspension.

The points E' and F' coincide only when the sides AB, AC of the triangle are equal.

EXAMPLES

Find the position of the centre of percussion in the following cases:

1. A uniform rod with one end fixed.

2. A uniform circular plate; axis a horizontal tangent.

3. A sector of a circle; axis in the plane of the sector, perpendicular to its symmetrical radius, and passing through the centre of the circle.

4. A uniform circular lamina rests on a smooth horizontal plane, shew that it will commence to turn about a point O on its circumference if it be struck a horizontal blow whose line of action is perpendicular to the diameter through O and at a distance from O equal to three-quarters of the diameter of the lamina.

5. A pendulum is constructed of a solid sphere, of mass M and radius a, which is attached to the end of a rod, of mass m and length b. Shew that there will be no strain on the axis if the pendulum be struck at a distance $\{M[\frac{2}{5}a^2+(a+b)^2]+\frac{1}{3}mb^2\}$ $\div[M(a+b)+\frac{1}{2}mb]$ from the axis.

6. Find how an equilateral triangular lamina must be struck that it may commence to rotate about a side.

7. A uniform beam AB can turn about its end A and is in equilibrium; find the points of its length where a blow must be applied to it so that the impulses at A may be in each case $\frac{1}{n}$th of that of the blow.

8. A uniform bar AB, of length 6 feet and mass 20 lbs., hangs vertically from a smooth horizontal axis at A; it is struck normally at a point 5 feet below A by a blow which would give a mass of 2 lbs. a velocity of 30 feet per second; find the impulse received by the axis and the angle through which the bar rises.

9. A rod, of mass m and length $2a$, which is capable of free motion about one end A, falls from a vertical position, and when it is horizontal strikes a fixed inelastic obstacle at a distance b from the end A. Shew that the impulse of the blow is $m.\frac{2a}{b}.\sqrt{\frac{2ga}{3}}$, and that the impulse of the reaction at A is $m\sqrt{\frac{3ga}{2}}\left[1-\frac{4a}{3b}\right]$ vertically upwards.

10. A rod, of mass nM, is lying on a horizontal table and has one end fixed; a particle, of mass M, is in contact with it. The rod receives a horizontal blow at its free end; find the position of the particle so that it may start moving with the maximum velocity.

In this case shew that the kinetic energies communicated to the rod and mass are equal.

11. A uniform inelastic beam can revolve about its centre of gravity in a vertical plane and is at rest inclined at an angle α to the vertical. A particle of given mass is let fall from a given height above the centre and hits the beam in a given point P; find the position of P so that the resulting angular velocity may be a maximum.

12. A rod, of mass M and length $2a$, is rotating in a vertical plane with angular velocity ω about its centre which is fixed. When the rod is horizontal its ascending end is struck by a ball of mass m which is falling with velocity u, and when it is next horizontal the same end is struck by a similar ball falling with the same velocity u; the coefficient of restitution being unity, find the subsequent motion of the rod and balls.

13. A uniform beam, of mass m and length $2l$, is horizontal and can turn freely about its centre which is fixed. A particle, of mass m' and moving with vertical velocity u, hits the beam at one end. If the coefficient of restitution for the impact be e, shew that the angular velocity of the beam immediately after the impact is

$$3m'(1+e)u/(m+3m')l,$$

and that the vertical velocity of the ball is then $u(em-3m')/(m+3m')$.

14. Two wheels on spindles in fixed bearings suddenly engage so that their angular velocities become inversely proportional to their radii and in opposite directions. One wheel, of radius a and moment of inertia I_1, has angular velocity ω initially; the other, of radius b and moment of inertia I_2, is initially at rest. Shew that their new angular velocities are

$$\frac{I_1 b^2}{I_1 b^2 + I_2 a^2}\omega \quad \text{and} \quad \frac{I_1 ab}{I_1 b^2 + I_2 a^2}\omega.$$

15. A rectangular parallelepiped, of edges $2a$, $2b$, $2c$, and weight W, is supported by hinges at the upper and lower ends of a vertical edge $2a$, and is rotating with uniform angular velocity ω about that edge. Find, in so far as they are determinate, the component pressures on the hinges.

16. A rod, of length $2a$, revolves with uniform angular velocity ω about a vertical axis through a smooth joint at one extremity of the rod so that it describes a cone of semi-vertical angle α; shew that

$$\omega^2 = \frac{3}{4}\frac{g}{a\cos\alpha}.$$

Prove also that the direction of the reaction at the hinge makes with the vertical the angle $\tan^{-1}(\frac{3}{4}\tan\alpha)$.

17. A door, l feet wide and of mass m lbs., swinging to with angular velocity ω is brought to rest in a small angle θ by a buffer-stop which applies a uniform force P at a distance $\frac{l}{6}$ from the axis of the hinges. Find the magnitude of P and the hinge reactions normal to the door when the buffer is placed in a horizontal plane half-way up.

If the door be $2l$ feet high and have two hinges disposed symmetrically and $2b$ apart, find the hinge-reactions when the buffer is placed at the top edge.

18. A uniform rod AB, of length c and mass m, hangs from a fixed point about which it can turn freely, and the wind blows horizontally with steady velocity v. Assuming the wind pressure on an element dr of the rod to be kv'^2dr, where v' is the normal relative velocity, shew that the inclination α of the rod to the vertical in the position of stable equilibrium is given by $mg \sin \alpha = ckv^2 \cos^2 \alpha$; and find the time in which the rod will fall to this inclination if it be given a slightly greater inclination and let fall against the wind.

19. A rod is supported by a stiff joint at one end which will just hold it at an angle θ with the vertical. If the rod be lifted through a small angle α and be let go, shew that it will come to rest after moving through an angle $2\alpha - \frac{1}{3}\alpha^2 \tan \theta$ nearly, the friction couple at the joint being supposed constant.

20. The door of a railway carriage stands open at right angles to the length of the train when the latter starts to move with an acceleration f; the door being supposed to be smoothly hinged to the carriage and to be uniform and of breadth $2a$, shew that its angular velocity, when it has turned through an angle θ is $\sqrt{\frac{3f}{2a}\sin \theta}$.

21. A Catherine wheel is constructed by rolling a thin casing of powder several times round the circumference of a circular disc of radius a. If the wheel burn for a time T and the powder be fired off at a uniform rate with relative velocity V along the circumference, shew that the angle turned through by the wheel in time T will be

$$\frac{V.T}{a}\left\{1 - c \log\left(1 + \frac{1}{c}\right)\right\},$$

where $2c$ is the ratio of the masses of the disc and powder.

The casing is supposed so thin that the distance of all the powder from the centre of the disc is a.

[If m be the whole mass of the powder and P the impulsive action at any time t, the equations of motion are

$$\left\{2cm.\frac{a^2}{2} + m\left(1 - \frac{t}{T}\right)a^2\right\}\ddot{\theta} = P.a, \text{ and } P.\delta t = \frac{m\delta t}{T}.V.]$$

22. A small weight is attached to a compound pendulum; show that the period of oscillation would be thereby increased or diminished according as the point of attachment is below or above the centre of oscillation.

23. A rigid body is movable about a horizontal axis and starts from rest with its centre of gravity in the horizontal plane through the axis. Show that the mean kinetic energy during one complete oscillation is to the greatest kinetic energy during the motion in the duplicate ratio of $2\Gamma(\frac{3}{4})$ to $\Gamma(\frac{1}{4})$, the mean being taken with regard to the time.

24. A fly-wheel, turning with average angular velocity p, is acted on by a driving couple $A \sin^2 pt$, and has a constant couple $\frac{1}{2}A$ opposing the motion. Find the least moment of inertia required to make the difference between the greatest and least angular velocities less than $\frac{p}{100}$.

25. A fly-wheel weighing 40 lbs. has a radius of gyration 9 inches; it is driven by a couple fluctuating during each revolution, so that the curve connecting this couple and the angular position during one revolution is a triangle, the couple becoming zero once per revolution and reaching a maximum of 2 ft.-lbs. There is also a constant resisting couple such that the motion is the same in each revolution. If the fly-wheel makes 60 revolutions per minute, shew that the difference between the greatest and least angular velocities is approximately 5·7 per cent. of either.

[cosh ·336 = 1·057.]

CHAPTER XIV

MOTION IN TWO DIMENSIONS. FINITE FORCES

187. The position of a lamina compelled to move in the plane of xy is clearly known when we are given the position of some definite point of it, say its centre of inertia, and also the position of some line fixed in the body, *i.e.* when we know the angle that a line fixed in the body makes with a line fixed in space. These quantities (say $\bar{x}$, $\bar{y}$, θ) are called the coordinates of the body, and, if we can determine them in terms of the time t, we have completely determined the motion of the body.

The motion of the centre of inertia is, by Art. 162, given by the equations

$$M\frac{d^2\bar{x}}{dt^2}=\Sigma X \quad \text{.............................(1)},$$

and

$$M\frac{d^2\bar{y}}{dt^2}=\Sigma Y \quad \text{.............................(2)}.$$

If (x', y') be the coordinates of any point of the body relative to the centre of inertia, then the motion about the centre of inertia G is, by equation (6) of Art. 162, given by

$$\Sigma m\left(x'\frac{d^2y'}{dt^2}-y'\frac{d^2x'}{dt^2}\right)=\Sigma(x'Y-y'X),$$

i.e.

$$\frac{d}{dt}\Sigma m\left(x'\frac{dy'}{dt}-y'\frac{dx'}{dt}\right)=\Sigma(x'Y-y'X) \quad \text{..............(3)}.$$

Now $x'\frac{dy'}{dt}-y'\frac{dx'}{dt}=$ the moment about G of the velocity of m relative to G.

Let ϕ be the angle that the line joining m to G makes with a line GB *fixed in space*; and θ the angle that a line GA, *fixed in the body*, makes with GB.

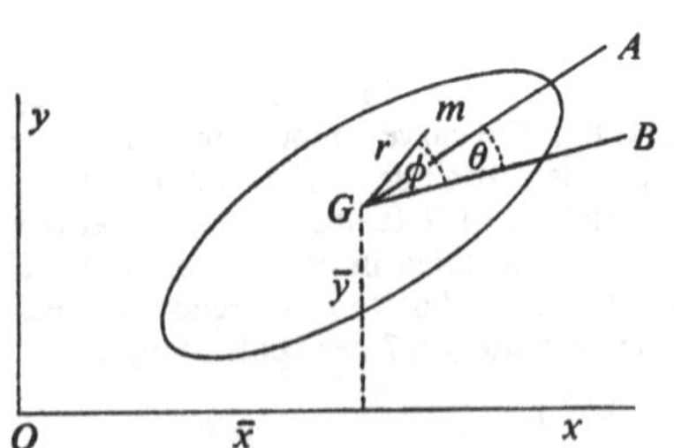

Then, as in Art. 168, since AGm is the same for all positions of the body, we have

$$\frac{d\phi}{dt}=\frac{d\theta}{dt}.$$

If $Gm=r$, the velocity of m relative to G is $r\,\dfrac{d\phi}{dt}$.

Hence its moment about G

$$=r\frac{d\phi}{dt}\times r=r^2\frac{d\phi}{dt}=r^2\frac{d\theta}{dt}.$$

Hence $\Sigma m\left(x'\frac{dy'}{dt}-y'\frac{dx'}{dt}\right)=$ the sum of the moments about G of the velocities of all the mass-elements such as m

$$=\Sigma m . r^2\frac{d\theta}{dt}=\frac{d\theta}{dt}\times\Sigma mr^2=\frac{d\theta}{dt}\times Mk^2,$$

where k is the radius of gyration of the body about an axis through G perpendicular to the plane of the motion.

Hence equation (3) becomes

$$\frac{d}{dt}\left[Mk^2\frac{d\theta}{dt}\right]=\Sigma(x'Y-y'X),$$

i.e. $Mk^2\dfrac{d^2\theta}{dt^2}=$ the moment about G of all the external forces acting on the system ....(4).

188. The equations (1), (2) and (4) of the previous article are the three dynamical equations for the motion of any body in one plane. In general there will be geometrical equations connecting x, y, and θ. These must be written down for any particular problem.

Often, as in the example of Art. 196, the moving body is in contact with fixed surfaces; for each such contact there will be a normal reaction R and corresponding to each such R there will be a geometrical relation expressing the condition that the velocity of the point of contact of the moving body resolved along the normal to the fixed surface is zero.

If, as in Art. 202, we have two moving bodies which are always in contact there is a normal reaction R at the point of contact, and a corresponding geometrical relation expressing that the velocity of the point of contact of each body resolved along the common normal is the same.

Similarly for other cases; it will be clear that for each reaction we have a forced connection and a corresponding geometrical equation so that the number of geometrical equations is the same as the number of reactions.

189. Friction. The same laws are assumed for friction, as in Statics, *viz.* that friction is a self-adjusting force, tending to stop the relative motion of the point at which it acts, but that it cannot exceed a fixed multiple (μ) of the corresponding normal reaction, where μ is a quantity depending on the substances which are in contact. This value of μ is assumed to be constant in dynamical problems, but in reality its value gets less as the relative velocity increases.

The fundamental axiom concerning friction is that it will keep the point of contact at which it acts at relative rest if it can, *i.e.* if the amount of friction required is not greater than the limiting friction. Hence the friction will, if it be possible, make a body roll.

In any practical problem therefore we assume a friction F in a direction opposite to what would be the direction of relative motion, and assume that the point of contact is at relative rest; there is a geometrical equation expressing this latter condition. So to each unknown friction there is a geometrical equation.

If, however, the value of F required to prevent sliding is greater than μR, then sliding follows, and there is discontinuity in our equations. We then have to write them down afresh, substituting μR for F and omitting the corresponding geometrical equation.

190. Kinetic energy of a body moving in two dimensions. Let $(\bar{x}, \bar{y})$ be the centre of inertia, G, of the body referred to fixed axes; let (x, y) be the coordinates of any element m whose coordinates referred to parallel axes through the centre of inertia (x', y').

Then $$x=\bar{x}+x' \quad \text{and} \quad y=\bar{y}+y'.$$

Hence the kinetic energy of the body

$$\frac{1}{2}\Sigma m\left[\left(\frac{dx}{dt}\right)^2+\left(\frac{dy}{dt}\right)^2\right]=\frac{1}{2}\Sigma m\left[\left(\frac{d\bar{x}}{dt}+\frac{dx'}{dt}\right)^2+\left(\frac{d\bar{y}}{dt}+\frac{dy'}{dt}\right)^2\right]$$

$$=\frac{1}{2}\Sigma m\left[\left(\frac{d\bar{x}}{dt}\right)^2+\left(\frac{d\bar{y}}{dt}\right)^2\right]+\frac{1}{2}\Sigma m\left[\left(\frac{dx'}{dt}\right)^2+\left(\frac{dy'}{dt}\right)^2\right]$$

$$+\Sigma m\frac{d\bar{x}}{dt}\frac{dx'}{dt}+\Sigma m\frac{d\bar{y}}{dt}\frac{dy'}{dt} \quad \text{.....................(1).}$$

Since x' is the x-coordinate of the point referred to the centre of inertia as origin, therefore, as in Art. 162.

$$\Sigma mx'=0 \quad \text{and} \quad \Sigma m\frac{dx'}{dt}=0.$$

$$\therefore\ \Sigma m\frac{d\bar{x}}{dt}\cdot\frac{dx'}{dt}=\frac{d\bar{x}}{dt}\cdot\Sigma m\frac{dx'}{dt}=0.$$

Hence the last two terms of (1) vanish, and the kinetic energy

$$=\frac{1}{2}M\left[\left(\frac{d\bar{x}}{dt}\right)^2+\left(\frac{d\bar{y}}{dt}\right)^2\right]+\frac{1}{2}\Sigma m\left[\left(\frac{dx'}{dt}\right)^2+\left(\frac{dy'}{dt}\right)^2\right]$$

= the kinetic energy of a particle of mass M placed at the centre of inertia and moving with it

+ the kinetic energy of the body relative to the centre of inertia.

Now the velocity of the particle m relative to G

$$=r\frac{d\phi}{dt}=r\frac{d\theta}{dt},$$

and therefore the kinetic energy of the body relative to G

$$=\frac{1}{2}\Sigma mr^2\left(\frac{d\theta}{dt}\right)^2=\frac{1}{2}\cdot\left(\frac{d\theta}{dt}\right)^2\Sigma mr^2=\frac{1}{2}\cdot\left(\frac{d\theta}{dt}\right)^2.Mk^2=\frac{1}{2}Mk^2\dot{\theta}^2.$$

Hence the required kinetic energy

$$=\tfrac{1}{2}Mv^2+\tfrac{1}{2}Mk^2\dot{\theta}^2,$$

where v is the velocity of the centre of inertia G, θ is the angle that any line fixed in the body makes with a line fixed in space, and k is the radius of gyration of the body about a line through G perpendicular to the plane of motion.

191. Moment of momentum about the origin O of a body moving in two dimensions.

With the notation of the last article, the moment of momentum of the body about the origin

$$=\Sigma m\left[x\frac{dy}{dt}-y\frac{dx}{dt}\right]$$

$$=\Sigma m\left[(\bar{x}+x')\left(\frac{d\bar{y}}{dt}+\frac{dy'}{dt}\right)-(\bar{y}+y')\left(\frac{d\bar{x}}{dt}+\frac{dx'}{dt}\right)\right]$$

$$=\Sigma m\left[\bar{x}\frac{d\bar{y}}{dt}-\bar{y}\frac{d\bar{x}}{dt}\right]+\Sigma m\left[x'\frac{dy'}{dt}-y'\frac{dx'}{dt}\right]$$

$$+\Sigma m\left[\bar{x}\frac{dy'}{dt}+x'\frac{d\bar{y}}{dt}-\bar{y}\frac{dx'}{dt}-y'\frac{d\bar{x}}{dt}\right]\quad\text{.................. (1.)}$$

But, as in the last article, $\Sigma mx'=0$ and $\Sigma m\frac{dx'}{dt}=0$.

$$\therefore\ \Sigma mx'\frac{d\bar{y}}{dt}=\frac{d\bar{y}}{dt}\Sigma mx'=0,$$

and

$$\Sigma m\bar{y}\frac{dx'}{dt}=\bar{y}\Sigma m\frac{dx'}{dt}=0.$$

So also for the corresponding y terms. Hence, from (1), the moment of momentum about O

$$=M\left[\bar{x}\frac{d\bar{y}}{dt}-\bar{y}\frac{d\bar{x}}{dt}\right]+\Sigma m\left[x'\frac{dy'}{dt}-y'\frac{dx'}{dt}\right]$$

=moment of momentum about O of a particle of mass M placed at the centre of inertia G and moving with it

+the moment of momentum of the body relative to G.

Now the velocity of the particle m relative to G

$$=r\frac{d\phi}{dt}=r\frac{d\theta}{dt},$$

and its moment of momentum about G

$$=r\times r\frac{d\theta}{dt}=r^2\frac{d\theta}{dt}.$$

Therefore the moment of momentum of the body relative to G

$$=\Sigma mr^2\frac{d\theta}{dt}=\frac{d\theta}{dt}\cdot\Sigma mr^2=Mk^2.\dot{\theta}.$$

Hence the total moment of momentum

$$=Mvp+Mk^2\dot{\theta} \quad\text{..........................(2)},$$

where p is the perpendicular from O upon the direction of the velocity v of the centre of inertia.

Or again, if the polar coordinates of the centre of inertia G referred to the fixed point O as origin be (R, ψ), this expression may be written

$$MR^2\frac{d\psi}{dt}+Mk^2\frac{d\theta}{dt} \quad\text{..........................(3).}$$

192. The origin O being a fixed point, the rate of change of the moment of momentum about an axis through it perpendicular to the plane of rotation (for brevity called the moment of momentum about O) is, by equation (3) of Art. 187, equal to the moment of the impressed forces about O. For the moment of momentum we may take either of the expressions (1), (2), or (3).

Thus taking (1) we have

$$\frac{d}{dt}\left[M\left(\bar{x}\frac{d\bar{y}}{dt}-\bar{y}\frac{d\bar{x}}{dt}\right)+Mk^2\dot{\theta}\right]=L,$$

the moment of the forces about O.

Hence $$M\left[\bar{x}\frac{d^2\bar{y}}{dt^2}-\bar{y}\frac{d^2\bar{x}}{dt^2}\right]+Mk^2\ddot{\theta}=L.$$

Similarly, if we took moments about the point (x_0, y_0), the equation is

$$M\left[(\bar{x}-x_0)\frac{d^2\bar{y}}{dt^2}-(\bar{y}-y_0)\frac{d^2\bar{x}}{dt^2}\right]+Mk^2\ddot{\theta}$$

$=$the moment of the impressed forces about (x_0, y_0).

The use of the expressions of this article often simplifies the solution of a problem; but the beginner is very liable to make mistakes, and, to begin with, at any rate, he would do well to confine himself to the formulæ of Art. 187.

193. Instead of the equations (1) and (2) of Art. 187 may be used any other equations which give the motion of a particle, *e.g.* we may use the expressions for the accelerations given in Art. 49 or in Art. 88.

The remainder of this Chapter will consist of examples illustrative of the foregoing principles.

194. *A uniform sphere rolls down an inclined plane, rough enough to prevent any sliding; to find the motion.*

Let O be the point of contact initially when the sphere was at rest. At time t, when the centre of the sphere has described a distance x, let A be the position of the point of the sphere which was originally at O, so that CA is a line fixed in the body. Let $\angle KCA$, being an angle that a line fixed in the body makes with a line fixed in space, be θ. Let R and F be the normal reaction and the friction.

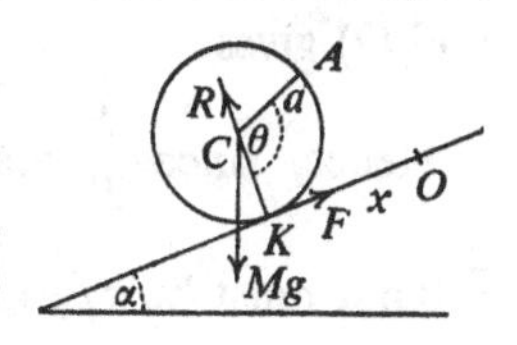

Then the equations of motion of Art. 187 are

$$M\frac{d^2x}{dt^2}=Mg\sin\alpha-F \qquad \text{.......................(1),}$$

$$0=Mg\cos\alpha-R \qquad \text{.......................(2),}$$

and $$Mk^2\frac{d^2\theta}{dt^2}=F.a \qquad \text{.......................(3).}$$

Since there is no sliding the arc $KA=$line KO, so that, throughout the motion, $$x=a\theta \qquad \text{.......................(4).}$$

(1) and (3) give $$\frac{d^2x}{dt^2}+\frac{k^2}{a}\frac{d^2\theta}{dt^2}=g\sin\alpha,$$

i.e. by (4), $$\left(1+\frac{k^2}{a^2}\right)\frac{d^2x}{dt^2}=g\sin\alpha,$$

i.e. $$\frac{d^2x}{dt^2}=\frac{a^2}{a^2+k^2}g\sin\alpha \qquad \text{.......................(5).}$$

Hence the centre of the sphere moves with constant acceleration $\frac{a^2}{a^2+k^2} g \sin \alpha$, and therefore its velocity $v=\frac{a^2}{a^2+k^2} g \sin \alpha . t$ and

$$x=\frac{1}{2}\frac{a^2}{a^2+k^2} g \sin \alpha . t^2.$$

In a sphere $k^2=\frac{2a^2}{5}$; hence the acceleration $=\frac{5}{7}g \sin \alpha$.

[If the body were a thin spherical shell, k^2 would $=\frac{2a^2}{3}$, and the acceleration be $\frac{3}{5}g \sin \alpha$.

If it were a uniform solid disc, k^2 would $=\frac{a^2}{2}$, and the acceleration be $\frac{2}{3}g \sin \alpha$.

If it were a uniform thin ring, k^2 would $=a^2$, and the acceleration be $\frac{1}{2}g \sin \alpha$.]

From (1) we have $F=Mg \sin \alpha - \frac{5}{7}Mg \sin \alpha = \frac{2}{7}Mg \sin \alpha$,

and (2) gives $R=Mg \cos \alpha$.

Hence, since $\frac{F}{R}$ must be $<$ the coefficient of friction μ, therefore $\frac{2}{7}\tan \alpha$ must be $<\mu$, in order that there may be no sliding in the case of a solid sphere.

Equation of Energy. On integrating equation (5), we have

$$\frac{1}{2}\dot{x}^2=\frac{a^2}{a^2+k^2} gx \sin \alpha,$$

the constant vanishing since the body started from rest.

Hence the kinetic energy at time t, by Art. 190,

$$=\frac{1}{2}M\dot{x}^2+\frac{1}{2}Mk^2\dot{\theta}^2=\frac{1}{2}M\dot{x}^2\left(1+\frac{k^2}{a^2}\right)$$

$$=Mg.x \sin \alpha = \text{the work done by gravity.}$$

Ex. 1. A uniform solid cylinder is placed with its axis horizontal on a plane, whose inclination to the horizon is α. Shew that the least coefficient of friction between it and the plane, so that it may roll and not slide, is $\frac{1}{3}\tan \alpha$.

If the cylinder be hollow, and of small thickness, the least value is $\frac{1}{2}\tan \alpha$.

Ex. 2. A hollow cylinder rolls down a perfectly rough inclined plane in one minute; shew that a solid cylinder will roll down the same distance in 52 seconds nearly, a hollow sphere in 55 seconds and a solid sphere in 50 seconds nearly.

Ex. 3. A uniform circular disc, 10 inches in diameter and weighing 5 lbs., is supported on a spindle, $\frac{1}{4}$ inch in diameter, which rolls down an inclined railway with a slope of 1 vertical in 30 horizontal. Find (1) the time it takes, starting from rest, to roll 4 feet, and (2) the linear and angular velocities at the end of that time.

Ex. 4. A cylinder rolls down a smooth plane whose inclination to the horizon is α, unwrapping, as it goes, a fine string fixed to the highest point of the plane; find its acceleration and the tension of the string.

Ex. 5. One end of a thread, which is wound on to a reel, is fixed, and the reel falls in a vertical line, its axis being horizontal and the unwound part of the thread being vertical. If the reel be a solid cylinder of radius a and weight W, shew that the acceleration of the centre of the reel is $\frac{2}{3}g$ and the tension of the thread $\frac{1}{3}W$.

Ex. 6. Two equal cylinders, each of mass m, are bound together by an elastic string, whose tension is T, and roll with their axes horizontal down a rough plane of inclination α. Shew that their acceleration is

$$\frac{2}{3}g \sin \alpha \left[1 - \frac{2\mu T}{mg \sin \alpha}\right],$$

where μ is the coefficient of friction between the cylinders.

Ex. 7. A circular cylinder, whose centre of inertia is at a distance c from its axis, rolls on a horizontal plane. If it be just started from a position of unstable equilibrium, shew that the normal reaction of the plane when the centre of mass is in its lowest position is $1 + \frac{4c^2}{(a-c)^2 + k^2}$ times its weight, where k is the radius of gyration about an axis through the centre of mass.

195. *A uniform rod is held in a vertical position with one end resting upon a perfectly rough table, and when released rotates about the end in contact with the table. Find the motion.*

Let R and F be the normal reaction and the friction when the rod is inclined at an angle θ to the vertical; let x and y be the coordinates of its centre, so that $x = a \sin \theta$ and $y = a \cos \theta$.

The equations of Art. 187 are then

$$F = M\frac{d^2x}{dt^2} = M[a\cos\theta\ddot{\theta} - a\sin\theta\dot{\theta}^2] \qquad \text{.............(1)},$$

$$R - Mg = M\frac{d^2y}{dt^2} = M[-a\sin\theta\ddot{\theta} - a\cos\theta.\dot{\theta}^2] \qquad \text{...(2)},$$

and $$M.\frac{a^2}{3}.\ddot{\theta} = Ra\sin\theta - Fa\cos\theta$$

$$= Mga\sin\theta - Ma^2\ddot{\theta}, \text{ by (1) and (2)},$$

so that $$M\frac{4a^2}{3}\ddot{\theta} = Mga\sin\theta \qquad \text{........................(3)}.$$

[This latter equation could have been written down at once by Art. 171, since the rod is rotating about A as a fixed point.]

(3) gives, on integration, $\dot{\theta}^2 = \frac{3g}{2a}(1-\cos\theta)$, since $\dot{\theta}$ is zero when $\theta = 0$.

Hence (1) and (2) give

$$F = M.\frac{3g}{4}\sin\theta(3\cos\theta - 2) \quad \text{and} \quad R = \frac{Mg}{4}(1 - 3\cos\theta)^2.$$

It will be noted that R vanishes, but does not change its sign, when $\cos\theta = \frac{1}{3}$. The end A does not therefore leave the plane.

The friction F changes its sign as θ passes through the value $\cos^{-1}\frac{2}{3}$; hence its direction is then reversed.

The ratio $\frac{F}{R}$ becomes infinite when $\cos\theta = \frac{1}{3}$; hence unless the plane be infinitely rough there must be sliding then.

In any practical case the end A of the rod will begin to slip for some value of θ less than $\cos^{-1}\frac{1}{3}$, and it will slip backwards or forwards according as the slipping occurs before or after the inclination of the rod is $\cos^{-1}\frac{2}{3}$.

196. *A uniform straight rod slides down in a vertical plane, its ends being in contact with two smooth planes, one horizontal and the other vertical. If it started from rest at an angle* α *with the horizontal, find the motion.*

Let R and S be the reactions of the two planes when the rod is inclined at θ to the horizon. Let x and y be the coordinates of the centre of gravity G. Then the equations of Art. 187 give

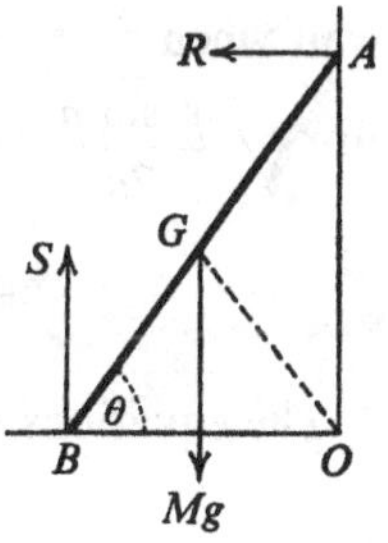

$$M\frac{d^2x}{dt^2}=R \qquad \text{......(1)},$$

$$M\frac{d^2y}{dt^2}=S-Mg \qquad \text{......(2)},$$

and
$$Mk^2\frac{d^2\theta}{dt^2}=R.a\sin\theta-Sa\cos\theta \qquad \text{......(3)}.$$

Since $k^2=\frac{a^2}{3}$, these give

$$\frac{a}{3}\frac{d^2\theta}{dt^2}=\sin\theta\frac{d^2x}{dt^2}-\cos\theta.g-\cos\theta\frac{d^2y}{dt^2} \qquad \text{......(4)}.$$

Now $x=a\cos\theta$ and $y=a\sin\theta$, so that

$$\frac{d^2x}{dt^2}=-a\cos\theta\dot{\theta}^2-a\sin\theta.\ddot{\theta},$$

and
$$\frac{d^2y}{dt^2}=-a\sin\theta.\dot{\theta}^2+a\cos\theta\ddot{\theta}.$$

Hence (4) gives
$$\frac{4}{3}a\frac{d^2\theta}{dt^2}=-g\cos\theta \qquad \text{......(5)}.$$

Hence, on integration, $\frac{1}{2}\left(\frac{d\theta}{dt}\right)^2=\frac{-3g}{4a}\sin\theta+C$, where

$$0=-\frac{3g}{4a}\sin\alpha+C.$$

$$\therefore \left(\frac{d\theta}{dt}\right)^2=\frac{3g}{2a}(\sin\alpha-\sin\theta) \qquad \text{......(6)}.$$

From (1), $\frac{R}{M}=-a\cos\theta.\dot{\theta}^2-a\sin\theta\ddot{\theta}$

$$=-\frac{3g}{2}(\sin\alpha-\sin\theta)\cos\theta+\frac{3g}{4}\sin\theta\cos\theta, \text{ from (5) and (6),}$$

$$=\frac{3g}{4}\cos\theta(3\sin\theta-2\sin\alpha) \qquad \text{......(7)}.$$

From (2),
$$\frac{S}{M}=g-a\sin\theta\dot{\theta}^2+a\cos\theta\ddot{\theta}$$

$$=\frac{g}{4}[1-6\sin\alpha\sin\theta+9\sin^2\theta]=\frac{g}{4}[(3\sin\theta-\sin\alpha)^2+\cos^2\alpha] \text{...(8)}.$$

From (7) it follows that R is zero when $\sin\theta=\frac{2}{3}\sin\alpha$, and, for a smaller value of θ, R becomes negative. Hence the end A leaves the wall when $\sin\theta=\frac{2}{3}\sin\alpha$, and its angular velocity is then, by (6), equal to $\sqrt{\frac{g\sin\alpha}{2a}}$. Also at this instant the horizontal velocity of G

$$=\frac{dx}{dt}=-a\sin\theta.\frac{d\theta}{dt}=\frac{1}{3}\sqrt{2ag\sin^3\alpha}.$$

The equations of motion now take a different form. They become

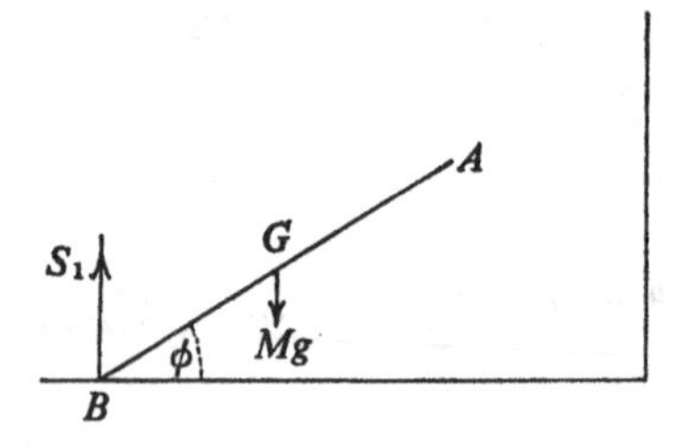

$$M\frac{d^2x}{dt^2}=0 \quad\text{.................(1')},$$

$$M\frac{d^2y}{dt^2}=S_1-Mg \quad\text{............(2')},$$

and $$M\cdot\frac{a^2}{3}\cdot\frac{d^2\phi}{dt^2}=-S_1.a\cos\phi \quad\text{...(3')}.$$

Also $y=a\sin\phi$, so that

$$\frac{d^2y}{dt^2}=-a\sin\phi.\dot{\phi}^2+a\cos\phi.\ddot{\phi}.$$

(2′) and (3′) now give

$$\frac{d^2\phi}{dt^2}\left(\frac{1}{3}+\cos^2\phi\right)-\sin\phi\cos\phi.\left(\frac{d\phi}{dt}\right)^2=-\frac{g}{a}\cos\phi \quad\text{......(4')}.$$

On integration we have

$$\left(\frac{d\phi}{dt}\right)^2\left[\frac{1}{3}+\cos^2\phi\right]=-\frac{2g}{a}\sin\phi+C_1 \quad\text{..............(5')}.$$

The constant is found from the fact that when the rod left the wall, *i.e.* when $\sin\phi=\frac{2}{3}\sin\alpha$, the value of $\frac{d\phi}{dt}$ was equal to $\sqrt{\frac{g\sin\alpha}{2a}}$.

Hence (5′) gives $$\frac{g\sin\alpha}{2a}\left[\frac{4}{3}-\frac{4\sin^2\alpha}{9}\right]=-\frac{2g}{a}.\frac{2\sin\alpha}{3}+C_1,$$

so that $$C_1=\frac{2g\sin\alpha}{a}\left[1-\frac{\sin^2\alpha}{9}\right].$$

Hence we have

$$\left(\frac{d\phi}{dt}\right)^2\left[\frac{1}{3}+\cos^2\phi\right]=\frac{2g\sin\alpha}{a}\left[1-\frac{\sin^2\alpha}{9}\right]-\frac{2g}{a}\sin\phi \quad\text{......(6')}.$$

When the rod reaches the horizontal plane, *i.e.* when $\phi=0$, the angular velocity is Ω, where $\Omega^2\left[\frac{1}{3}+1\right]=\frac{2g\sin\alpha}{a}\left(1-\frac{\sin^2\alpha}{9}\right)$,

i.e. $$\Omega^2=\frac{3g\sin\alpha}{2a}\left[1-\frac{\sin^2\alpha}{9}\right] \qquad \text{.....................(7').}$$

The equation (1′) shews that during the second part of the motion $\frac{dx}{dt}$ is constant and equal to its value $\frac{1}{3}\sqrt{2ag\sin^3\alpha}$ at the end of the first part of the motion.

Energy and Work. The equation (6) may be deduced from the principle of the Conservation of Energy. For as long as the rod is in contact with the wall it is clear that $GO=a$, and $GOB=\theta$, so that G is turning round O as centre with velocity $a\dot\theta$. Hence, by Art. 190, the kinetic energy of the rod is $\frac{1}{2}Ma^2\dot\theta^2+\frac{1}{2}M\frac{a^2}{3}\dot\theta^2$, *i.e.* $\frac{2Ma^2}{3}\dot\theta^2$.

Equating this to the work done by gravity, *viz.* $Mga(\sin\alpha-\sin\theta)$, we have equation (6).

Ex. 1. A uniform rod is held in a vertical position with one end resting upon a horizontal table and, when released, rotates about the end in contact with the table. Shew that, when it is inclined at an angle of 30° to the horizontal, the force of friction that must be exerted to prevent slipping is approximately 0·32 of the weight.

Ex. 2. A uniform rod is placed with one end in contact with a horizontal table, and is then at an inclination α to the horizon and is allowed to fall. When it becomes horizontal, shew that its angular velocity is $\sqrt{\frac{3g}{2a}\sin\alpha}$, whether the plane be perfectly smooth or perfectly rough. Shew also that the end of the rod will not leave the plane in either case.

Ex. 3. A rough uniform rod, of length $2a$, is placed on a rough table at right angles to its edge; if its centre of gravity be initially at a distance b beyond the edge, shew that the rod will begin to slide when it has turned through an angle $\tan^{-1}\frac{\mu a^2}{a^2+9b^2}$, where μ is the coefficient of friction.

Ex. 4. A uniform rod is held at an inclination α to the horizon with one end in contact with a horizontal table whose coefficient of friction is μ. If it be then released, shew that it will commence to slide if

$$\mu<\frac{3\sin\alpha\cos\alpha}{1+3\sin^2\alpha}.$$

Ex. 5. The lower end of a uniform rod, inclined initially at an angle α to the horizon, is placed on a smooth horizontal table. A horizontal force is applied to its lower end of such a magnitude that the rod rotates in a vertical plane with constant angular velocity ω. Shew that when the rod is inclined at an angle θ to the horizon the magnitude of the force is

$$mg \cot \theta - ma\omega^2 \cos \theta,$$

where m is the mass of the rod.

Ex. 6. A heavy rod, of length $2a$, is placed in a vertical plane with its ends in contact with a rough vertical wall and an equally rough horizontal plane, the coefficient of friction being $\tan \epsilon$. Shew that it will begin to slip down if its initial inclination to the vertical is greater than 2ϵ. Prove also that the inclination θ of the rod to the vertical at any time is given by

$$\ddot{\theta}(k^2 + a^2 \cos 2\epsilon) - a^2\dot{\theta}^2 \sin 2\epsilon = ag \sin (\theta - 2\epsilon).$$

Ex. 7. At the ends of a uniform beam are two small rings which slide on two equally rough rods, which are respectively horizontal and vertical; obtain the value of the angular velocity of the beam when it is inclined at any angle to the vertical, the initial inclination being α.

197. *A solid homogeneous sphere, resting on the top of another fixed sphere, is slightly displaced and begins to roll down it. Shew that it will slip when the common normal makes with the vertical an angle θ given by the equation*

$$2 \sin (\theta - \lambda) = 5 \sin \lambda (3 \cos \theta - 2),$$

where λ is the angle of friction.

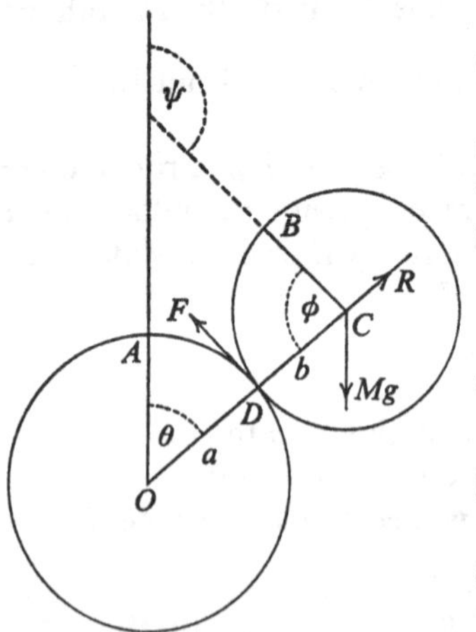

Assume that the motion continues to be one of pure rolling.

Let CB be the position, at time t, of the radius of the upper sphere which was originally vertical, so that if D be the point of contact and A the highest point of the sphere, then

$$\text{arc } AD = \text{arc } BD,$$

i.e. $\quad a\theta = b\phi$ ....(1).

Let R, F be the normal reaction and friction acting on the upper sphere.

Since C describes a circle of radius $a+b$ about O, its accelerations are $(a+b)\dot{\theta}^2$ and $(a+b)\ddot{\theta}$ along and perpendicular to CO.

Hence $$M(a+b)\dot{\theta}^2 = Mg\cos\theta - R \qquad\qquad (2),$$
$$M(a+b)\ddot{\theta} = Mg\sin\theta - F \qquad\qquad (3).$$

Also if ψ be the angle that CB, a line fixed in the moving body, makes with a line fixed in space, *viz.* the vertical, then,

$$Mk^2\ddot{\psi} = Fb.$$

But $$\psi = \theta + \phi = \frac{a+b}{b}\theta \quad \text{and} \quad k^2 = \frac{2b^2}{5},$$

so that $$M.\frac{2(a+b)}{5}\ddot{\theta} = F \qquad\qquad (4).$$

(3) and (4) give $$\ddot{\theta} = \frac{5}{7}\frac{g}{a+b}\sin\theta.$$

$\therefore\ \dot{\theta}^2 = \dfrac{10}{7}\dfrac{g}{a+b}(1-\cos\theta)$, since the sphere started from rest when $\theta = 0$.

[This equation could be directly obtained from the principle of energy.]

(2) and (3) then give

$$F = \frac{2M}{7}g\sin\theta \text{ and } R = \frac{Mg}{7}(17\cos\theta - 10) \qquad\qquad (5).$$

The sphere will slip when the friction becomes limiting, *i.e.* when $F = R\tan\lambda$, *i.e.* when

$$\cos\lambda . 2\sin\theta = \sin\lambda\,(17\cos\theta - 10),$$

i.e. $$2\sin(\theta - \lambda) = 5\sin\lambda\,(3\cos\theta - 2).$$

If the sphere were rough enough to prevent any slipping, then by (5) R would be zero, and change its sign, *i.e.* the upper sphere would leave the lower, when $\cos\theta = \frac{10}{17}$.

[If the spheres were both smooth it could be shewn that the upper sphere would leave the lower when $\cos\theta = \frac{2}{3}$.]

198. *A hollow cylinder, of radius a, is fixed with its axis horizontal; inside it moves a solid cylinder, of radius b, whose velocity in its lowest position is given; if the friction between the cylinders be sufficient to prevent any sliding, find the motion.*

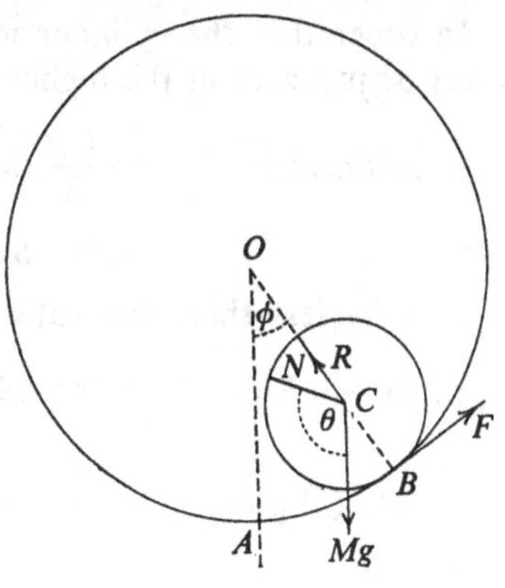

Let O be the centre of the fixed cylinder, and C that of the movable one at time t; let CN be the radius of the movable cylinder which was vertical when it was in its lowest position. Since there has been no sliding the arcs BA and BN are equal; therefore

$$a\phi = b \times \angle BCN.$$

Hence if θ be the angle which CN, a line fixed in the body, makes with the vertical, *i.e.* a line fixed in space, then

$$\theta = \angle BCN - \phi = \frac{a-b}{b}\phi \quad \text{.....................(1).}$$

The accelerations of C are $(a-b)\dot{\phi}^2$ and $(a-b)\ddot{\phi}$ along and perpendicular to CO. Since the motion of the centre of inertia of the cylinder is the same as if all the forces were applied at it, therefore

$$M(a-b)\dot{\phi}^2 = R - Mg\cos\phi \quad \text{.....................(2),}$$

and

$$M(a-b)\ddot{\phi} = F - Mg\sin\phi \quad \text{.....................(3),}$$

where R is the normal reaction, and F is the friction at B as marked.

Also for the motion relative to the centre of inertia, we have

$$Mk^2\ddot{\theta} = \text{moment of the forces about } C = -F.b,$$

i.e.

$$M.\frac{b^2}{2}.\frac{a-b}{b}\ddot{\phi} = -Fb \quad \text{.....................(4).}$$

These equations are sufficient to determine the motion.

Eliminating F between (3) and (4), we have

$$\ddot{\phi} = -\frac{2g}{3(a-b)}\sin\phi \quad \text{.....................(5).}$$

Integrating this equation, we have

$$\dot{\phi}^2 = \frac{4g}{3(a-b)}\cos\phi + \text{const.} = \Omega^2 - \frac{4g}{3(a-b)}(1-\cos\phi) \quad \text{...(6),}$$

where Ω is the value of $\dot{\phi}$ when the cylinder is in its lowest position.

This equation cannot in general be integrated further.

(2) and (6) give

$$\frac{R}{M} = (a-b)\Omega^2 + \frac{g}{3}[7\cos\phi - 4] \quad \text{.................(7).}$$

In order that the cylinder may just make complete revolutions, R must be just zero at the highest point, where $\phi = \pi$.

In this case $(a-b)\Omega^2 = \frac{11g}{3}$, and hence the velocity of projection

$$= (a-b)\Omega = \sqrt{\tfrac{11}{3}g(a-b)}.$$

If Ω be less than this value, R will be zero, and hence the inner cylinder will leave the outer, when $\cos\phi = \frac{1}{7}\left[4 - \frac{3(a-b)\Omega^2}{g}\right]$.

(4) and (5) give

$$F = \frac{Mg}{3}\sin\phi \quad \text{.............................(8).}$$

The friction is therefore zero when the cylinder is in its lowest position, and for any other position F is positive, and therefore acts in the direction marked in the figure.

Equation of Energy. The equation (6) may be deduced at once by assuming that the change in the kinetic energy is equal to the work done.

When the centre is at C the energy (by Art. 190)

$$=\frac{1}{2}M(a-b)^2\dot{\phi}^2+\frac{1}{2}M.\frac{b^2}{2}.\dot{\theta}^2=\frac{3}{4}M(a-b)^2\dot{\phi}^2.$$

Hence the loss in the kinetic energy as the cylinder moves from its lowest position $=\frac{3}{4}M(a-b)^2(\Omega^2-\dot{\phi}^2)$. This equated to the work done against gravity, *viz.* $Mg(a-b)(1-\cos\phi)$, gives equation (6).

Small oscillations. Suppose the cylinder to make small oscillations about the lowest position so that ϕ is always small. Equation (5) then gives $\ddot{\phi}=-\frac{2g}{3(a-b)}\phi$, so that the time of a small oscillation

$$=2\pi\sqrt{\frac{3(a-b)}{2g}}.$$

Ex. 1. A disc rolls on the inside of a fixed hollow circular cylinder whose axis is horizontal, the plane of the disc being vertical and perpendicular to the axis of the cylinder; if, when in its lowest position, its centre is moving with a velocity $\sqrt{\frac{8}{3}g(a-b)}$, shew that the centre of the disc will describe an angle ϕ about the centre of the cylinder in time

$$\sqrt{\frac{3(a-b)}{2g}}\log\tan\left(\frac{\pi}{4}+\frac{\phi}{4}\right].$$

Ex. 2. A solid homogeneous sphere is rolling on the inside of a fixed hollow sphere, the two centres being always in the same vertical plane. Shew that the smaller sphere will make complete revolutions if, when it is in its lowest position, the pressure on it is greater than $\frac{34}{7}$ times its own weight.

Ex. 3. A circular plate rolls down the inner circumference of a rough circle under the action of gravity, the planes of both the plate and circle being vertical. When the line joining their centres is inclined at an angle θ to the vertical, shew that the friction between the bodies is $\frac{\sin\theta}{3}\times$ the weight of the plate.

Ex. 4. A cylinder, of radius a, lies within a rough fixed cylindrical cavity of radius $2a$. The centre of gravity of the cylinder is at a distance

c from its axis, and the initial state is that of stable equilibrium at the lowest point of the cavity. Shew that the smallest angular velocity with which the cylinder must be started that it may roll right round the cavity is given by

$$\Omega^2(a+c)=g\left\{1+\frac{4(a+c)^2}{(a-c)^2+k^2}\right\},$$

where k is the radius of gyration about the centre of gravity.

Find also the normal reaction between the cylinders in any position.

199. *An imperfectly rough sphere moves from rest down a plane inclined at an angle α to the horizon; to determine the motion.*

Let the centre C have described a distance x in time t, and the sphere have rolled through an angle θ; so that θ is the angle between the normal CB to the plane at time t and the radius CA which was normal at zero time. Let us assume that the friction was not enough to produce pure rolling, and hence that the sphere slides as well as turns; in this case the friction will be the maximum that the plane can exert, *viz.* μR, where μ is the coefficient of friction.

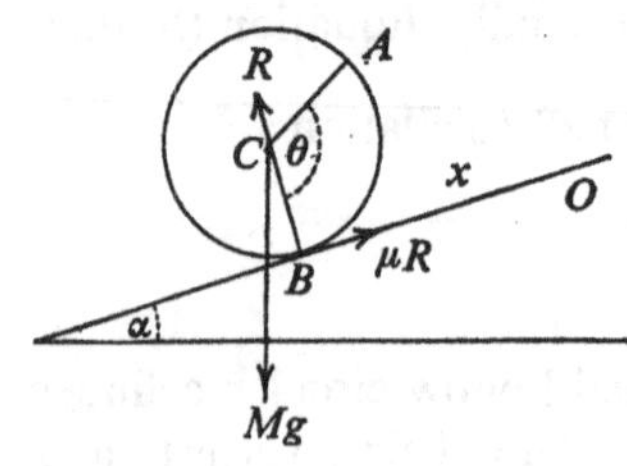

Since the sphere remains in contact with the plane, its centre is always at the same distance a from the plane, so that $\dot{y}$ and $\ddot{y}$ are both zero.

Hence the equations of motions are

$$M\frac{d^2x}{dt^2}=Mg\sin\alpha-\mu R \quad\text{.......................(1),}$$

$$0=R-Mg\cos\alpha \quad\text{.......................(2),}$$

and

$$M.k^2.\frac{d^2\theta}{dt^2}=\mu R.a \quad\text{...................................(3).}$$

Since $k^2\equiv\frac{2a^2}{5}$, (2) and (3) give $\frac{d^2\theta}{dt^2}=\frac{5\mu g}{2a}\cos\alpha$.

$$\therefore\ \frac{d\theta}{dt}=\frac{5\mu g}{2a}\cos\alpha.t \quad\text{.......................(4),}$$

and

$$\theta=\frac{5\mu g}{2a}\cos\alpha.\frac{t^2}{2} \quad\text{.......................(5),}$$

the constants of integration vanishing since θ and $\dot{\theta}$ are both zero initially.

So (1) and (2) give $\frac{d^2x}{dt^2}=g(\sin\alpha-\mu\cos\alpha)$.

$$\therefore\ \frac{dx}{dt}=g(\sin\alpha-\mu\cos\alpha)t \quad\text{.....................(6),}$$

and
$$x=g(\sin\alpha-\mu\cos\alpha)\frac{t^2}{2} \quad\text{.....................(7),}$$

the constants vanishing as before.

The velocity of the point B down the plane = the velocity of C + the velocity of B relative to $C=\frac{dx}{dt}-a\frac{d\theta}{dt}=g\left(\sin\alpha-\frac{7}{2}\mu\cos\alpha\right)t$.

First, suppose $\sin\alpha-\frac{7}{2}\mu\cos\alpha$ to be positive, *i.e.* $\mu<\frac{2}{7}\tan\alpha$.

In this case the velocity of B is always positive and never vanishes; so that the point of contact B has always a velocity of sliding, and the sphere therefore never rolls; the motion is then entirely given by equations (4), (5), (6), and (7).

Secondly, suppose $\sin\alpha-\frac{7}{2}\mu\cos\alpha$ to be zero, so that $\mu=\frac{2}{7}\tan\alpha$.

In this case the velocity of B vanishes at the start and is always zero. The motion is then throughout one of pure rolling and the maximum friction μR is always being exerted.

Thirdly, suppose $\sin\alpha-\frac{7}{2}\mu\cos\alpha$ to be negative, so that $\mu>\frac{2}{7}\tan\alpha$.

In this case the velocity of B appears to be negative which is impossible; for friction only acts with force sufficient at the most to reduce the point on which it acts to rest; and then is only sufficient to keep this point at rest. In this case then pure rolling takes place from the start, and the maximum friction μR is not always exerted.

The equations (1), (2), (3) should then be replaced by

$$M\frac{d^2x}{dt^2}=Mg\sin\alpha-F \quad\text{.........................(8),}$$

$$0=R-Mg\cos\alpha \quad\text{.........................(9),}$$

and
$$Mk^2\frac{d^2\theta}{dt^2}=F.a \quad\text{.................................(10).}$$

Also, since the point of contact is at rest, we have

$$\frac{dx}{dt}-a\frac{d\theta}{dt}=0 \quad\text{..........................(11),}$$

(8) and (10) now give $\frac{d^2x}{dt^2}+\frac{2a}{5}\frac{d^2\theta}{dt^2}=g\sin\alpha$.

Therefore, by (11), $\ddot{x}=a\ddot{\theta}=\frac{5}{7}g\sin\alpha$.

$$\therefore\ \dot{x}=a\dot{\theta}=\tfrac{5}{7}g\sin\alpha . t \quad \ldots\ldots\ldots\ldots\ldots\ldots\ldots(12),$$

and
$$x=a\theta=\frac{5}{7}g\sin\alpha.\frac{t^2}{2} \quad \ldots\ldots\ldots\ldots\ldots\ldots\ldots(13),$$

the constants of integration vanishing as before.

Equation of Energy. The work done by gravity when the centre has described a distance $x=Mg.x\sin\alpha$, and the kinetic energy then

$$=\frac{1}{2}M\dot{x}^2+\frac{1}{2}Mk^2.\dot{\theta}^2=\frac{1}{2}M\left[\dot{x}^2+\frac{2a^2}{5}\dot{\theta}^2\right].$$

In the first case the energy, by (4) and (6),

$$=\tfrac{1}{2}Mg^2t^2[(\sin\alpha-\mu\cos\alpha)^2+\tfrac{5}{2}\mu^2\cos^2\alpha] \quad \ldots\ldots\ldots(14),$$

and the work done by gravity, by (7),

$$=Mg.x\sin\alpha=\tfrac{1}{2}Mg^2t^2\sin\alpha\,(\sin\alpha-\mu\cos\alpha) \quad \ldots\ldots(15).$$

It is easily seen that (14) is less than (15) so long as $\mu<\frac{2}{7}\tan\alpha$, *i.e.* so long as there is any sliding. In this case then there is work lost on account of the friction, and the equation of work and energy does not hold.

In the third case the kinetic energy, by (12) and (13),

$$=\tfrac{1}{2}M\dot{x}^2+\tfrac{1}{2}M.\tfrac{2}{5}a^2.\dot{\theta}^2=\tfrac{1}{2}M.\tfrac{7}{5}a^2.\dot{\theta}^2=\tfrac{1}{2}M.\tfrac{5}{7}g^2\sin^2\alpha t^2,$$

and the work done, by (13),

$$=Mg.x\sin\alpha=Mg\sin\alpha.\frac{5}{7}g\sin\alpha\frac{t^2}{2}=\frac{1}{2}M.\frac{5}{7}g^2\sin^2\alpha t^2.$$

In this case, and similarly in the second case, the kinetic energy acquired is equal to the work done and the equation of work and energy holds.

This is a simple example of a general principle, *viz.* that where there is no friction, *i.e.* where there is pure sliding, or where there is pure rolling, there is no loss of kinetic energy; but where there is not pure rolling, but sliding and rolling combined, energy is lost.

Ex. 1. A homogeneous sphere, of radius a, rotating with angular velocity ω about a horizontal diameter, is gently placed on a table whose coefficient of friction is μ. Shew that there will be slipping at the point of contact for a time $\frac{2\omega a}{7\mu g}$, and that then the sphere will roll with angular velocity $\frac{2\omega}{7}$.

Ex. 2. A solid circular cylinder rotating about its axis is placed gently with its axis horizontal on a rough plane, whose inclination to the horizon is α. Initially the friction acts up the plane and the coefficient of friction is μ. Shew that the cylinder will move upwards if $\mu > \tan\alpha$, and find the time that elapses before rolling takes place.

Ex. 3. A sphere is projected with an underhand twist down a rough inclined plane; shew that it will turn back in the course of its motion is $2a\omega(\mu - \tan\alpha) > 5u\mu$, where u, ω are the initial linear and angular velocities of the sphere, μ is the coefficient of friction, and α is the inclination of the plane.

Ex. 4. A sphere, of radius a, is projected up an inclined plane with velocity V and angular velocity Ω in the sense which would cause it to roll up; if $V > a\Omega$ and the coefficient of friction $> \frac{2}{7}\tan\alpha$, shew that the sphere will cease to ascend at the end of a time $\dfrac{5V + 2a\Omega}{5g\sin\alpha}$, where α is the inclination of the plane.

Ex. 5. If a sphere be projected up an inclined plane, for which $\mu = \frac{1}{7}\tan\alpha$, with velocity V and an initial angular velocity Ω (in the direction in which it would roll up), and if $V > a\Omega$, shew that friction acts downwards at first, and upwards afterwards, and prove that the whole time during which the sphere rises is $\dfrac{17V + 4a\Omega}{18g\sin\alpha}$.

Ex. 6. A hoop is projected with velocity V down a plane of inclination α, the coefficient of friction being $\mu(>\tan\alpha)$. It has initially such a backward spin Ω that after a time t_1 it starts moving uphill and continues to do so for a time t_2 after which it once more descends. The motion being in a vertical plane at right angles to the given inclined plane, shew that

$$(t_1 + t_2)g\sin\alpha = a\Omega - V.$$

Ex. 7. A uniform sphere, of radius a, is rotating about a horizontal diameter with angular velocity Ω and is gently placed on a rough plane which is inclined at an angle a to the horizontal, the sense of the rotation being such as to tend to cause the sphere to move up the plane along the line of greatest slope. Shew that, if the coefficient of friction be $\tan\alpha$, the centre of the sphere will remain at rest for a time $\dfrac{2a\Omega}{5g\sin\alpha}$, and will then move downwards with acceleration $\frac{5}{7}g\sin\alpha$.

If the body be a thin circular hoop instead of a sphere, shew that the time is $\dfrac{a\Omega}{g\sin\alpha}$ and the acceleration $\dfrac{1}{2}g\sin\alpha$.

200. *A sphere, of radius a, whose centre of gravity G is at a distance c from its centre C is placed on a rough plane so that CG is horizontal; shew that it will begin to roll or slide according as* μ *is* $\gtrless \frac{ac}{k^2+a^2}$, *where k is the radius of gyration about a horizontal axis through G. If* μ *is equal to this value, what happens?*

When CG is inclined at an angle θ to the horizontal let A, the point of contact, have moved through a horizontal distance x from its initial position O, and let $OA=x$.

Assume that the sphere rolls so that the friction is F; since the point of contact A is at rest,

$$\therefore \dot{x}=a\dot{\theta} \quad \text{.................................(1).}$$

The equations of motion of Art. 187 are

$$F=M\frac{d^2}{dt^2}[x+c\cos\theta]$$

$$=M[(a-c\sin\theta)\ddot{\theta}-c\cos\theta\dot{\theta}^2] \quad \text{..................(2),}$$

$$R-Mg=M\frac{d^2}{dt^2}[a-c\sin\theta]=M[-c\cos\theta\ddot{\theta}+c\sin\theta\dot{\theta}^2] \quad \text{...(3),}$$

and
$$Rc\cos\theta-F(a-c\sin\theta)=Mk^2\ddot{\theta} \quad \text{..................(4).}$$

We only want the initial motion when $\theta=0$, and then $\dot{\theta}$ is zero but $\ddot{\theta}$ is not zero. The equations (2), (3), (4) then give

$$\left.\begin{aligned} F&=Ma\ddot{\theta},\\ R&=Mg-Mc\ddot{\theta},\\ \text{and}\quad Rc-Fa&=Mk^2\ddot{\theta},\end{aligned}\right\}\text{for the } \textit{initial} \text{ values.}$$

Hence we have
$$\ddot{\theta}=\frac{gc}{k^2+a^2+c^2},$$

$$\frac{R}{M}=\frac{k^2+a^2}{k^2+a^2+c^2}g, \text{ and } \frac{F}{M}=\frac{gac}{k^2+a^2+c^2}.$$

In order that the initial motion may be really one of rolling, we must have $F<\mu R$, *i.e.* $\mu>\frac{ac}{k^2+a^2}$.

If μ be $<$ this value, the sphere will not roll, since the friction is not sufficient.

Critical Case. If $\mu=\frac{ac}{k^2+a^2}$ it will be necessary to consider whether, when θ is small but not absolutely zero, the value of $\frac{F}{R}$ is a little greater or a little less than μ.

The question must therefore be solved from the beginning, keeping in the work first powers of θ and neglecting θ^2, θ^3, ... etc.

(2), (3), and (4) then give, on eliminating F and R,

$$\ddot{\theta}[k^2+a^2+c^2-2ac\sin\theta]-ac\cos\theta\dot{\theta}^2=gc\cos\theta \quad \text{......(5).}$$

Hence, on integration,

$$\dot{\theta}^2[k^2+a^2+c^2-2ac\sin\theta]=2gc\sin\theta \quad \text{..............(6).}$$

If $K^2\equiv k^2+a^2+c^2$ these give, neglecting squares of θ,

$$\dot{\theta}^2=\frac{2gc\theta}{K^2} \text{ and } \ddot{\theta}[K^2-2ac\theta]=gc+\frac{2gac^2\theta}{K^2},$$

i.e.
$$K^2\ddot{\theta}=\left[gc+\frac{2gac^2\theta}{K^2}\right]\left[1+\frac{2ac\theta}{K^2}\right]=gc\left[1+\frac{4ac\theta}{K^2}\right].$$

Hence, to the first power of θ, we have from (2) and (3),

$$\frac{F}{R}=\frac{(a-c\theta)\ddot{\theta}-c\dot{\theta}^2}{g-c\ddot{\theta}}=\frac{ac}{k^2+a^2}\left[1-c\frac{3k^2-a^2}{a(k^2+a^2)}\theta\right],$$

on substitution and simplification.

If $k^2>\frac{a^2}{3}$, then $\frac{F}{R}$ is less than $\frac{ac}{k^2+a^2}$, *i.e.* $\frac{F}{R}$ is less than the coefficient of friction and the sphere rolls.

If $k^2<\frac{a^2}{3}$, then $\frac{F}{R}>$ the coefficient of friction and the sphere slides.

Ex. 1. A homogeneous sphere, of mass M, is placed on an imperfectly rough table, and a particle, of mass m, is attached to the end of a horizontal diameter. Shew that the sphere will begin to roll or slide according as μ is greater or less than $\frac{5(M+m)m}{7M^2+17Mm+5m^2}$. If μ be equal to this value, shew that the sphere will begin to roll.

Ex. 2. A homogeneous solid hemisphere, of mass M and radius a, rests with its vertex in contact with a rough horizontal plane, and a particle, of mass m, is placed on its base, which is smooth, at a distance c from the centre. Shew that the hemisphere will commence to roll or slide according as the coefficient of friction $\gtrless\frac{25mac}{26(M+m)a^2+40mc^2}$.

Ex. 3. A sphere, of radius a, whose centre of gravity G is not at its centre O, is placed on a rough table so that OG is inclined at an angle α to the upward drawn vertical; shew that it will commence to slide along the table if the coefficient of friction be $<\frac{c\sin\alpha(a+c\cos\alpha)}{k^2+(a+c\cos\alpha)^2}$, where $OG = c$ and k is the radius of gyration about a horizontal axis through G.

Ex. 4. If a uniform semi-circular wire be placed in a vertical plane with one extremity on a rough horizontal plane, and the diameter through that extremity vertical, shew that the semi-circle will begin to roll or slide according as $\mu \gtrless \frac{\pi}{\pi^2 - 2}$. If μ has this value, prove that the wire will roll.

Ex. 5. A heavy uniform sphere, of mass M, is resting on a perfectly rough horizontal plane, and a particle, of mass m, is gently placed on it at an angular distance α from its highest point. Shew that the particle will at once slip on the sphere if $\mu < \frac{\sin \alpha \{7M + 5m(1 + \cos \alpha)\}}{7M \cos \alpha + 5m(1 + \cos \alpha)^2}$, where μ is the coefficient of friction between the sphere and the particle.

201. *A uniform circular disc is projected, with its plane vertical, along a rough horizontal plane with a velocity v of translation, and an angular velocity ω about the centre. Find the motion.*

Case I. $v \rightarrow$, $\omega \curvearrowright$ and $v > a\omega$.

In this case the initial velocity of the point of contact is $v - a\omega$ in the direction $\rightarrow$ and the friction is $\mu Mg \leftarrow$.

When the centre has described a distance x, and the disc has turned through an angle θ, the equations of motion are

$$M\ddot{x} = -\mu Mg, \text{ and } M.\frac{a^2}{2}.\ddot{\theta} = \mu Mga,$$

$$\therefore\ \dot{x} = v - \mu gt \text{ and } \frac{a}{2}\dot{\theta} = \frac{a}{2}\omega + \mu gt \qquad \text{.................(1).}$$

Hence the velocity of the point of contact P

$$= \dot{x} - a\dot{\theta} = v - a\omega - 3\mu gt.$$

Sliding therefore continues until $t = \frac{v - a\omega}{3\mu g}$ and pure rolling then begins.

Also at this time the velocity of the centre $= \frac{2v + a\omega}{3}$ ....(2).

The equations of motion then become

$$\left.\begin{aligned} M\ddot{x} &= -F \\ \text{and} \qquad M\frac{a^2}{2}\ddot{\phi} &= F.a \end{aligned}\right\}, \text{ where } F \text{ is the friction } \leftarrow.$$

Also $\dot{x}=a\dot{\phi}$, since the point of contact is now at rest; $\therefore\ \ddot{x}=a\ddot{\phi}$. These three equations give $F=0$, *i.e.* no friction is now required. Also $a\dot{\phi}=\dot{x}=\text{constant}=$ the velocity at the commencement of the rolling $=\dfrac{2v+a\omega}{3}$, by (2).

The disc therefore continues to roll with a constant velocity which is less than its initial velocity.

Case II. $v\rightarrow$, $\omega\circlearrowleft$ and $v<a\omega$.

Here the initial velocity of the point of contact is $\leftarrow$ and hence the friction is $\mu Mg\rightarrow$. The equations of motion are then

$$M\ddot{x}=\mu Mg, \text{ and } M.\frac{a^2}{2}\ddot{\theta}=-\mu Mga,$$

giving

$$\dot{x}=v+\mu gt \text{ and } \frac{a}{2}\dot{\theta}=\frac{a}{2}\omega-\mu gt.$$

Hence pure rolling begins when $\dot{x}=a\dot{\theta}$, *i.e.* when $t=\dfrac{a\omega-v}{3\mu g}$.

The velocity, $\dot{x}$, of the centre then $=\dfrac{a\omega+2v}{3}$, and, as in Case I, the disc rolls on with constant velocity which is greater than the initial velocity of the centre.

Case III. $v\rightarrow$, $\omega\circlearrowright$.

Initially the velocity of the point of contact is $v+a\omega\rightarrow$, so that the friction is $\mu Mg\leftarrow$. The equations of motion are

$$M\ddot{x}=-\mu Mg, \text{ and } M\frac{a^2}{2}\ddot{\theta}=\mu Mga,$$

$$\therefore\ \dot{x}=v-\mu gt \text{ and } \frac{a}{2}\dot{\theta}=\mu gt-\frac{a}{2}\omega.$$

Pure rolling begins when $\dot{x}=a\dot{\theta}$, *i.e.* when $t=\dfrac{v+a\omega}{3\mu g}$, and the velocity of the centre then $=\dfrac{2v-a\omega}{3}$.

If $2v>a\omega$, this velocity is $\rightarrow$, and the motion during pure rolling is $\rightarrow$ with constant velocity as before.

If however $2v<a\omega$, the velocity of the centre when pure rolling commences is $\leftarrow$ and the disc rolls back towards O. In this particular case the velocity of the centre vanishes when $t=\dfrac{v}{\mu g}$ which is less than

$\frac{v+aw}{3\mu g}$, if $2v<aw$; hence the disc begins to move in the direction $\leftarrow$ before pure rolling commences.

[In this last case the motion is of the same kind as that in the well-known experiment of a napkin-ring projected along the table with a velocity $v\rightarrow$ and a sufficient angular velocity ω in the direction ↻.]

Ex. A napkin-ring, of radius a, is propelled forward on a rough horizontal table with a linear velocity u and a backward spin ω, which is $>u/a$. Find the motion and shew that the ring will return to the point of projection in time $\frac{(u+a\omega)^2}{4\mu g(a\omega-u)}$, where μ is the coefficient of friction.

What happens if $u>a\omega$?

202. *Two unequal smooth spheres are placed one on the top of the other in unstable equilibrium, the lower sphere resting on a smooth table. The system is slightly disturbed; shew that the spheres will separate when the line joining their centres makes an angle θ with the vertical given by the equation* $\frac{m}{M+m}\cos^3\theta-3\cos\theta+2=0$, *where M is the mass of the lower, and m of the upper, sphere.*

Let the radii of the two spheres be a and b, and G their centre of gravity, so that

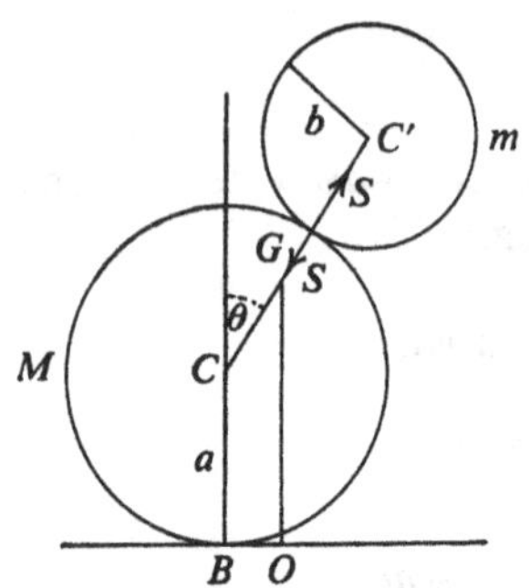

$$\frac{CG}{m}=\frac{C'G}{M}=\frac{a+b}{M+m}.$$

There being no friction at the table the resultant horizontal force on *the system consisting of the two spheres* is zero.

Hence, by Art. 162, the horizontal velocity of the centre of gravity is constant, and equal to its value at the commencement of the motion, *i.e.* it is always zero. Hence the only velocity of G is vertical, and it therefore describes a vertical straight line GO, where O was the initial position of the point of contact B, so that O is a fixed point.

For the horizontal motion of the lower sphere, we thus have

$$S\sin\theta=M\frac{d^2}{dt^2}[CG.\sin\theta]=\frac{Mm(a+b)}{M+m}[\cos\theta\ddot{\theta}-\sin\theta\dot{\theta}^2]\ \ldots(1).$$

For the vertical motion of the upper sphere

$$S\cos\theta-mg=m\frac{d^2}{dt^2}[a+(a+b)\cos\theta]$$

$$=m(a+b)[-\sin\theta\ddot{\theta}-\cos\theta\dot{\theta}^2]\ldots\ldots(2).$$

Eliminating S, we have

$$\ddot{\theta}[M+m\sin^2\theta]+m\sin\theta\cos\theta\dot{\theta}^2=\frac{(M+m)g}{a+b}\sin\theta \quad \ldots\ldots(3).$$

Hence, by integration,

$$\dot{\theta}^2[M+m\sin^2\theta]=\frac{2g}{a+b}(M+m)(1-\cos\theta) \quad \ldots\ldots\ldots(4),$$

since the motion started from rest at the highest point.

By (1), S vanishes, *i.e.* the spheres separate, when

$$\cos\theta\ddot{\theta}=\sin\theta\,.\,\dot{\theta}^2 \quad \ldots\ldots\ldots\ldots\ldots\ldots\ldots\ldots\ldots(5).$$

(3) and (5) give $\dot{\theta}^2=\frac{g\cos\theta}{a+b}$ at this instant and then (4) gives, on substitution, $m\cos^3\theta=(M+m)(3\cos\theta-2)$.

There are no forces acting so as to turn either sphere about its centre, so that neither of them has any rotatory motion.

Work and Energy, Equation (4) may be obtained thus, by assuming the principle of work. The horizontal velocity of the lower sphere

$$=\frac{d}{dt}(CG\sin\theta)=\frac{m(a+b)}{M+m}\cos\theta\dot{\theta},$$

so that its kinetic energy is

$$\tfrac{1}{2}M\frac{m^2(a+b)^2}{(M+m)^2}\cos^2\theta\dot{\theta}^2.$$

The horizontal and vertical velocities of the upper sphere are

$$\frac{d}{dt}[GC'\sin\theta] \text{ and } \frac{d}{dt}[a+(a+b)\cos\theta],$$

i.e.
$$\frac{M(a+b)}{M+m}\cos\theta\dot{\theta} \text{ and } -(a+b)\sin\theta\dot{\theta},$$

so that its kinetic energy is

$$\tfrac{1}{2}m\,.\,(a+b)^2\,.\,\dot{\theta}^2\left[\frac{M^2}{(M+m)^2}\cos^2\theta+\sin^2\theta\right].$$

Equating the sum of these two energies to the work done, *viz.* $mg(a+b)(1-\cos\theta)$, we obtain equation (4).

203. Varying mass. In obtaining the equations of Art. 161 we assumed the mass of the body to remain constant. If the mass m of a particle is not constant, the component effective force is $\frac{d}{dt}\left(m\frac{dx}{dt}\right)$ and not $m\frac{d^2x}{dt^2}$.

The equation (1) of Art. 161 is then

$$\Sigma\left[X-\frac{d}{dt}\left(m\frac{dx}{dt}\right)\right]=0,$$

i.e.

$$\Sigma X=\Sigma\frac{d}{dt}\left(m\frac{dx}{dt}\right)=\frac{d}{dt}\Sigma m\frac{dx}{dt}=\frac{d}{dt}\left(M\frac{d\bar{x}}{dt}\right).$$

Also the equation (6) of the same article is

$$\Sigma(xY-yX)=\Sigma\left[x\frac{d}{dt}\left(m\frac{dy}{dt}\right)-y\frac{d}{dt}\left(m\frac{dx}{dt}\right)\right]$$

$$=\Sigma\frac{d}{dt}\left[x.m\frac{dy}{dt}-y.m\frac{dx}{dt}\right]=\frac{d}{dt}\Sigma\left[m\left(x\frac{dy}{dt}-y\frac{dx}{dt}\right)\right]$$

$$=\frac{d}{dt}\left[Mk^2\frac{d\theta}{dt}\right], \text{ as in Art. 187.}$$

Ex. A cylindrical mass of snow rolls down an inclined plane covered with snow of uniform depth E, gathering up all the snow it rolls over and always remaining circular; find the motion of the snow, and shew that it will move with an acceleration $\frac{1}{3}g\sin\alpha$, *if initially, when its radius is a, it be started with velocity* $a\sqrt{\dfrac{2\pi g\sin\alpha}{5E}}$, *where* α *is the inclination of the plane.*

At time t from the start, let x be the distance described down the plane, and r be the radius, so that

$$\pi(r^2-a^2)=\text{the amount of snow picked up}$$
$$=E.x \quad\text{..(1).}$$

If F be the friction up the plane, and θ the angle turned through by the snow-ball, we have

$$\frac{d}{dt}[\pi r^2\rho.\dot{x}]=\pi r^2 g\rho\sin\alpha-F \quad\text{.....................(2),}$$

and

$$\frac{d}{dt}[\pi r^2\rho.k^2\dot{\theta}]=F.r \quad\text{.........................(3),}$$

where ρ is the density of the snow-ball.

Also $$\dot{x}-r\dot{\theta}=0 \quad\text{............................(4),}$$

since there is no sliding.

Since $k^2=\frac{r^2}{2}$, the equations (2) and (3) give

$$r\frac{d}{dt}(r^2\dot{x})+\frac{d}{dt}\left(\frac{r^3}{2}\dot{x}\right)=r^3 g\sin\alpha,$$

i.e. $$3\ddot{x}+7\frac{\dot{r}}{r}\dot{x}=2g\sin\alpha,$$

or, from (1), $$\ddot{x}+\frac{7}{6}\frac{E}{\pi a^2+Ex}\dot{x}^2=\frac{2}{3}g\sin\alpha.$$

On putting $\dot{x}^2=u$ and hence $2\ddot{x}=\frac{du}{dx}$, this equation becomes linear, and its solution is

$$\dot{x}^2(\pi a^2+Ex)^{\frac{7}{3}}=\frac{4}{3}g\sin\alpha.\frac{3}{10E}(\pi a^2+Ex)^{\frac{10}{3}}+C,$$

i.e. $$\dot{x}^2=\frac{2g\sin\alpha}{5E}(\pi a^2+Ex)+\frac{C}{(\pi a^2+Ex)^{\frac{7}{3}}}.$$

This equation cannot in general be integrated further.

If, however, $\dot{x}=a\sqrt{\frac{2\pi g\sin\alpha}{5E}}$, when $x=0$, we have $C=0$, and then

$$\dot{x}^2=\frac{2g\sin\alpha}{5}x+\frac{2\pi ga^2\sin\alpha}{5E}=[\dot{x}^2]_{t=0}+\frac{2g\sin\alpha}{5}x,$$

so that the acceleration is $\frac{g\sin\alpha}{5}$.

MISCELLANEOUS EXAMPLES ON CHAPTER XIV

1. A uniform stick, of length $2a$, hangs freely by one end, the other being close to the ground. An angular velocity ω is then given to the stick, and when it has turned through a right angle the fixed end is let go. Shew that on first touching the ground it will be in an upright position if $\omega^2=\frac{g}{2a}\left[3+\frac{p^2}{p+1}\right]$, where p is any odd multiple of $\frac{\pi}{2}$.

2. A circular disc rolls in one plane upon a fixed plane and its centre describes a straight line with uniform acceleration f; find the magnitude and line of action of the impressed forces.

3. A spindle of radius a carries a wheel of radius b, the mass of the combination being M and the moment of inertia I; the spindle rolls down a fixed track at inclination α to the horizon, and a string, wound round the wheel and leaving it at its under side, passes over a light pulley and has a mass m attached to the end which hangs vertically, the string between the wheel and pulley being parallel to the track. Shew that the acceleration of the weight is

$$g(b-a)\{Ma\sin\alpha+m(b-a)\}\div[I+Ma^2+m(b-a)^2].$$

4. Three uniform spheres, each of radius a and of mass m, attract one another according to the law of the inverse square of the distance. Initially they are placed on a perfectly rough horizontal plane with their centres forming a triangle whose sides are each of length $4a$. Shew that the velocity of their centres when they collide is $\sqrt{\gamma\frac{5m}{14a}}$, where γ is the constant of gravitation.

5. A uniform sphere, of mass m and radius a, rolls on a horizontal plane. If the resistance of the air be represented by a horizontal force acting at the centre of the sphere equal to $\frac{m\lambda}{a}v^2$ and a couple about it equal to $m\mu v^2$, where v is the velocity of the sphere at any instant, and if V be the velocity at zero time, shew that the distance described by the centre in time t is $\frac{1}{A}\log[1+AVt]$, where $A=\frac{5}{7}\frac{\lambda+\mu}{a}$.

6. A uniform sphere rolls in a straight line on a rough horizontal plane and is acted upon by a horizontal force X at its centre in a direction opposite to the motion of the centre. Shew that the centre of the sphere moves as it would if its mass were collected there and the force reduced to $\frac{5}{7}X$, and that the friction is equal to $\frac{2}{7}X$ and is in a direction opposite to that of X.

7. A man walks on a rough sphere so as to make it roll straight up a plane inclined at an angle α to the horizon, always keeping himself at an angle β from the highest point of the sphere; if the masses of the sphere and man be respectively M and m, shew that the acceleration of the sphere

is
$$\frac{5g\{m\sin\beta-(M+m)\sin\alpha\}}{7M+5m\{1+\cos(\alpha+\beta)\}}.$$

8. A circular cylinder, of radius a and radius of gyration k, rolls inside a fixed horizontal cylinder of radius b. Shew that the plane through the axes will move like a simple circular pendulum of length

$$(b-a)\left(1+\frac{k^2}{a^2}\right).$$

If the fixed cylinder be instead free to move about its axis, and have its centre of gravity in its axis, the corresponding pendulum will be of length $(b-a)(1+n)$, where

$$n=\frac{\frac{k^2}{a^2}}{1+\frac{b^2}{a^2}\frac{mk^2}{MK^2}},$$

m and M are respectively the masses of the inner and outer cylinders, and K is the radius of gyration of the outer cylinder about its axis.

[In the second case, if the outer cylinder has at time t turned through an angle ψ, the equations of motion are, as in Art. 198,

$$m(b-a)\dot{\phi}^2=R-mg\cos\phi;\quad m(b-a)\ddot{\phi}=F-mg\sin\phi;$$
$$mk^2\ddot{\theta}=-F.a;\quad\text{and}\quad MK^2\ddot{\psi}=-Fb.$$

Also the geometrical equation is $a(\theta+\phi)=b(\phi-\psi)$.]

9. A uniform circular hoop has a fine string wound round it. The hoop is placed upright on a horizontal plane, and the string, leaving the hoop at its highest point, passes over a smooth pulley at a height above the plane equal to the diameter of the hoop and has a particle attached to its other end. Find the motion of the system, supposed to be all in one vertical plane; and shew that whether the plane be smooth or rough the hoop will roll without slipping.

10. A disc rolls upon a straight line on a horizontal table, the flat surface of the disc being in contact with the table. If v be the velocity of the centre of the disc at any instant, shew that it will be at rest after a time $\frac{27\pi v}{64\mu g}$, where μ is the coefficient of friction between the disc and table.

11. A perfectly rough cylindrical grindstone, of radius a, is rotating with uniform acceleration about its axis, which is horizontal. If a sphere in contact with its edge can remain with its centre at rest, shew that the angular acceleration of the grindstone must not exceed $\frac{5g}{2a}$.

12. A perfectly rough ball is at rest within a hollow cylindrical garden roller, and the roller is then drawn along a level path with uniform velocity V. If $V^2 > \frac{27}{7}g(b-a)$, shew that the ball will roll completely round the inside of the roller, a and b being the radii of the ball and roller.

13. A solid uniform disc, of radius a, can turn freely about a horizontal axis through its centre, and an insect, of mass $\frac{1}{n}$ that of the disc, starts from its lowest point and moves along the rim with constant velocity relative to the rim; shew that it will never get to the highest point of the disc if this constant velocity is less than $\frac{2}{n}\sqrt{2ga(n+2)}$.

14. Inside a rough hollow cylinder, of radius a and mass M, which is free to turn about its horizontal axis, is placed an insect of mass m; if the insect starts from the lowest generator and walks in a plane perpendicular to the axis of the cylinder at a uniform rate v relatively to the cylinder, shew that the plane containing it and the axis never makes with the upward drawn vertical an angle $< 2\cos^{-1}\left[\frac{v}{2a}\frac{Mk^2}{\sqrt{mga(Mk^2+ma^2)}}\right]$, where Mk^2 is the moment of inertia of the cylinder about its axis.

15. A rough lamina, of mass M, can turn freely about a horizontal axis passing through its centre of gravity, the moment of inertia about this axis being Mk^2. Initially the lamina was horizontal and a particle of mass m was placed on it at a distance c from the axis and then motion was allowed to ensue. Shew that the particle will begin to slide on the lamina when the latter has turned through an angle $\tan^{-1}\frac{\mu . Mk^2}{Mk^2+3mc^2}$, where μ is the coefficient of friction.

16. A uniform beam, of mass M and length l, stands upright on perfectly rough ground; on the top of it, which is flat, rests a weight of mass m, the coefficient of friction between the beam and the weight being μ. If the beam is allowed to fall to the ground, its inclination θ to the vertical when the weight slips is given by $\left(\frac{4M}{3}+3m\right)\cos\theta-\frac{M}{6\mu}\sin\theta=M+2m$.

17. A rough cylinder, of mass M, is capable of motion about its axis, which is horizontal; a particle of mass m is placed on it vertically above the axis and the system is slightly disturbed. Shew that the particle will slip on the cylinder when it has moved through an angle θ given by $\mu(M+6m)\cos\theta-M\sin\theta=4m\mu$, where μ is the coefficient of friction.

18. A hemisphere rests with its base on a smooth horizontal plane; a perfectly rough sphere is placed at rest on its highest point and is slightly displaced. Shew that in the subsequent motion the angular velocity of the line joining the centres, when its inclination to the vertical is θ, is $2\sin\frac{\theta}{2}\left[\frac{5ng}{c(7n-5\cos^2\theta)}\right]^{\frac{1}{2}}$, and shew also that the sphere will leave the hemisphere when θ satisfies the equation

$$5\left(3-\frac{5}{n}\right)\cos^3\theta+20\cos^2\theta+7(15-17n)\cos\theta+70(n-1)=0,$$

where c is the sum of the radii and n the ratio of the sum of the masses of the sphere and hemisphere to that of the sphere.

[Use the Principles of Linear Momentum and Energy.]

19. A thin hollow cylinder, of radius a and mass M, is free to turn about its axis, which is horizontal, and a similar cylinder, of radius b and mass m, rolls inside it without slipping, the axes of the two cylinders being parallel. Shew that, when the plane of the two axes is inclined at an angle ϕ to the vertical, the angular velocity Ω of the larger is given by

$$a^2 (M+m) (2M+m) \Omega^2 = 2gm^2 (a-b) (\cos \phi - \cos \alpha),$$

provided both cylinders are at rest when $\phi = \alpha$.

20. A perfectly rough solid cylinder, of mass m and radius r, rests symmetrically on another solid cylinder, of mass M and radius R, which is free to turn about its axis which is horizontal. If m rolls down, shew that at any time during the contact the angle ϕ which the line joining the centres makes with the vertical is given by

$$(R+r)\ddot{\phi} = \frac{2(M+m)}{3M+2m} . g \sin \phi.$$

Find also the value of ϕ when the cylinders separate.

21. A locomotive engine, of mass M, has two pairs of wheels, of radius a, the moment of inertia of each pair about its axis being Mk^2; and the engine exerts a couple L on the forward axle. If both pairs of wheels commence to roll without sliding when the engine starts, shew that the friction between each of the front wheels and the line capable of being called into action must be not less than $\dfrac{L}{2a} \dfrac{k^2+a^2}{2k^2+a^2}$.

22. A rod, of mass m, is moving in the direction of its length on a smooth horizontal plane with velocity u. A second perfectly rough rod, of the same mass and length $2a$, which is in the same vertical plane as the first rod, is gently placed with one end on the first rod; if the initial inclination of the second rod to the horizontal be α, shew that it will just rise into a vertical position if $3u^2 \sin^2 \alpha = 4ga(1-\sin \alpha)(5+3 \cos^2 \alpha)$.

23. A rough wedge, of mass M and inclination α, is free to move on a smooth horizontal plane; on the inclined face is placed a uniform cylinder, of mass m; shew that the acceleration of the centre of the cylinder down the face, and relative to it, is $2g \sin \alpha . \dfrac{M+m}{3M+m+2m \sin^2 \alpha}$.

24. A uniform circular ring moves on a rough curve under the action of no forces, the curvature of the curve being everywhere less than that of the ring. If the ring be projected from a point A of the curve without rotation and begin to roll at B, then the angle between the normals at A and B is $\dfrac{\log 2}{\mu}$.

25. A uniform rod has one end fastened by a pivot to the centre of a wheel which rolls on a rough horizontal plane, the other extremity resting against a smooth vertical wall at right angles to the plane containing the rod and wheel; shew that the inclination θ of the rod to the vertical, when it leaves the wall, is given by the equation

$$9M \cos^3 \theta + 6m \cos \theta - 4m \cos \alpha = 0,$$

where M and m are the masses of the wheel and rod and α is the initial inclination to the vertical when the system was at rest.

26. Rope is coiled round a drum of a feet radius. Two wheels each of radius b are fitted to the ends of the drum, and the wheels and drum form a rigid body having a common axis. The system stands on level ground and a free end of the rope, after passing under the drum, is inclined at an angle of 60° to the horizon. If a force P be applied to the rope, shew that the drum starts to roll in the opposite direction, its centre having acceleration $\frac{P(2a-b)b}{2M(b^2+k^2)}$, where M is the mass of the system and k its radius of gyration about the axis.

27. A thin circular cylinder, of mass M and radius b, rests on a perfectly rough horizontal plane and inside it is placed a perfectly rough sphere, of mass m and radius a. If the system be disturbed in a plane perpendicular to the generators of the cylinder, obtain the equations of finite motion and two first integrals of them; if the motion be small, shew that the length of the simple equivalent pendulum is $\frac{14M(b-a)}{10M+7m}$.

28. A uniform sphere, of mass M, rests on a rough plank of mass m which rests on a rough horizontal plane, and the plank is suddenly set in motion with velocity u in the direction of its length. Shew that the sphere will first slide, and then roll, on the plank, and that the whole system will come to rest in time $\frac{mu}{\mu g(M+m)}$, where μ is the coefficient of friction at each of the points of contact.

29. A board, of mass M, whose upper surface is rough and under surface smooth, rests on a smooth horizontal plane. A sphere of mass m is placed on the board and the board is suddenly given a velocity V in the direction of its length. Shew that the sphere will begin to roll after a time $\frac{V}{\left(\frac{7}{2}+\frac{m}{M}\right)\mu g}$.

30. On a smooth table there is placed a board, of mass M, whose upper surface is rough and whose lower surface is smooth. Along the upper surface of the board is projected a uniform sphere, of mass m, so that the vertical plane through the direction of projection passes through the centre of inertia of the board. If the velocity of projection be u and the initial angular velocity of the sphere be ω about a horizontal axis perpendicular to the initial direction of projection, shew that the motion will become uniform at the end of time $\frac{2M}{7M+2m}\,\frac{u-a\omega}{\mu g}$, and that the velocity of the board will then be $\frac{2m}{7M+2m}(u-a\omega)$.

31. A perfectly rough plane turns with uniform angular velocity ω about a horizontal axis lying in its plane; initially when the plane was horizontal a homogeneous sphere was in contact with it, and at rest relative to it at a distance a from the axis of rotation; shew that at time t the distance of the point of contact from the axis of rotation was

$$a\cosh\left(\sqrt{\frac{5}{7}}\omega t\right)+\frac{\sqrt{35}g}{12\omega^2}\sinh\left[\sqrt{\frac{5}{7}}\omega t\right]-\frac{5g}{12\omega^2}\sin\omega t.$$

Find also when the sphere leaves the plane.

[For the motion of the centre of gravity use revolving axes, as in Art. 51.]

32. In the previous question the plane turns about an axis parallel to itself and at a distance c from it; when the plane is horizontal and above the axis the sphere,

of radius b, is gently placed on the plane so that its centre is vertically over the axis; shew that in time t the centre of the sphere moves through a distance

$$\sqrt{35}\left[\frac{g}{12\omega^2}-\frac{b}{5}-\frac{c}{7}\right]\sinh\left[\sqrt{\frac{5}{7}}\omega t\right]-\frac{5g}{12\omega^2}\sin\omega t.$$

33. A six-foot gymnast makes a somersault dive into a net by standing stiff and erect on the edge of a platform and allowing himself to overbalance. He loses foothold without slipping, when the component pressure along his legs becomes zero and preserves his rigidity during his fall. With any reasonable and necessary assumptions, shew that he will fall flat on his back if the drop from the platform to the net is about 43 feet.

34. Two railway wagons are each of mass M and each has four wheels of radius a and the moment of inertia for each pair of wheels is I. The two wagons are coupled together and run down a slope inclined to the horizontal at an angle α. If the first wagon carries a load of mass M' and the second wagon is unloaded, shew that there is a pull P in the coupling given by $P\left[2M+M'+\frac{4I}{a^2}\right]=\frac{2g(\sin\alpha-n)M'I}{a^2}$, where the frictional resistances are equivalent to a back pull of n times the weight both for the loaded and unloaded wagon, and a is the radius of each wheel.

If $M=5$ tons, $M'=10$ tons, $a=18$ inches and the moment of inertia, I, is that of half a ton at a distance of one foot, shew that, when $\alpha=\sin^{-1}\frac{1}{10}$ and the back pull is 10 lbs. wt. per ton, the pull is about $45\frac{1}{2}$ lbs. wt.

35. If a uniform heavy right circular cylinder, of radius a, be rotated about its axis and laid gently on two rough horizontal rails, at the same level and distant $2a\sin\alpha$ apart, so that the axis of the cylinder is parallel to the rails, shew that the cylinder will remain in contact with both rails if $\mu<\tan\alpha$, but will initially rise on one rail if $\mu>\tan\alpha$.

36. A billiard ball, struck low, slides and rotates until uniform motion presently ensues. Investigate the motion, and shew that the point of the ball, which is three-tenths of the diameter below the top, has a velocity which remains of the same magnitude all the time.

CHAPTER XV

MOTION IN TWO DIMENSIONS. IMPULSIVE FORCES

204. In the case of impulsive forces the equations of Art. 187 can be easily transformed. For if T be the time during which the impulsive forces act we have, on integrating (1),

$$\left[M\frac{d\bar{x}}{dt}\right]_0^T = \int_0^T \Sigma X\,dt = \Sigma X',$$

where X' is the impulse of the force acting at any point (x, y). Let u and v be the velocities of the centre of inertia parallel to the axes just before the impulsive forces act, and u' and v' the corresponding velocities just after their action.

Then this equation gives

$$M(u'-u) = \Sigma X' \quad \text{.........................(1).}$$

So

$$M(v'-v) = \Sigma Y' \quad \text{.........................(2).}$$

These equations state that the change in the momentum of the mass M, supposed collected at the centre of inertia, in any direction is equal to the sum of the impulses in that direction.

So, on integrating equation (4), we have

$$\left[Mk^2\frac{d\theta}{dt}\right]_0^T = \Sigma\left[x'\int_0^T Y\,dt - y'\int_0^T X\,dt\right],$$

i.e. if ω and ω' be the angular velocities of the body before and after the action of the impulsive forces, we have

$$Mk^2(\omega'-\omega) = \Sigma(x'Y' - y'X').$$

Hence the change produced in the moment of momentum about the centre of inertia is equal to the moment about the centre of inertia of the impulses of the forces.

205. *Ex.* 1. *A uniform rod AB, of length 2a, is lying on a smooth horizontal plane and is struck by a horizontal blow, of impulse P, in a direction perpendicular to the rod at a point distant b from its centre; to find the motion.*

Let u' be the velocity of the centre of inertia perpendicular to the rod after the blow, and ω' the corresponding angular velocity about the centre. Then the equations of the last article give

$$Mu' = P, \quad \text{and} \quad M\frac{a^2}{3}\omega' = P.b.$$

Hence we have u' and ω'.

Ex. 2. *A uniform rod at rest is struck by a blow at right angles to its length at a distance x from its centre. Find the point about which it will begin to turn.*

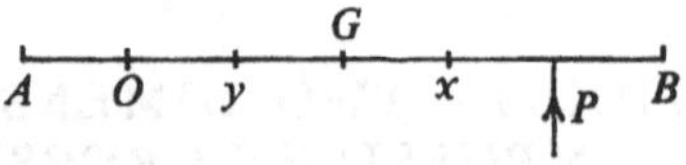

Let O be the required centre of motion, $GO=y$, where G is the centre of inertia, and $GA=GB=a$.

Let the impulse of the blow be P, and the resulting angular velocity about O be ω. The velocity acquired by G is $y\omega$. Hence, from Art. 204, we have

$$My\omega=P \quad\text{.............................(1)},$$

and

$$M\left[y^2+\frac{a^2}{3}\right]\omega=P(y+x) \quad\text{......................(2)}.$$

Solving, $\omega=\frac{3P.x}{Ma^2}$ and $y=\frac{a^2}{3x}$, giving the resulting angular velocity and the position of O. The velocity of the centre of inertia $G=y\omega=\frac{P}{M}$.

The kinetic energy acquired, by Art. 190,

$$=\frac{1}{2}M\left[y^2+\frac{a^2}{3}\right]\omega^2=\frac{P^2}{2M}\,\frac{a^2+3x^2}{a^2} \quad\text{................(3)}.$$

If the end A were fixed, the resulting angular velocity ω_1 would be given by the equation $M\left[a^2+\frac{a^2}{3}\right]\omega_1=P(a+x)$, so that $\omega_1=\frac{3P}{4M}\cdot\frac{a+x}{a^2}$, and the kinetic energy generated would

$$=\frac{1}{2}.M.\frac{4a^2}{3}\omega_1{}^2=\frac{3P^2}{8M}\,\frac{(a+x)^2}{a^2} \quad\text{................(4)}.$$

The ratio of the energies given by (3) and (4) $=\frac{4}{3}\,\frac{a^2+3x^2}{(a+x)^2}$.

The least value of this ratio is easily seen to be unity, when $x=\frac{a}{3}$.

Hence the kinetic energy generated when the rod is free is always greater than that when the end A is fixed, except when $x=\frac{a}{3}$, in which case A is the centre of rotation.

Ex. 3. *Two uniform rods AB, BC are freely jointed at B and laid on a horizontal table; AB is struck by a horizontal blow of impulse P in a direction perpendicular to AB at a distance c from its centre ; the lengths of AB, BC being 2a and 2b and their masses M and M', find the motion immediately after the blow.*

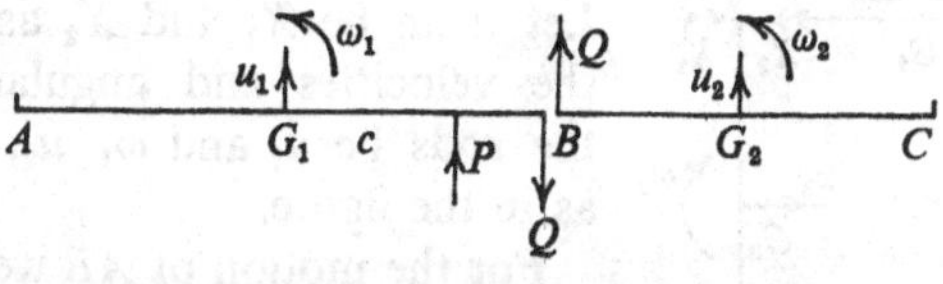

Let u_1 and ω_1 be the velocity of the centre of inertia of AB and its angular velocity just after the blow; u_2 and ω_2 similar quantities for BC. There will be an impulsive action between the two rods at B when the blow is struck; let its impulse be Q, in opposite directions on the two rods.

Then for the rod AB, since it was at rest before the blow, we have

$$Mu_1 = P - Q \quad \text{..............................(1)},$$

and

$$M\frac{a^2}{3}.\omega_1 = P.c - Q.a \quad \text{......................(2)}.$$

So, for BC, we have

$$M'u^2 = Q \quad \text{..............................(3)},$$

and

$$M'\frac{b^2}{3}.\omega_2 = -Q.b \quad \text{..........................(4)}.$$

Also, since the rods are connected at B, the motion of B, as deducted from each rod, must be the same.

$$\therefore \; u_1 + a\omega_1 = u_2 - b\omega_2 \quad \text{......................(5)}.$$

These five simple equations give u_1, ω_1, u_2, ω_2, and Q. On solving them, we obtain

$$Q = \frac{1}{4}P.\frac{M'}{M+M'}\left(1+\frac{3c}{a}\right), \quad u_1 = \frac{P}{M}\left[1 - \frac{1}{4}\frac{M'}{M+M'}\left(1+\frac{3c}{a}\right)\right],$$

$$\omega_1 = \frac{3P}{Ma}\left[\frac{c}{a} - \frac{1}{4}\frac{M'}{M+M'}\left(1+\frac{3c}{a}\right)\right], \quad u_2 = \frac{1}{4}.\frac{P}{M+M'}\left(1+\frac{3c}{a}\right),$$

and

$$\omega_2 = -\frac{3}{4}\frac{P}{(M+M')b}\left(1+\frac{3c}{a}\right).$$

Ex. 4. *Three equal uniform rods AB, BC, CD are hinged freely at their ends, B and C, so as to form three sides of a square and are laid on a smooth table ; the end A is struck by a horizontal blow P at right angles to AB. Shew that the initial velocity of A is nineteen times that of D, and that the impulsive actions at B and C are respectively* $\frac{5P}{12}$ *and* $\frac{P}{12}$.

The initial motion of the point B must be perpendicular to AB, so that the action at B must be along BC; similarly the action at C must be along CB. Let them be X_1 and X_2 as marked. Let the velocities and angular velocities of the rods be u_1 and ω_1, u_2, and u_3 and ω_3 as in the figure.

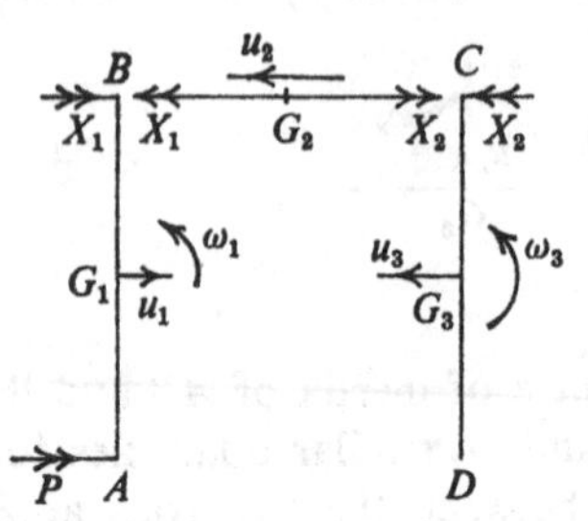

For the motion of AB we have

$$mu_1 = P + X_1 \quad \text{.................(1)},$$

and

$$m\frac{a^2}{3}\omega_1 = (P - X_1)a \quad \text{...........(2)},$$

where m is the mass and $2a$ the length of each rod.

For BC, we have $$mu_2 = X_1 - X_2 \quad \text{...........(3)}.$$
For CD, we have $$mu_3 = X_2 \quad \text{...........(4)},$$

and

$$m\frac{a^2}{3}\omega_3 = X_2 . a \quad \text{...........(5)}.$$

Also the motion of the point B of the rod AB is the same as that of the same point B of the rod BC.

$$\therefore \; u_1 - a\omega_1 = -u_2 \quad \text{...........(6)}.$$

So, for the point C,

$$u_3 + a\omega_3 = u_2 \quad \text{...........(7)}.$$

On substituting from (1)...(5) in (6) and (7), we obtain

$$5X_1 - X_2 = 2P \quad \text{and} \quad X_1 = 5X_2,$$

giving

$$X_1 = \frac{5P}{12} \quad \text{and} \quad X_2 = \frac{P}{12}.$$

Hence we have

$$u_1 = \frac{17P}{12m}; \; a\omega_1 = \frac{7P}{4m}; \; u_2 = \frac{P}{3m}; \; u_3 = \frac{P}{12m} \quad \text{and} \quad a\omega_3 = \frac{P}{4m}.$$

$$\therefore \; \frac{\text{velocity of the point } A}{\text{velocity of the point } D} = \frac{u_1 + a\omega_1}{a\omega_3 - u_3} = 19.$$

EXAMPLES

1. AB, BC are two equal similar rods freely hinged at B and lie in a straight line on a smooth table. The end A is struck by a blow perpendicular to AB; shew that the resulting velocity of A is $3\frac{1}{2}$ times that of B.

2. Two uniform rods, AB and BC, are smoothly jointed at B and placed in a horizontal line; the rod BC is struck at G by a blow at right angles to it; find the position of G so that the angular velocities of AB and BC may be equal in magnitude.

3. Two equal uniform rods, AB and AC, are freely hinged at A and rest in a straight line on a smooth table. A blow is struck at B perpendicular to the rods; shew that the kinetic energy generated is $\frac{7}{4}$ times what it would be if the rods were rigidly fastened together at A.

4. Two equal uniform rods, AB and BC, are freely jointed at B and turn about a smooth joint at A. When the rods are in a straight line, ω being the angular velocity of AB and u the velocity of the centre of mass of BC, BC impinges on a fixed inelastic obstacle at a point D; shew that the rods are instantaneously brought to rest if $BD = 2a\dfrac{2u - a\omega}{3u + 2a\omega}$, where $2a$ is the length of either rod.

5. Two rods, AB and BC, of length $2a$ and $2b$ and of masses proportional to their lengths, are freely jointed at B and are lying in a straight line. A blow is communicated to the end A; shew that the resulting kinetic energy when the system is free is to the energy when C is fixed as $(4a+3b)(3a+4b) : 12(a+b)^2$.

6. Three equal rods, AB, BC, CD, are freely jointed and placed in a straight line on a smooth table. The rod AB is struck at its end A by a blow which is perpendicular to its length; find the resulting motion, and shew that the velocity of the centre of AB is 19 times that of CD, and its angular velocity 11 times that of CD.

7. Three equal uniform rods placed in a straight line are freely jointed and move with a velocity v perpendicular to their lengths. If the middle point of the middle rod be suddenly fixed, shew that the ends of the other two rods will meet in time $\dfrac{4\pi a}{9v}$, where a is the length of each rod.

8. Two equal uniform rods, AB and AC, are freely jointed at A, and are placed on a smooth table so as to be at right angles. The rod AC is struck by a blow at C in a direction perpendicular to itself; shew that the resulting velocities of the middle points of AB and AC are in the ratio $2:7$.

9. Two uniform rods, AB, AC, are freely jointed at A and laid on a smooth horizontal table so that the angle BAC is a right angle. The rod AB is struck by a blow P at B in a direction perpendicular to AB; shew that the initial velocity of A is $\dfrac{2P}{4m'+m}$, where m and m' are the masses of AB, AC respectively.

10. AB and CD are two equal and similar rods connected by a string BC; AB, BC, and CD form three sides of a square. The point A of the rod AB is struck a blow in a direction perpendicular to the rod; shew that the initial velocity of A is seven times that of D.

11. Three particles of equal mass are attached to the ends, A and C, and the middle point B of a light rigid rod ABC, and the system is at rest on a smooth table. The particle C is struck a blow at right angles to the rod; shew that the energy communicated to the system when A is fixed, is to the energy communicated when the system is free as 24 to 25.

12. A uniform straight rod, of length 2 ft. and mass 2 lbs., has at each end a mass of 1 lb., and at its middle point a mass of 4 lbs. One of the 1 lb. masses is struck a blow at right angles to the rod and this end starts off with a velocity of 5 ft. per second; shew that the other end of the rod begins to move in the opposite direction with a velocity of 2·5 ft. per sec.

206. *A uniform sphere, rotating with an angular velocity ω about an axis perpendicular to the plane of motion of its centre, impinges on a horizontal plane; find the resulting change in its motion.*

First, suppose the plane rough enough to prevent any sliding.

Let u and v be the components of its velocity before impact as marked in the figure; u' and v' the components, and ω' the angular velocity, just after the impact.

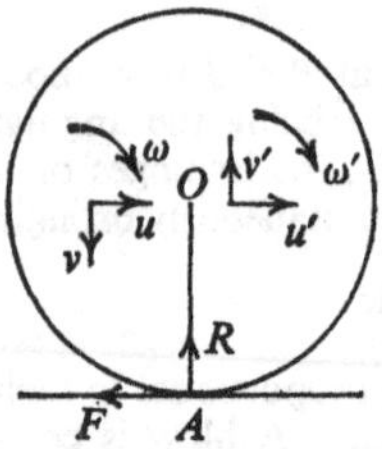

Let R be the normal impulsive reaction and F the impulsive friction.

Then the equations of Art. 204 give

$$M(u'-u)=-F \quad\text{.........................(1)},$$

$$M(v'+v)=R \quad\text{.........................(2)},$$

and $$Mk^2(\omega'-\omega)=Fa \quad\text{.........................(3)}.$$

Also since the point A is instantaneously reduced to rest, there being no sliding,

$$u'-a\omega'=0 \quad\text{.............................(4)}.$$

Also, if e be the coefficient of restitution,

$$v'=ev \quad\text{...............................(5)}.$$

Solving (1), (3), and (4), we have

$$u'=a\omega'=\frac{5u+2a\omega}{7} \quad\text{.........................(6)},$$

and $$F=M.\tfrac{2}{7}(u-a\omega) \quad\text{.........................(7)}.$$

Case I. $u=a\omega$.

There is no friction called into play, and u and ω are unaltered.

Case II. $u<a\omega$.

Then F acts $\rightarrow$; $\omega'<\omega$, and $u'>u$. Hence when the point of contact A before impact is moving $\leftarrow$, the angular velocity is decreased by the impact, the horizontal velocity is increased, and the direction of motion of the sphere after impact makes a smaller angle with the plane than it would if there were no friction.

Case III. $u>a\omega$.

Then F acts $\leftarrow$; $\omega'>\omega$ and $u'<u$. Hence when the point of contact A before impact is moving $\rightarrow$, the angular velocity is increased, the horizontal velocity is diminished, and the direction of motion after impact makes a greater angle with the plane than it would if there were no friction.

Case IV. Let the angular velocity before the impact be $\circlearrowleft$.

We must now change the sign of ω, and have

$$u'=a\omega'=\frac{5u-2a\omega}{7} \quad\text{.........................(8)},$$

and $$F=M.\tfrac{2}{7}(u+a\omega) \quad\text{.........................(9)}.$$

If $u=\frac{2a\omega}{5}$ then u' and ω' are both zero, and the sphere rebounds from the plane vertically with no spin.

If $u<\frac{2a\omega}{5}$, then u' is negative and the sphere after the impact rebounds towards the direction from which it came.

[Compare the motion of a tennis ball on hitting the ground when it has been given sufficient " under-cut."]

In each case the vertical velocity after the impact is ev and $R=M(1+e)v$.

In Cases I, II, and III, in order that the point of contact may be instantaneously brought to rest, we must have $\frac{F}{R}<\mu$, the coefficient of friction, *i.e.* from (2), (5), and (7),

$$\tfrac{2}{7}(u-a\omega)<\mu(1+e)v.$$

If $\frac{2}{7}(u-a\omega)>\mu(1+e)v$, the friction is not sufficient to bring the point of contact A to instantaneous rest, equation (4) will not hold and for equations (1), (2), (3), we must have

$$M(u'-u)=-\mu R \qquad \text{(1')},$$

$$M(v'+v)=R \qquad \text{(2')},$$

and

$$Mk^2(\omega'-\omega)=\mu R.a \qquad \text{(3')}.$$

These with equation (5) give

$$u'=u-\mu v(1+e), \quad v'=ev,$$

and

$$\omega'=\omega+\frac{5\mu}{2a}v(1+e) \qquad \text{(4')}.$$

In Case IV in order that the friction may bring the point of contact to rest we must have $F<\mu R$.

Hence from (2), (5), and (9) we must have

$$\tfrac{2}{7}(u+a\omega)<\mu v(1+e).$$

If however

$$\tfrac{2}{7}(u+a\omega)>\mu v(1+e) \qquad \text{(5')},$$

the friction is not sufficient, and we have equations similar to (1'), (2'), and (3'), but with the sign of ω changed. They will give

$$u'=u-\mu v(1+e), \quad v'=ev,$$

and

$$\omega'=\frac{5\mu}{2a}v(1+e)-\omega.$$

In this case it will, from (5'), be possible for u to be less than $\mu v(1+e)$, if ω be large enough; hence, if the ball has sufficiently large enough under-cut, u' can be negative, *i.e.* the ball can rebound backwards.

[Compare again the motion of a tennis ball.]

207. *Ex.* 1. *A rod, of length* $2a$, *is held in a position inclined at an angle* α *to the vertical, and is then let fall on to a smooth inelastic horizontal plane. Shew that the end which hits the plane will leave it immediately after the impact if the height through which the rod falls is greater than*

$$\tfrac{1}{18}a \sec \alpha \operatorname{cosec}^2 \alpha (1+3 \sin^2 \alpha)^2.$$

If u and ω be the vertical and angular velocity just after the impact, V, the vertical velocity before the impact and R the impulse of the reaction of the plane, then

$$m(V-u)=R, \quad mk^2\omega=Ra \sin \alpha, \quad \text{and} \quad u-a\omega \sin \alpha$$
$$=\text{vertical velocity of the end in contact with the plane}=0.$$

Hence $$\omega=\frac{u}{a \sin \alpha}=\frac{3V \sin \alpha}{a(1+3 \sin^2 \alpha)} \quad \text{.....................(1).}$$

Assuming the end to remain in contact with the plane, and that S is the normal reaction when the rod is inclined at θ to the vertical, we have

$$S-mg=m\frac{d^2}{dt^2}(a \cos \theta), \quad \text{and} \quad S.a \sin \theta=m\frac{a^2}{3}\ddot{\theta} \quad \text{......(2).}$$

Eliminating S, we have

$$\ddot{\theta}(1+3 \sin^2 \theta)+3 \sin \theta \cos \theta\dot{\theta}^2=\frac{3g}{a} \sin \theta \quad \text{...........(3).}$$

Now S is negative when $\theta=\alpha$ if $\ddot{\theta}$ is negative then, so that equation (3) then gives $3 \sin \alpha \cos \alpha\omega^2>\frac{3g}{a} \sin \alpha$, *i.e.* $\omega^2>\frac{g}{a \cos \alpha}$.

Hence, from (1),

$$V^2=\frac{a^2}{9}\frac{(1+3 \sin^2 \alpha)^2}{\sin^2 \alpha}\omega^2>\frac{ga(1+3 \sin^2 \alpha)^2}{9 \cos \alpha \sin^2 \alpha}.$$

Hence the given answer.

Ex. 2. *Four equal rods, each of mass m and length* $2a$, *are freely jointed at their ends so as to form a rhombus. The rhombus falls with a diagonal vertical, and is moving with velocity* V *when it hits a fixed horizontal inelastic plane. Find the motion of the rods immediately after the impact, and shew that their angular velocities are each equal to* $\frac{3}{2}\frac{V \sin \alpha}{a(1+3 \sin^2 \alpha)}$, *where* α *is the angle each rod makes with the vertical.*

Shew also that the impact destroys a fraction $\frac{1}{1+3 \sin^2 \alpha}$ *of the kinetic energy just before the impact.*

After the impact it is clear that AB is moving with some angular velocity ω_1) about A, and BC with an angular velocity ω_2) about B.

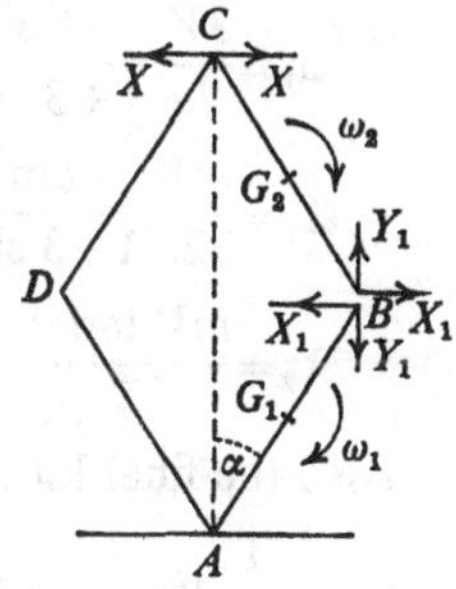

Since C is, by symmetry, moving vertically after the impact, its horizontal velocity is zero.

$\therefore 0 =$ horizontal velocity of B + horizontal velocity of C relative to B

$$= 2a\omega_1 \cos\alpha + 2a\omega_2 \cos\alpha,$$

i.e. $$\omega_2 = -\omega_1 \qquad\text{.............................(1).}$$

The horizontal velocity of G_2 similarly

$$= 2a\omega_1 \cos\alpha + a\omega_2 \cos\alpha = a\omega_1 \cos\alpha \rightarrow,$$

and its vertical velocity

$$= 2a\omega_1 \sin\alpha - a\omega_2 \sin\alpha = 3a\omega_1 \sin\alpha \downarrow.$$

If X be the horizontal impulse at C as marked (there being no vertical impulse there by symmetry) we have, as in Art. 192, on taking moments about A for the two rods AB, BC

$$m\frac{4a^2}{3}\omega_1 + m\left[a\omega_1 \cos\alpha . 3a\cos\alpha + 3a\omega_1 \sin\alpha . a\sin\alpha + \frac{a^2}{3}\omega_2\right]$$
$$-2mVa\sin\alpha = X . 4a\cos\alpha,$$

i.e. $$2\omega_1 = \frac{V}{a}\sin\alpha + \frac{2X}{ma}\cos\alpha \qquad\text{.....................(2).}$$

Similarly, taking moments about B for the rod BC, we have

$$m\left[a\omega_1 \cos\alpha . a\cos\alpha - 3a\omega_1 \sin\alpha . a\sin\alpha + \frac{a^2}{3}\omega_2\right] - m[-V] . a\sin\alpha$$
$$= X . 2a\cos\alpha,$$

i.e. $$\omega_1\left(\frac{2}{3} - 4\sin^2\alpha\right) = -\frac{V}{a}\sin\alpha + \frac{2X}{ma}\cos\alpha \qquad\text{...........(3).}$$

Solving (2) and (3), we have ω_1 and X, and the results given are obtained.

The impulsive actions $X_1 \rightarrow$ and $Y_1 \uparrow$ at B on the rod BC are clearly given by

$X_1 + X = m$. horizontal velocity communicated to $G_2 = m . a\omega_1 \cos\alpha$,

and $Y_1 = m \times$ vertical velocity communicated to G_2

$$= m(-3a\omega_1 \sin\alpha) - m[-V] = m[V - 3a\omega_1 \sin\alpha].$$

Also the impulsive action $X_2 \rightarrow$ at A on AB is given by

$X_2 - X_1 = m$. horizontal velocity communicated to $G_1 = m . a\omega_1 \cos\alpha$.

The total action $Y_2 \uparrow$ at A = total change in the vertical momentum

$$= 4mV - 8ma\omega_1 \sin\alpha.$$

On solving these equations, we have

$$\omega_1 = \frac{3V}{2a} \cdot \frac{\sin\alpha}{1+3\sin^2\alpha}; \qquad X = \frac{mV\tan\alpha}{2} \cdot \frac{3\cos^2\alpha - 1}{1+3\sin^2\alpha};$$

$$X_1 = \frac{mV}{2} \cdot \frac{\tan\alpha}{1+3\sin^2\alpha}; \qquad Y_1 = \frac{mV}{2} \cdot \frac{3\cos^2\alpha - 1}{1+3\sin^2\alpha};$$

$$X_2 = \frac{mV\tan\alpha}{2} \cdot \frac{1+3\cos^2\alpha}{1+3\sin^2\alpha}; \text{ and } Y_2 = \frac{4mV}{1+3\sin^2\alpha}.$$

Also the final kinetic energy

$$= \frac{1}{2} \cdot 2m \cdot \frac{4a^2}{3}\omega_1^2 + \frac{1}{2} \cdot 2m \left[a^2\omega_1^2 \cos^2\alpha + 9a^2\omega_1^2 \sin^2\alpha + \frac{a^2}{3}\omega_2^2 \right]$$

$$= \frac{3\sin^2\alpha}{1+3\sin^2\alpha} \times \text{original kinetic energy.}$$

It will be noted that, since we are considering only the change in the motion produced by the blow, the finite external forces (the weights of the rods in this case) do not come into our equations. For these finite forces produce no effect during the very short time that the blow lasts.

Ex. 3. *A body, whose mass is m, is acted upon at a given point P by a blow of impulse X. If V and V' be the velocities of P in the direction of X just before and just after the action of X, shew that the change in the kinetic energy of the body, i.e. the work done on it by the impulse, is* $\frac{1}{2}(V+V')X$.

Take the axis of x parallel to the direction of X. Let u and v be the velocities of the centre of inertia G parallel to Ox and Oy, and ω the angular velocity round G just before the action of X. Let u', v', and ω' be the same quantities just after the blow. The equations of Art. 204 then become

$$m(u'-u) = X; \quad m(v'-v) = 0, \quad \text{and} \quad mk^2(\omega'-\omega) = -y' \cdot X \ldots(1),$$

where (x', y') are the coordinates of P relative to G.

By Art. 190, the change in the kinetic energy

$$= \tfrac{1}{2}m(u'^2 + v'^2 + k^2\omega'^2) - \tfrac{1}{2}m(u^2 + v^2 + k^2\omega^2) = \tfrac{1}{2}m(u'^2 - u^2) + \tfrac{1}{2}mk^2(\omega'^2 - \omega^2)$$

$$= \tfrac{1}{2}X(u'+u) - \tfrac{1}{2}y' \cdot X(\omega'+\omega), \text{ by (1)}, = \tfrac{1}{2}X[(u'-y'\omega') + (u - y'\omega)].$$

Now V = the velocity of G parallel to Ox + the velocity of P relative to G

$$= u - \omega . GP \sin GPx = u - y'\omega,$$

and similarly $\qquad V' = u' - y'\omega'.$

Hence the change in the kinetic energy $= \frac{1}{2}X(V'+V)$.

EXAMPLES ON CHAPTER XV

1. A uniform inelastic rod falls without rotation, being inclined at any angle to the horizon, and hits a smooth fixed peg at a distance from its upper end equal to one-third of its length. Shew that the lower end begins to descend vertically.

2. A light string is wound round the circumference of a uniform reel, of radius a and radius of gyration k about its axis. The free end of the string being tied to a fixed point, the reel is lifted up and let fall so that, at the moment when the string becomes tight, the velocity of the centre of the reel is u and the string is vertical. Find the change in the motion and show that the impulsive tension is $mu.\frac{k^2}{a^2+k^2}$.

3. A square plate, of side $2a$, is falling with velocity u, a diagonal being vertical, when an inelastic string attached to the middle point of an upper edge becomes tight in a vertical position. Shew that the impulsive tension of the string is $\frac{4}{7}Mu$, where M is the mass of the plate.

Verify the theorem of Art. 207, Ex. 3.

4. If a hollow lawn tennis ball of elasticity e has on striking the ground, supposed perfectly rough, a vertical velocity u and an angular velocity ω about a horizontal axis, find its angular velocity after impact and prove that the range of the rebound will be $\frac{4}{5}\frac{a\omega}{g}eu$.

5. An imperfectly elastic sphere descending vertically comes in contact with a fixed rough point, the impact taking place at a point distant α from the lowest point, and the coefficient of elasticity being e. Find the motion, and shew that the sphere will start moving horizontally after the impact if $\alpha=\tan^{-1}\sqrt{\frac{7e}{5}}$.

6. A billiard ball is at rest on a horizontal table and is struck by a horizontal blow in a vertical plane passing through the centre of the ball; if the initial motion is one of pure rolling, find the height of the point struck above the table.

[There is no impulsive friction.]

7. A rough imperfectly elastic ball is dropped vertically, and, when its velocity is V, a man suddenly moves his racket forward in its own plane with velocity U, and thus subjects the ball to pure cut in a downward direction making an angle α with the horizon. Shew that, on striking the rough ground, the ball will not proceed beyond the point of impact, provided $(U-V\sin\alpha)(1-\cos\alpha)>(1+e)\left(1+\frac{a^2}{k^2}\right)V\sin\alpha\cos\alpha$.

8. An inelastic sphere, of radius a, rolls down a flight of perfectly rough steps; shew that if the velocity of the centre on the first step exceeds $\sqrt{ga}$, its velocity will be the same on every step, the steps being such that, in its flight, the sphere never impinges on an edge.

[The sphere leaves each edge immediately.]

9. An equilateral triangle, formed of uniform rods freely hinged at their ends, is falling freely with one side horizontal and uppermost. If the middle point of this side be suddenly stopped, shew that the impulsive actions at the upper and lower hinges are in the ratio $\sqrt{13}:1$.

10. A lamina in the form of an equilateral triangle ABC lies on a smooth horizontal plane. Suddenly it receives a blow at A in a direction parallel to BC, which causes A to move with velocity V. Determine the instantaneous velocities of B and C and describe the subsequent motion of the lamina.

11. A rectangular lamina, whose sides are of length $2a$ and $2b$, is at rest when one corner is caught and suddenly made to move with prescribed speed V in the plane of the lamina. Shew that the greatest angular velocity which can thus be imparted to the lamina is $\dfrac{3V}{4\sqrt{a^2+b^2}}$.

12. Four freely-jointed rods, of the same material and thickness, form a rectangle of sides $2a$ and $2b$ and of mass M'. When lying in this form on a horizontal plane an inelastic particle of mass M moving with velocity V in a direction perpendicular to the rod of length $2a$ impinges on it at a distance c from its centre. Shew that the kinetic energy lost in the impact is $\dfrac{1}{2}V^2 \div \left[\dfrac{1}{M}+\dfrac{1}{M'}\left(1+\dfrac{3a+3b}{a+3b}\dfrac{c^2}{a^2}\right)\right]$.

13. Four equal uniform rods, AB, BC, CD, and DE, are freely jointed at B, C, and D and lie on a smooth table in the form of a square. The rod AB is struck by a blow at A at right angles to AB from the inside of the square; shew that the initial velocity of A is 79 times that of E.

14. A rectangle formed of four uniform rods freely jointed at their ends is moving on a smooth horizontal plane with velocity V in a direction along one of its diagonals which is perpendicular to a smooth inelastic vertical wall on which it impinges; shew that the loss of energy due to the impact is

$$V^2\Big/\left\{\frac{1}{m_1+m_2}+\frac{3\cos^2\alpha}{3m_1+m_2}+\frac{3\sin^2\alpha}{m_1+3m_2}\right\},$$

where m_1 and m_2 are the masses of the rods and α is the angle the above diagonal makes with the side of mass m_1.

15. Of two inelastic circular discs with milled edges, each of mass m and radius a, one is rotating with angular velocity ω round its centre O which is fixed on a smooth plane, and the other is moving without spin in the plane with velocity v directed towards O. Find the motion immediately afterwards, and shew that the energy lost by the impact is $\dfrac{1}{2}m\left(v^2+\dfrac{a^2\omega^2}{5}\right)$.

16. A uniform circular disc, of mass M and radius a, is rotating with uniform angular velocity ω on a smooth plane and impinges normally with any velocity u upon a rough rod, of mass m, resting on the plane. Find the resulting motion of the rod and disc, and shew that the angular velocity of the latter is immediately reduced to $\dfrac{M+m}{M+3m}\omega$.

17. An elliptic disc, of mass m, is dropped in a vertical plane with velocity V on a perfectly rough horizontal plane; shew that the loss of kinetic energy by the impact is $\dfrac{1}{2}(1-e^2)\,mV^2.\dfrac{k^2+p^2}{k^2+r^2}$, where r is the distance of the centre of the disc from the point of contact, p is the central perpendicular on the tangent, and e is the coefficient of elasticity.

18. Two similar ladders, of mass m and length $2a$, smoothly hinged together at the top, are placed on a smooth floor and released from rest when their inclination to the horizontal is α. When their inclination to the horizontal is θ they are brought to rest by the tightening of a string of length l which joins similarly situated rungs. Shew that the jerk in the string is

$$4m.\frac{a}{l}\cot\theta\sqrt{\frac{2}{3}ga(\sin\alpha-\sin\theta)}.$$

19. A sphere of mass m falls with velocity V on a perfectly rough inclined plane of mass M and angle α which rests on a smooth horizontal plane. Shew that the vertical velocity of the centre of the sphere immediately after the impact is $\dfrac{5(M+m)V\sin^2\alpha}{7M+2m+5m\sin^2\alpha}$, the bodies being all supposed perfectly inelastic.

20. A sphere, of mass m, is resting on a perfectly rough horizontal plane. A second sphere, of mass m', falling vertically with velocity V strikes the first; both spheres are inelastic and perfectly rough and the common normal at the point of impact makes an angle γ with the horizon. Shew that the vertical velocity of the falling sphere will be instantaneously reduced to $V(m+m')\div\left[\frac{7}{5}m\sec^2\gamma+m' + \frac{2}{7}m'\tan^2\left(\frac{\pi}{4}+\frac{\gamma}{2}\right)\right]$.

Shew also that the lower sphere will not be set in motion if $\sin\gamma=\frac{2}{5}$, but that the upper sphere will be set spinning in any case.

CHAPTER XVI

INSTANTANEOUS CENTRE. ANGULAR VELOCITIES. MOTION IN THREE DIMENSIONS

208. To fix the position of a point in space we must know its three coordinates; this may be otherwise expressed by saying that it has three degrees of freedom.

If one condition be given (*e.g.* a relation between its coordinates, so that it must lie on a fixed surface) it is said to have two degrees of freedom and one of constraint.

If two conditions are given (*e.g.* two relations between its coordinates so that it must be on a line, straight or curved) it is said to have one degree of freedom and two of constraint.

A rigid body, free to move, has six degrees of freedom. For its position is fully determined when three points of it are given. The nine coordinates of these three points are connected by three relations expressing the invariable lengths of the three lines joining them. Hence, in all, the body has 6 degrees of freedom.

A rigid body with one point fixed has $6-3$, *i.e.* three degrees of freedom, and therefore three of constraint.

A rigid body with two of its points fixed, *i.e.* free to move about an axis, has one degree of freedom. For the six coordinates of these two points are equivalent to five constraining conditions, since the distance between the two points is constant.

209. A rigid body has its position determined when we know the three coordinates of any given point G of it, and also the angles which any two lines, GA and GB, fixed in the body make with the axes of coordinates.

[If G and GA only were given the body might revolve round GA.]

Since there are the three relations (1) $l^2+m^2+n^2=1$, (2) $l'^2+m'^2+n'^2=1$, and (3) $ll'+mm'+nn'=$the cosine of the given angle AGB, between the direction cosines (l, m, n) and (l', m', n') of the two lines, it follows that, as before, six quantities, viz. three coordinates and three angles must be known to fix the position of the body.

210. Uniplanar motion.

At any instant there is always an axis of pure rotation, *i.e.* a body can be moved from one position into any other by a rotation about some point without any translation.

During any motion let three points A, B, C fixed in the body move into the positions A', B', and C' respectively. Bisect AA', BB' at M and N and erect perpendiculars to meet in O, so that $OA=OA'$ and $OB=OB'$.

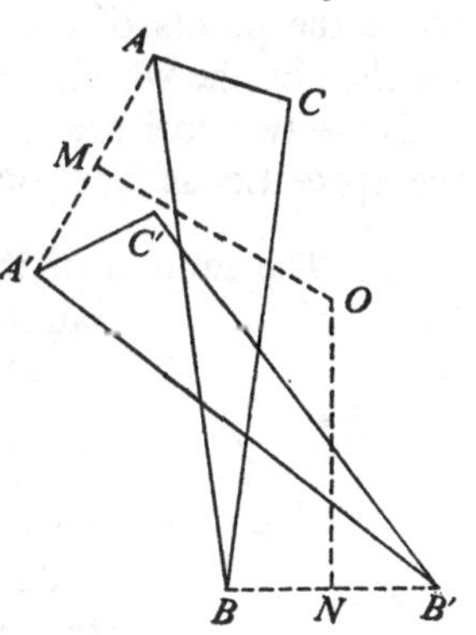

Then the triangles AOB, $A'OB'$ are equal in all respects, so that $\angle AOB=\angle A'OB'$,

and $\therefore\ \angle AOA'=\angle BOB'$.........(1),

and $\angle OBA=\angle OB'A'$.

But $\angle CBA=\angle C'B'A'$.

$\therefore$ by subtraction $\angle OBC=\angle OB'C'$.

Also $OB=OB'$ and $BC=B'C'$.

Hence the triangles OBC, $OB'C'$ are equal in all respects,

and hence $OC=OC'$ ....(2),

and $\angle COB=\angle C'OB'$,

i.e. $\angle COC'=\angle BOB'=\angle AOA'$ ....(3).

Hence the same rotation about O, which brings A to A' and B to B', brings any point C to its new position, *i.e.* O is the required centre of rotation.

The point O always exists unless AA' and BB' are parallel, in which case the motion is one of simple translation and the corresponding point O is at infinity.

Since the proposition is true for all finite displacements, it is true for very small displacements. Hence a body, in uniplanar motion, may be moved into the successive positions it occupies by successive instantaneous rotations about some centre or centres.

To obtain the position of the point O at any instant let A and A' be successive positions of one point, and B and B' successive positions of another point, of the body.

Erect perpendiculars to AA' and BB'; these meet in O.

211. The centre, or axis, of rotation may be either *permanent*, as in the case of the axis of rotation of an ordinary pendulum, or *instantaneous*, as in the case of a wheel rolling in a straight line on the ground, where the point of contact of the wheel with the ground is, for the moment, the centre of rotation.

The instantaneous centre has two loci according to whether we consider its position with regard to the body, or in space. Thus in the case of the cart-wheel the successive points of contact are the points on the edge of the wheel; their locus with regard to the body is the edge itself, *i.e.* a circle whose centre is that of the wheel. In space the points of contact are the successive points on the ground touched by the wheel, *i.e.* a straight line on the ground.

These two loci are called the Body-Locus, or Body-Centrode, and the Space-Locus, or Space-Centrode.

212. The motion of the body is given by the rolling of the body-centrode, carrying the body with it, upon the space-centrode.

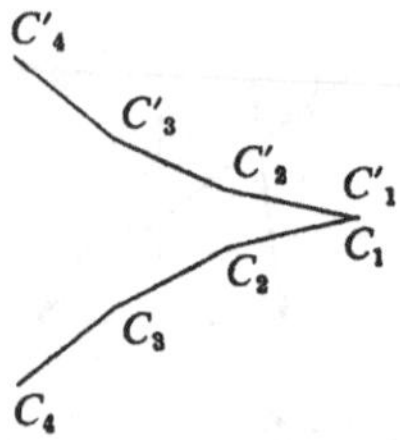

Let C_1', C_2', C_3', C_4', ... be successive points of the body-centrode, and C_1, C_2, C_3, C_4, ... successive points of the space-centrode.

At any instant let C_1 and C_1' coincide so that the body is for the instant moving about C_1 as centre. When the body has turned through the small angle θ the point C_2' coincides with C_2 and becomes the new centre of rotation; a rotation about C_2 through a small angle brings C_3' to C_3 and then a small rotation about C_3 brings C_4' to C_4 and so on.

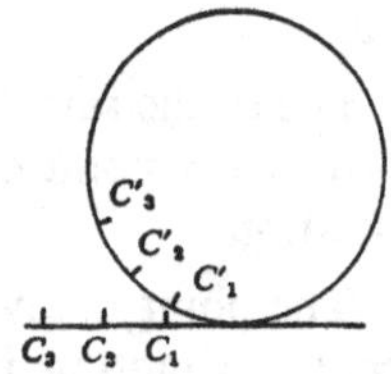

In the case of the wheel the points C_1', C_2' ... lie on the wheel and the points C_1, C_2 ... on the ground.

Ex. 1. *Rod sliding on a plane with its ends on two perpendicular straight lines CX and CY.*

At A and B draw perpendiculars to CX and CY and let them meet in O. The motions of A and B are instantaneously along AX and BC, so that O is the instantaneous centre of rotation.

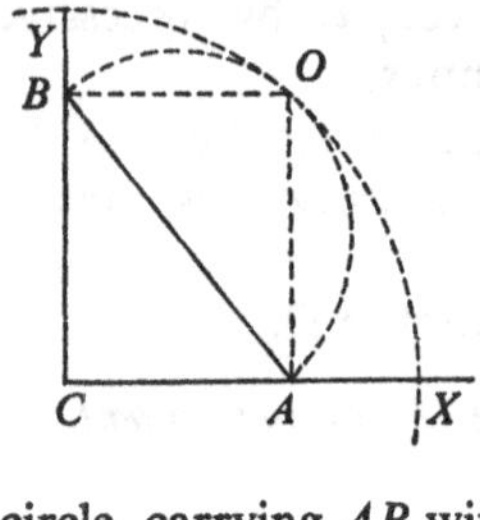

Since BOA is a right angle, the locus of O with respect to the body is a circle on AB as diameter, and thus the body-centrode is a circle of radius $\frac{1}{2}.AB$.

Since $CO=AB$, the locus of O in space is a circle of centre C and radius AB. Hence the motion is given by the rolling of the smaller circle, carrying AB with it, upon the outer circle of double its size, the point of contact of the two circles being the instantaneous centre.

Ex. 2. *The end A of a given rod is compelled to move on a given straight line CY, whilst the rod itself always passes through a fixed point B.*

Draw BC $(=a)$ perpendicular to CY. The instantaneous motion of A is along CY, so that the instantaneous centre O lies on the perpendicular AO.

The point B of the rod is for the moment moving in the direction AB, so that O lies on the perpendicular OB to AB.

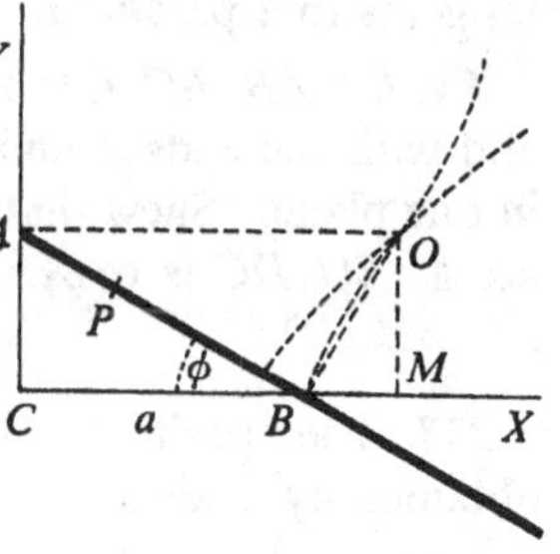

Body-Centrode. By similar triangles OAB, ABC we have

$$\frac{AB}{AO}=\frac{a}{AB}; \therefore AO=\frac{a}{\cos^2 OAB},$$

so that with respect to the body the locus of O is the curve

$$r=\frac{a}{\cos^2\theta} \quad \text{..............................(1).}$$

Space-Centrode. If OM be perpendicular to CB, and $CM=x$, $MO=y$, then

$$x=a+y\cot OBM=a+y\tan\phi, \quad \text{and} \quad y=CA=a\tan\phi.$$

Therefore the locus of O in space is the parabola $y^2=a(x-a)$.

The motion is therefore given by the rolling of the curve (1), carrying the rod with it, upon the parabola.

This motion is sometimes known as *Conchoidal Motion*; for every fixed point P on the rod clearly describes a conchoid whose pole is B.

Ex. 3. Obtain the position of the centre of instantaneous motion, and the body- and space-centrodes in the following cases:

(i) A rod AB moves with its ends upon two fixed straight lines not at right angles.

(ii) A rod AB moves so that its end A describes a circle, of centre O and radius a, whilst B is compelled to move on a fixed straight line passing through O. [Connecting rod motion.] Compare the velocities of A and B.

(iii) Two rods AB, BD are hinged at B; AB is hinged to a fixed point at A and revolves round A; BD always passes through a small fixed ring at C which is free to rotate about C. [Oscillating cylinder motion.]

(iv) The middle point G of a rod AB is forced to move on a given circle whilst at the same time the rod passes through a small ring at a fixed point C of the circle, the ring being free to rotate.

[Hence show that in a limaçon the locus of the intersection of normals at the ends of a focal chord is a circle.]

Ex. 4. The arms AC, CB of a wire bent at right angles slide upon two fixed circles in a plane. Shew that the locus of the instantaneous centre in space is a circle, and that its locus in the body is a circle of double the radius of the space-centrode.

Ex. 5. A straight thin rod moves in any manner in a plane; shew that, at any instant, the directions of motion of all its particles are tangents to a parabola.

Ex. 6. AB, BC, CD are three bars connected by joints at B and C, and with the ends A and D fixed, and the bars are capable of motion in one plane. Shew that the angular velocities of the rods AB and CD are as $BO.DC$ is to $AB.CO$, where O is the point of intersection of AB and CD.

213. The position of the instantaneous centre may be easily obtained by analysis.

Let u and v be the velocities parallel to the axes of the centre of mass G of the body and ω the angular velocity about G. Then the velocities of any point P, whose coordinates referred to G are x and y and such that PG is inclined at θ to the axis of x, are

$$u - PG.\sin\theta.\omega \quad \text{and} \quad v + PG.\cos\theta.\omega \text{ parallel to the axes,}$$

i.e.
$$u - y\omega \quad \text{and} \quad v + x\omega.$$

These are zero if $x = -\frac{v}{\omega}$ and $y = \frac{u}{\omega}$.

The coordinates of the centre of no acceleration are also easily found. For the accelerations of any point P relative to G are $PG.\omega^2$ along PG and $PG.\dot{\omega}$ perpendicular to PG.

Therefore the acceleration of P parallel to OX

$$= \dot{u} - PG.\omega^2.\cos\theta - PG.\dot{\omega}\sin\theta = \dot{u} - \omega^2 x - \dot{\omega} y,$$

and its acceleration parallel to Oy

$$= \dot{v} - PG.\omega^2.\sin\theta + PG.\dot{\omega}\cos\theta = \dot{v} - \omega^2 y + \dot{\omega} x.$$

These vanish at the point

$$\frac{x}{\dot{u}\omega^2 - \dot{v}\dot{\omega}} = \frac{y}{\dot{v}\omega^2 + \dot{u}\dot{\omega}} = \frac{1}{\omega^4 + \dot{\omega}^2}.$$

214. The point P, whose coordinates referred to G are (x, y), being the instantaneous centre and L the moment of the forces about it, the equation of Art. 192 gives

$$L = M[k^2\dot{\omega} + y\dot{u} - x\dot{v}],$$

where Mk^2 is the moment of inertia about G,

$$= M\left[k^2\dot{\omega} + \frac{u\dot{u} + v\dot{v}}{\omega}\right] = \frac{M}{2\omega}\frac{d}{dt}[k^2\omega^2 + u^2 + v^2].$$

Now, since P is the instantaneous centre,

$$u^2+v^2=PG^2.\omega^2.$$

$$\therefore\ L=\frac{M}{2\omega}\frac{d}{dt}[k^2\omega^2+PG^2\omega^2]=\frac{M}{2\omega}\frac{d}{dt}[k_1{}^2\omega^2]\text{ ....(1)},$$

where k_1 is the radius of gyration about the instantaneous centre.

(1) If the instantaneous centre be fixed in the body, so that $k_1{}^2$ is constant, this quantity $=Mk_1{}^2\dot{\omega}$.

(2) If $PG\ (=r)$ be not constant, the quantity (1)

$$=\frac{M}{2\omega}\frac{d}{dt}\{(k^2+r^2)\omega^2\}=M(k^2+r^2)\dot{\omega}+\frac{M}{2\omega}.2r\dot{r}\omega^2$$

$$=Mk_1{}^2\dot{\omega}+Mr\dot{r}\omega.$$

Now if, as in the case of a small oscillation, the quantities $\dot{r}$ and ω are such that their squares and products can be neglected, this quantity becomes $Mk_1{}^2\dot{\omega}$, so that in the case of a small oscillation the equation of moments of momentum about the instantaneous centre reduces to

$$\dot{\omega}=\frac{L}{Mk_1{}^2}=\frac{\text{moment of momentum about the instantaneous centre } I}{\text{moment of inertia about } I},$$

the squares of small quantities being neglected,

i.e. as far as small oscillations are concerned we may treat the instantaneous centre as if it were fixed in space.

215. Motion in three dimensions.

One point O of a rigid body being fixed, to shew that the body may be transferred from one position into any other position by a rotation about a suitable axis.

Let the radii from O to any two given points α, β of a body meet any spherical surface, of centre O, in the points A and B, and when the body has been moved into a second position let A and B go to A' and B' respectively.

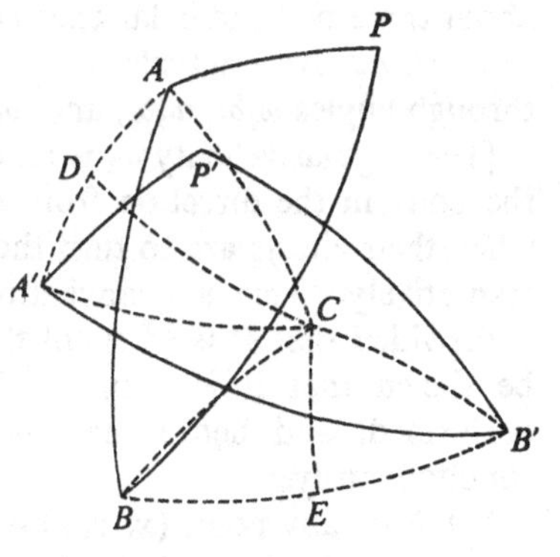

Bisect AA' and BB' in D and E and let great circles through D and E perpendicular to AA' and BB' meet in C.

Then

$CA=CA'$, $CB=CB'$, and $AB=A'B'$.

$\therefore\ \angle ACB=\angle A'CB'$.

$\therefore\ \angle ACA'=\angle BCB'$, so that the same rotation about OC which brings A to A' will bring B to B'.

Now the position of any rigid body is given when three points of it are given, and as the three points O, A, B have been brought into their second positions O, A', B' by the same rotation about OC, it follows that *any* other point P will be brought into its second position by the same rotation.

216. Next, remove the restriction that O is to be fixed, and take the most general motion of the body. Let O' be the position of O in the second position of the body.

Give to the whole body the translation, without any rotation, which brings O to O'. O' being now kept fixed, the same rotation about some axis $O'C$, which brings A and B into their final positions, will bring any other point of the body into its final position.

Hence, generally, every displacement of a rigid body is compounded of, and is equivalent to, (1) some motion of translation whereby every particle has the same translation as any assumed point O, and (2) some motion of rotation about some axis passing through O.

These motions are clearly independent, and can take place in either order or simultaneously.

217. *Angular velocities of a body about more than one axis. Indefinitely small rotations.*

A body has an angular velocity ω about an axis when every point of the body can be brought from its position at time t to its position at time $t+\delta t$ by a rotation round the axis through an angle $\omega\delta t$.

When a body is said to have three angular velocities ω_1, ω_2, and ω_3 about three perpendicular axes Ox, Oy, and Oz it is meant that during three successive intervals of time δt the body is turned in succession through angles $\omega_1\delta t$, $\omega_2\delta t$, and $\omega_3\delta t$ about these axes.

[The angular velocity ω_1 is taken as positive when its effect is to turn the body in the direction from Oy to Oz; so ω_2 and ω_3 are positive when their effects are to turn the body from Oz to Ox, and Ox to Oy respectively. This is a convention always adopted.]

Provided that δt is so small that its square may be neglected it can be shewn that it is immaterial in what order these rotations are performed, and hence that they can be considered to take place simultaneously.

Let P be any point (x, y, z) of a body; draw PM perpendicular to Ox and let PM be inclined at an angle θ to the plane xOy so that

$$y = MP\cos\theta,\ z = MP\sin\theta.$$

Let a rotation $\omega_1\delta t$ be made about Ox so that P goes to P' whose coordinates are

$$x,\ y+\delta y,\ z+\delta z.$$

Then $y+\delta y = MP\cos(\theta+\omega_1\delta t)$

$$= MP(\cos\theta - \sin\theta\,.\,\omega_1\delta t) = y - z\omega_1\delta t,$$

powers of δt above the first being neglected.

So $z+\delta z = MP\sin(\theta+\omega_1\delta t)$

$$= MP(\sin\theta + \cos\theta\,.\,\omega_1\delta t) = z + y\omega_1\delta t.$$

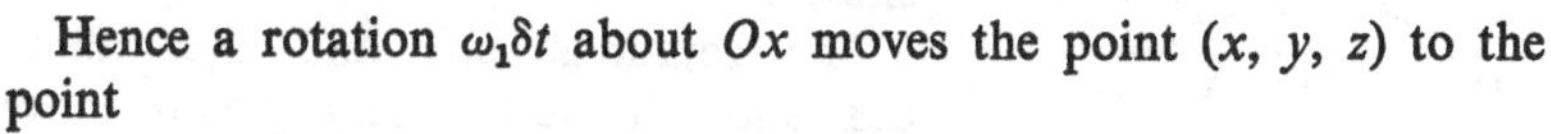

Hence a rotation $\omega_1\delta t$ about Ox moves the point (x, y, z) to the point

$$(x,\ y - z\omega_1\delta t,\ z + y\omega_1\delta t)\ \ldots\ldots\ldots\ldots\ldots\ldots(1).$$

So a rotation $\omega_2\delta t$ about Oy would move the point (x, y, z) to the point

$$(x + z\omega_2\delta t,\ y,\ z - x\omega_2\delta t)\ \ldots\ldots\ldots\ldots\ldots\ldots(2).$$

Also a rotation $\omega_3\delta t$ about Oz would move the point (x, y, z) to the point

$$(x - y\omega_3\delta t,\ y + x\omega_3\delta t,\ z)\ \ldots\ldots\ldots\ldots\ldots\ldots(3).$$

218. Now perform the three rotations, about the perpendicular axes Ox, Oy, Oz, of magnitudes $\omega_1\delta t$, $\omega_2\delta t$, $\omega_3\delta t$ respectively in succession.

By (1) the rotation $\omega_1\delta t$ takes the point $P\,(x, y, z)$ to the point P_1, *viz.*

$$(x,\ y - z\omega_1\delta t,\ z + y\omega_1\delta t).$$

By (2) the rotation $\omega_2\delta t$ takes P_1 to the point P_2, *viz.*

$$[x + (z + y\omega_1\delta t)\omega_2\delta t,\ y - z\omega_1\delta t,\ z + y\omega_1\delta t - x\omega_2\delta t],$$

i.e. $$[x + z\omega_2\delta t,\ y - z\omega_1\delta t,\ z + (y\omega_1 - x\omega_2)\delta t],$$

on neglecting squares of δt.

Finally the rotation $\omega_3\delta t$ about Oz takes P_2 to the point P_3, *viz.*

$$[x + z\omega_2\delta t - (y - z\omega_1\delta t)\omega_3\delta t,\ y - z\omega_1\delta t + (x + z\omega_2\delta t)\omega_3\delta t,\ z + (y\omega_1 - x\omega_2)\delta t],$$

i.e. P_3 is the point

$$[x + (z\omega_2 - y\omega_3)\delta t,\ y + (x\omega_3 - z\omega_1)\delta t,\ z + (y\omega_1 - x\omega_2)\delta t],$$

on again neglecting squares of δt.

The symmetry of the final result shews, that, if the squares of δt be neglected, the rotations about the axes might have been made in any order.

Hence *when a body has three instantaneous angular velocities the rotations may be treated as taking place in any order and therefore as taking place simultaneously.*

If the rotations are of finite magnitude, this statement is not correct, as will be seen in Art. 225.

219. *If a body possesses two angular velocities ω_1 and ω_2, about two given lines which are represented in magnitude by distances OA and OB measured along these two lines, then the resultant angular velocity is about a line OC, where OACB is a parallelogram, and will be represented in magnitude by OC.*

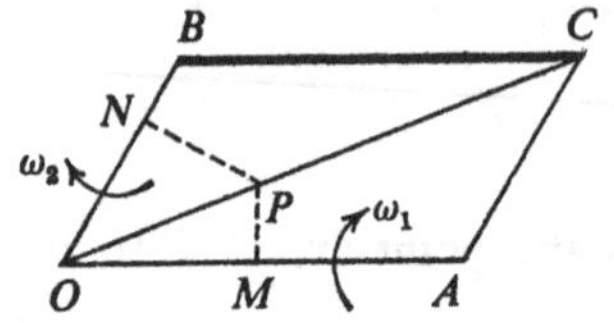

Consider any point P lying on OC and draw PM and PN perpendicular to OA and OB.

The rotations $\omega_1 \delta t$ and $\omega_2 \delta t$ about OA and OB respectively would move P through a small distance perpendicular to the plane of the paper which

$$= -PM.\omega_1 \delta t + PN.\omega_2 \delta t$$

$$= \lambda[-PM.OA + PN.OB]\delta t = 2\lambda[-\triangle POA + \triangle POB]\delta t = 0.$$

Hence P, and similarly any point on OC, is at rest.

Hence OC must be the resultant axis of rotation; for we know, by Art. 215, that there is always one definite axis of rotation for any motion.

If ω be the resultant angular velocity about OC, then the motion of any point, A say, will be the same whether we consider it due to the motion about OC, or about OA and OB together.

Hence $\omega \times$ perpendicular from A on $OC = \omega_2 \times$ perpendicular from A on OB.

$$\therefore \omega \times OA \sin AOC = \omega_2 \times OA \sin AOB.$$

$$\therefore \frac{\omega}{\omega_2} = \frac{\sin AOB}{\sin AOC} = \frac{\sin OAC}{\sin AOC} = \frac{OC}{AC} = \frac{OC}{OB}.$$

Hence on the same scale that ω_2 is represented by OB, or ω_1 by OA, the resultant angular velocity ω about OC is represented by OC.

Angular velocities are therefore compounded by the same rules as forces or linear velocities, *i.e.* they follow the Parallelogram Law.

Similarly, as in Statics and Elementary Dynamics, the Parallelogram of Angular Accelerations and the Parallelepiped of Angular Velocities and Accelerations would follow.

Hence an angular velocity ω about a line OP is equivalent to an angular velocity $\omega \cos \alpha$ about Ox, where $xOP = \alpha$, and an angular velocity $\omega \sin \alpha$ about a perpendicular line.

Also angular velocities ω_1, ω_2, and ω_3 about three rectangular axes Ox, Oy, and Oz are equivalent to an angular velocity $\omega(=\sqrt{\omega_1^2+\omega_2^2+\omega_3^2})$ about a line whose direction-cosines are $\frac{\omega_1}{\omega}$, $\frac{\omega_2}{\omega}$, and $\frac{\omega_3}{\omega}$.

220. *A body has angular velocities, ω_1 and ω_2, about two parallel axes; to find the motion.*

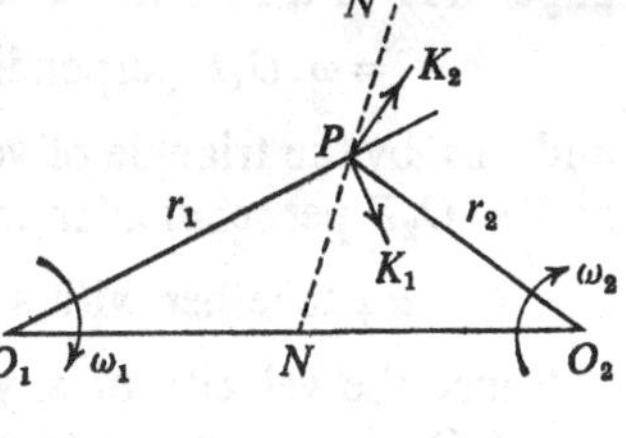

Take the plane of the paper through any point P of the body perpendicular to the two axes, meeting them in O_1 and O_2.

Then the velocities of P are $r_1\omega_1$ and $r_2\omega_2$ along PK_1 and PK_2 perpendicular to O_1P and O_2P respectively.

Take N on O_1O_2, such that $\omega_1 . O_1N=\omega_2 . NO_2$.

The velocities of P are $\omega_1 . PO_1$ and $\omega_2 . PO_2$ perpendicular to PO_1 and PO_2 respectively.

Hence by the ordinary rule their resultant is $(\omega_1+\omega_2)PN$ perpendicular to PN, *i.e.* P moves as it would if it had an angular velocity $(\omega_1+\omega_2)$ about N.

Hence two angular velocities ω_1 and ω_2 about two parallel axes O_1 and O_2 are equivalent to an angular velocity $\omega_1+\omega_2$ about an axis which divides the distance O_1O_2 inversely in the ratio of ω_1 to ω_2.

221. If the angular velocities are unlike and $\omega_1>\omega_2$ numerically, then N divides O_1O_2 externally so that $\omega_1 . O_1N=\omega_2 . O_2N$, and the resultant angular velocity is $\omega_1-\omega_2$.

Exceptional case. If the angular velocities are unlike and numerically equal, N is at infinity and the resultant angular velocity is zero.

The resultant motion is then a linear velocity. For, in this case, the velocities of P are perpendicular and proportional to O_1P and PO_2, and hence its resultant velocity is perpendicular and proportional to O_1O_2, *i.e.* it is

$$\omega_1 . O_1O_2 \downarrow .$$

Aliter. The velocity of P parallel to O_1O_2

$$=\omega_1 . O_1P \sin PO_1O_2-\omega_1 . O_2P . \sin PO_2O_1=0,$$

and its velocity perpendicular to O_1O_2

$$=\omega_1 . O_1P \cos PO_1O_2+\omega_1 . O_2P . \cos PO_2O_1=\omega_1 . O_1O_2 \downarrow .$$

222. *An angular velocity ω about an axis is equivalent to an angular velocity ω about a parallel axis distant a from the former together with a linear velocity $\omega . a$.*

Let the two axes meet the plane of the paper in O_1 and O_2 and be perpendicular to it.

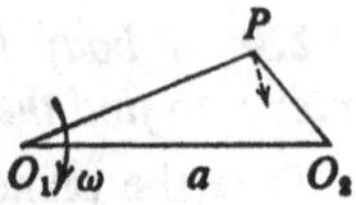

The velocity of any point P in the plane of the paper due to a rotation ω about O_1

$$= \omega . O_1P \text{ perpendicular to } O_1P,$$

and this, by the triangle of velocities, is equivalent to velocities $\omega . O_1O_2$ and $\omega . O_2P$ perpendicular to O_1O_2 and O_2P in the same sense

$$= \omega . a \downarrow \text{ together with a velocity } \omega . O_2P \text{ perpendicular to } O_2P.$$

Hence the velocity of any point P, given by an angular velocity ω about O_1, is equivalent to that given by an equal angular velocity ω about O_2, together with a linear velocity $\omega . O_1O_2$ perpendicular to O_1O_2.

223. In practice the results of Arts. 220—222 are remembered most easily by taking the point P on O_1O_2;

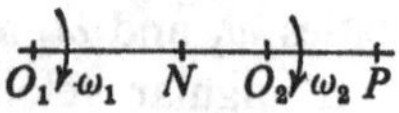

Thus (1) the velocity of $P = \omega_1 . O_1P + \omega_2 . O_2P$

$$= \omega_1(O_1O_2 + O_2P) + \omega_2 . O_2P = (\omega_1 + \omega_2)\left(O_2P + \frac{\omega_1}{\omega_1 + \omega_2} . O_1O_2\right)$$

$$= (\omega_1 + \omega_2) . NP, \text{ where } NO_2 = \frac{\omega_1}{\omega_1 + \omega_2} . O_1O_2.$$

(2) The velocity of $P = \omega_1 . O_1P - \omega_2 . O_2P$

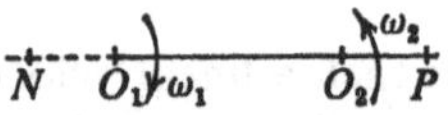

$$= \omega_1(O_1O_2 + O_2P) - \omega_2 . O_2P = (\omega_1 - \omega_2)\left[O_2P + \frac{\omega_1}{\omega_1 - \omega_2} . O_1O_2\right]$$

$$= (\omega_1 - \omega_2) . NP, \text{ where } O_2N = \frac{\omega_1}{\omega_1 - \omega_2} . O_1O_2.$$

(3) The velocity of $P = \omega . O_1P - \omega . O_2P$

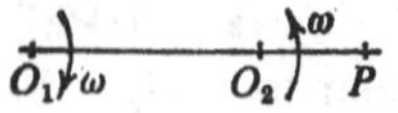

$= \omega . O_1O_2 =$ a constant velocity perpendicular to O_1O_2.

(4) The velocity of $P=\omega.O_1P=\omega.O_1O_2+\omega.O_2P$, and is therefore equivalent to a

O_1 ω O_2 P

linear velocity $\omega.O_1O_2$ perpendicular to O_1O_2 together with an angular velocity ω about O_2.

224. ***To shew that the instantaneous motion of a body may be reduced to a twist, i.e. to a linear velocity along a certain line together with an angular velocity about the line.***

By Art. 216 the instantaneous motion of a rigid body is equivalent to a translational velocity of any point O together with an angular velocity about a straight line passing through O.

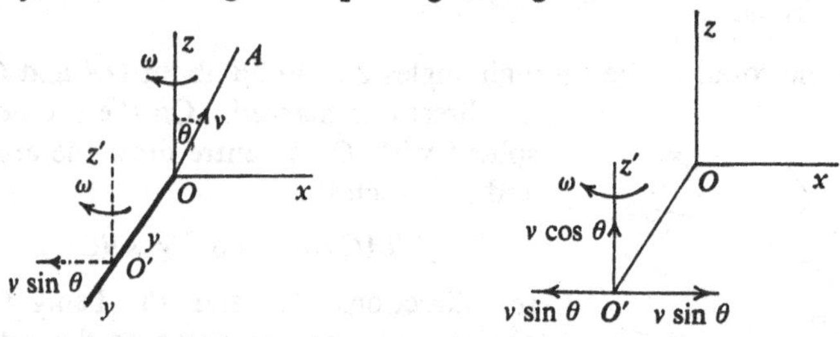

Let OA be the direction of this linear velocity v, and Oz the axis of the angular velocity ω.

In the plane zOA draw Ox at right angles to Oz and draw Oy at right angles to the plane zOx. Let $\angle zOA=\theta$.

On Oy take OO' such that $OO'.\omega=v\sin\theta$.

Then, by Art. 222, the angular velocity ω about Oz is equivalent to ω about a parallel axis $O'z'$ to Oz together with a linear velocity $\omega.OO'$, *i.e.* $v\sin\theta$, through O perpendicular to the plane zOO'.

Also a linear velocity v may be transferred to a parallel linear velocity through O', and then resolved into two velocities $v\cos\theta$ and $v\sin\theta$. We thus obtain the second figure.

In it the two linear velocities $v\sin\theta$ destroy one another, and we have left the motion consisting of a linear velocity $v\cos\theta$ along $O'z'$ and an angular velocity ω about it.

This construction is clearly similar to that for Poinsot's Central Axis in Statics; and properties similar to those for Poinsot's Central Axis follow.

It will be noticed that, in the preceding constructions, an angular velocity corresponds to a force in Statics, and a linear velocity corresponds to a couple.

225. Finite Rotations. If the rotations are through finite angles it is easily seen that the order of the rotations about the axes is important. As a simple case suppose the body to be rotated through a right angle about each of two perpendicular axes Ox and Oy.

The rotation through a right angle about Ox would bring any point P on Oz to a position on the negative axis of y, and a second rotation about Oy would not further alter its position.

A rotation, first about the axis of y, would have brought P to a position on the axis of x, and then a second rotation about Ox would not have had any effect on its position.

Thus in the case of finite rotations their order is clearly material.

226. *To find the effect of two finite rotations about axes OA and OB in succession.*

Let the rotations be through angles 2α and 2β about OA and OB in the directions marked. On the geometrical sphere with O as centre draw the arcs AC and BC, such that

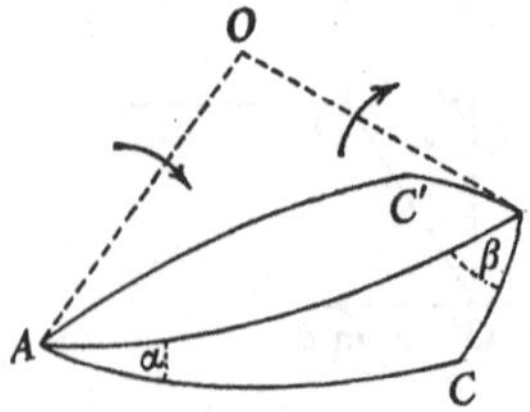

$$\angle BAC = \alpha \quad \text{and} \quad \angle ABC = \beta.$$

the directions AC and BC being taken, one in the same direction as the rotation about OA, and the second in the opposite direction to the rotation about OB.

Take C' on the other side of AB, symmetrical with C, so that $\angle CAC' = 2\alpha$ and $\angle CBC' = 2\beta$.

A rotation of the body through an angle 2α about OA would bring OC' into the position OC, and a second rotation 2β about OB would bring OC back again into the position OC'.

Hence the effect of the two component rotations would be that the position of OC' is unaltered, *i.e.* OC' is the resultant axis of rotation.

[If the rotations had been first about OB and secondly about OA it is clear, similarly, that OC would have been the resultant axis of rotation.]

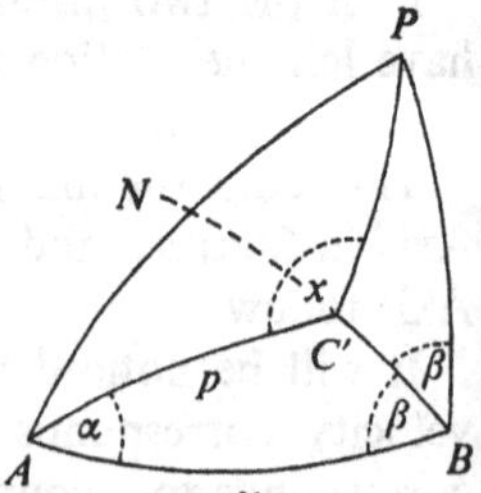

Magnitude of the resultant rotation.

The point A is unaltered by a rotation about OA; the rotation 2β about OB takes it to the point P, where $\angle ABP = 2\beta$ and the arc BP = the arc BA, and therefore

$$\angle BAP = \angle BPA.$$

Hence the resultant rotation is through an angle $AC'P(=x)$ about C', and $C'A=C'P$.

If BC' meets AP in N, then N is the middle point of the arc AP and $AC'N=NC'P=\frac{x}{2}$.

If the axes OA and OB meet at an angle γ, then $AB=\gamma$. Let AC' be p.

Then
$$\sin\gamma\sin\beta=\sin AN=\sin p\sin\frac{x}{2}\text{.................(1).}$$

Also, from the triangle ABC', we have
$$\cos\gamma\cos\alpha=\sin\gamma\cot p-\sin\alpha\cot\beta,$$
which gives
$$\sin p=\frac{1}{\sqrt{1+\cot^2 p}}=\frac{\sin\gamma}{\sqrt{\sin^2\gamma+(\cos\alpha\cos\gamma+\sin\alpha\cot\beta)^2}}.$$

Hence (1) gives
$$\sin\frac{x}{2}=\sin\beta\sqrt{\sin^2\gamma+(\cos\gamma\cos\alpha+\sin\alpha\cot\beta)^2}.$$

Hence the position of the resultant axis OC', and the magnitude of the resultant rotation, are given for any case.

Ex. 1. If a plane figure be rotated through 90° about a fixed point A, and then through 90° (in the same sense) about a fixed point B, the result is equivalent to a rotation of 180° about a certain fixed point C; find the position of C.

Ex. 2. Find the resultant rotation when a body revolves through a right angle in succession about two axes which are inclined to one another at an angle of 60°.

Ex. 3. When the rotations are each through two right angles, shew that the resultant axis of rotation is perpendicular to the plane through the two component axes, and that the resultant angle of rotation is equal to twice the angle between them.

227. *Velocity of any point of a body parallel to fixed axes in terms of the instantaneous angular velocities of the body about the axes.*

Let P be any point (x, y, z) of the body. Draw PM perpendicular to the plane of xy, MN perpendicular to the axis of x, and PT perpendicular to NP in the plane NPM to meet NM in T.

The angular velocity ω_1 about Ox gives to P a velocity along TP equal to $\omega_1 . PN$ which is equivalent to a velocity $-\omega_1 . PN \cos PTN$, *i.e.* $-\omega_1 . PN . \sin PNT$, *i.e.* $-\omega_1 . z$ along NT, and a velocity $\omega_1 . PN \sin PTN$, *i.e.* $\omega_1 . PN \cos PNM$, *i.e.* $\omega_1 . y$ along MP.

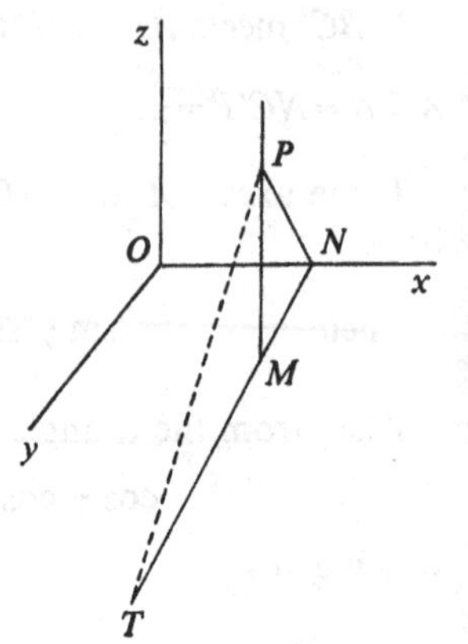

Hence the ω_1-rotation about Ox gives a component velocity $-\omega_1 . z$ parallel to Oy and $\omega_1 . y$ parallel to Oz.

So the rotation about Oy by symmetry gives component velocities $-\omega_2 . x$ parallel to Oz and $\omega_2 . z$ parallel to Ox.

Finally the rotation about Oz gives $-\omega_3 . y$ parallel to Ox and $\omega_3 . x$ parallel to Oy.

Summing up, the component velocities are

$$\omega_2 . z - \omega_3 . y \text{ parallel to } Ox,$$
$$\omega_3 . x - \omega_1 . z \quad ,, \quad ,, \; Oy,$$
and
$$\omega_1 . y - \omega_2 . x \quad ,, \quad ,, \; Oz.$$

If O be at rest, these are the component velocities of P parallel to the axes.

If O be in motion and u, v, w are the components of its velocity parallel to the fixed axes of coordinates, then the component velocities of P in space are

$$u + \omega_2 . z - \omega_3 . y \text{ parallel to } Ox,$$
$$v + \omega_3 . x - \omega_1 . z \quad ,, \quad ,, \; Oy,$$
and
$$w + \omega_1 . y - \omega_2 . x \quad ,, \quad ,, \; Oz.$$

228. *A rigid body is moving about a fixed point O; to find* (1) *the moments of momentum about any axes through O fixed in space, and* (2) *the kinetic energy of the body.*

The moment of momentum of the body about the axis of x

$$= \Sigma m \left(y \frac{dz}{dt} - z \frac{dy}{dt} \right).$$

But, by the previous article, since O is fixed,

$$\frac{dy}{dt} = \omega_z . x - \omega_x . z \quad \text{and} \quad \frac{dz}{dt} = \omega_x . y - \omega_y . x$$

where ω_x, ω_y, and ω_z are the angular velocities of the body about the axes.

On substitution, the moment of momentum about Ox

$$= \Sigma m[(y^2 + z^2)\omega_x - xy\omega_y - zx\omega_z] = A . \omega_x - F . \omega_y - E . \omega_z.$$

Similarly the moment of momentum about Oy

$$=B\omega_y-D\omega_z-F\omega_x,$$

and that about Oz

$$=C\omega_z-E\omega_x-D\omega_y.$$

(2) The kinetic energy

$$=\tfrac{1}{2}\Sigma m[\dot{x}^2+\dot{y}^2+\dot{z}^2]$$
$$=\tfrac{1}{2}\Sigma m[(\omega_y.z-\omega_z.y)^2+(\omega_z.x-\omega_x.z)^2+(\omega_x.y-\omega_y.x)^2]$$
$$=\tfrac{1}{2}\Sigma m[\omega_x^2(y^2+z^2)+\ldots+\ldots-2\omega_y\omega_z.yz-\ldots-\ldots]$$
$$=\tfrac{1}{2}(A\omega_x^2+B\omega_y^2+C\omega_z^2-2D\omega_y\omega_z-2E\omega_z\omega_x-2F\omega_x\omega_y).$$

229. In the previous article the axes are fixed in space, and therefore since the body moves with respect to them, the moments and products of inertia A, B, C ... are in general variable.

Other formulæ, more suitable for many cases, may be obtained as follows.

Let Ox', Oy', and Oz' be three axes *fixed in the body*, and therefore not in general fixed in space, passing through O and let ω_1, ω_2, ω_3 be the angular velocities of the body about them.

The fixed axes Ox, Oy, and Oz are any whatever, but let them be so chosen that at the instant under consideration the moving axes Ox', Oy', and Oz' coincide with them. Then $\omega_x=\omega_1$, $\omega_y=\omega_2$, $\omega_z=\omega_3$.

The expression for the moments of momentum of the last article are now $A\omega_1-F\omega_2-E\omega_3$ and two similar expressions, and the kinetic energy is

$$\tfrac{1}{2}(A\omega_1^2+B\omega_2^2+C\omega_3^2-2D\omega_2\omega_3-2E\omega_3\omega_1-2F\omega_1\omega_2),$$

where A, B, C are now the moments of inertia and D, E, F the products of inertia about axes fixed in the body and moving with it.

If these latter axes are the principal axes at O, then D, E, and F vanish and the expressions for the component moments of momentum are $A\omega_1$, $B\omega_2$, and $C\omega_3$, and that for the kinetic energy is

$$\tfrac{1}{2}(A\omega_1^2+B\omega_2^2+C\omega_3^2).$$

230. *General equations of motion of a body with one point fixed which is acted upon by given blows.*

The fixed point being the origin, let the axes be three rectangular axes through it.

Let ω_x, ω_y, ω_z be the angular velocities of the body about the axes just before the action of the blows, and ω_x', ω_y', ω_x' the corresponding quantities just after.

The moment of momentum of the body, just before, about the axis of x, is $A\omega_x-F\omega_y-E\omega_z$ and just after it is

$$A\omega_x'-F\omega_y'-E\omega_z'.$$

Hence the change in the moment of momentum about the axis of x is $A(\omega_x' - \omega_x) - F(\omega_y' - \omega_y) - E(\omega_z' - \omega_z)$.

But, by Art. 166, the change in the moment of momentum about any axis is equal to the moments of the blows about that axis.

If then L, M, N are the moments of the blows about the axes of x, y, and z, we have

$$A(\omega_x' - \omega_x) - F(\omega_y' - \omega_y) - E(\omega_z' - \omega_z) = L,$$

and similarly

$$B(\omega_y' - \omega_y) - D(\omega_z' - \omega_z) - F(\omega_x' - \omega_x) = M,$$

and $$C(\omega_z' - \omega_z) - E(\omega_x' - \omega_x) - D(\omega_y' - \omega_y) = N.$$

These three equations determine ω_x', ω_y', and ω_z'.

The axes of reference should be chosen so that A, B, C, D, E, and F may be most easily found. In general the principal axes are the most suitable. If they be taken as the axes of reference the equations become

$$A(\omega_x' - \omega_x) = L,\ B(\omega_y' - \omega_y) = M \text{ and } C(\omega_z' - \omega_z) = N.$$

231. If the body start from rest, so that ω_x, ω_y, and ω_z are zero, we have

$$\omega_x' = \frac{L}{A},\quad \omega_y' = \frac{M}{B},\quad \text{and } \omega_z' = \frac{N}{C}.$$

Hence the direction cosines of the instantaneous axis are proportional to

$$\left(\frac{L}{A}, \frac{M}{B}, \frac{N}{C}\right) \quad \text{.............................(1).}$$

The direction cosines of the axis of the impulsive couple are proportional to

$$(L, M, N) \quad \text{.............................(2).}$$

In general it is therefore clear that (1) and (2) are not the same, *i.e.* in general the body does not start to rotate about a perpendicular to the plane of the impulsive couple.

(1) and (2) coincide if $A = B = C$, in which case the momental ellipsoid at the fixed point becomes a sphere.

Again if $M = N = 0$, *i.e.* if the axis of the impulsive couple coincides with the axis of x, one of the principal axes at the fixed point, then the direction cosines (1) become proportional to (1, 0, 0), and the instantaneous axis also coincides with the axis of x, *i.e.* with the direction of the impulsive couple. Similarly if the axis of the impulsive couple coincides with either of the other two principal axes at the fixed point.

In the general case the instantaneous axis may be found geometrically. For the plane of the impulsive couple is

$$Lx+My+Nz=0.$$

Its conjugate diameter with respect to the momental ellipsoid

$$Ax^2+By^2+Cz^2=k$$

is easily seen to be $\dfrac{x}{\frac{L}{A}}=\dfrac{y}{\frac{M}{B}}=\dfrac{z}{\frac{N}{C}}$,

i.e. it is the instantaneous axis.

Hence *if an impulse couple act on a body, fixed at a point O and initially at rest, the body begins to turn about the diameter of the momental ellipsoid at O which is conjugate to the plane of the impulsive couple.*

232. *Ex.* 1. *A lamina in the form of a quadrant of a circle OHO′, whose centre is H, has one extremity O of its arc fixed and is struck by a blow P at the other extremity O′ perpendicular to its plane ; find the resulting motion.*

Take OH as the axis of x, the tangent at O as the axis of y, and a perpendicular to the plane at O as the axis of z.

Let G be the centre of gravity, GL perpendicular to OH, so that

$$HL=LG=\frac{4a}{3\pi}.$$

Then
$$A=M\frac{a^2}{4};$$

$$B\text{ (by Art. 147)}=M\frac{a^2}{4}-M.HL^2+M.OL^2=M\left(\frac{5}{4}-\frac{8}{3\pi}\right)a^2;$$

$$C=A+B;\quad D=E=0;$$

$$F=\int_0^{\frac{\pi}{2}}\int_0^a mr\,d\theta\,dr(a-r\cos\theta)r\sin\theta=M.\frac{5}{6\pi}a^2$$

The equations of Art. 230 then give

and
$$\left.\begin{aligned}A\omega_x'-F\omega_y'&=\ \ Pa\\ B\omega_y'-F\omega_x'&=-Pa\\ C\omega_z'&=\ \ 0\end{aligned}\right\}.$$

These give $\dfrac{\omega_x'}{B-F}=\dfrac{\omega_y'}{-(A-F)}=\dfrac{Pa}{AB-F^2}$, and $\omega_z'=0$, and the solution can be completed.

If ϕ be the inclination to Ox of the instantaneous axis, we have

$$\tan\phi=\frac{\omega_y'}{\omega_x'}=\frac{A-F}{F-B}=\frac{10-3\pi}{15\pi-42}.$$

Ex. 2. A uniform cube has its centre fixed and is free to turn about it; it is struck by a blow along one of its edges; find the instantaneous axis.

Ex. 3. A uniform solid ellipsoid is fixed at its centre and is free to turn about it. It is struck at a given point of its surface by a blow whose direction is normal to the ellipsoid. Find the equation to its instantaneous axis.

Ex. 4. A disc, in the form of a portion of a parabola bounded by its latus-rectum and its axis, has its vertex A fixed, and is struck by a blow through the end of its latus-rectum perpendicular to its plane. Shew that the disc starts revolving about a line through A inclined at $\tan^{-1}\frac{14}{25}$ to the axis.

Ex. 5. A uniform triangular lamina ABC is free to turn in any way about A which is fixed. A blow is given to it at B perpendicular to its plane. Shew that the lamina begins to turn about AD where D is a point on BC such that $CD=\frac{1}{3}CB$.

233. *General equations of motion of a body in three dimensions, referred to axes whose directions are fixed.*

If (x, y, z) be the coordinates of the centre of gravity of the body we have, by Art. 162, $M\frac{d^2x}{dt^2}=$ sum of the components of the impressed forces parallel to Ox, and similar equations for the motion parallel to the other axes.

If ω_x, ω_y, ω_z be the angular velocities at any instant about axes through the centre of inertia parallel to the axes of coordinates then, by Arts. 164 and 228, we have

$$\frac{d}{dt}[A\omega_x-F\omega_y-E\omega_z]$$

$=$ moment about a line parallel to Ox through G of the effective forces $=L$.

So $$\frac{d}{dt}[B\omega_y-D\omega_z-F\omega_x]=M,$$

and $$\frac{d}{dt}[C\omega_z-E\omega_x-D\omega_y]=N.$$

[If the body be a uniform sphere, of mass M_1, then $D=E=F=0$, and $A=B=C=M.\frac{2a^2}{5}$; these equations then become

$$M_1.\frac{2a^2}{5}\frac{d\omega_x}{dt}=L,\ M_1.\frac{2a^2}{5}\frac{d\omega_y}{dt}=M,\ \text{and}\ M_1.\frac{2a^2}{5}\frac{d\omega_z}{dt}=N.]$$

Impulsive forces. If u, v, w, ω_x, ω_y, ω_z be the component velocities of the centre of inertia G and the component angular velocities about lines through G parallel to the fixed axes of coordinates just before the action of the impulsive forces, and u', v', w', ω_x', ω_y', ω_z' similar quantities just after them, by Arts. 166 and 228 the dynamical equations are $M(u'-u)=X_1$, etc., and $A(\omega_x'-\omega_x)-F(\omega_y'-\omega_y)-E(\omega_z'-\omega_z)$

$$=\text{the moment of the impulsive forces about } Ox,$$

and two similar equations.

234. *Ex. A homogeneous billiard ball, spinning about any axis, moves on a billiard table which is not rough enough to always prevent sliding; to shew that the path of the centre is at first an arc of a parabola and then a straight line.*

Take as origin the initial position of the centre, and as axis of x the initial direction of sliding of the point of contact. If u and v be the initial velocities of the centre parallel to the axes, and Ω_x, Ω_y, and Ω_z the initial angular velocities about the axes, then since the initial velocity of the point of contact parallel to the axis of y is zero, we have

$$v+a\Omega_x=0 \quad\ldots\ldots\ldots\ldots\ldots(1).$$

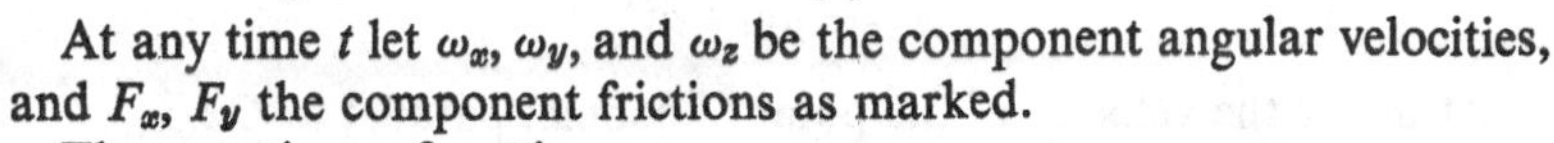

At any time t let ω_x, ω_y, and ω_z be the component angular velocities, and F_x, F_y the component frictions as marked.

The equations of motion are

$$\left.\begin{aligned} M\ddot{x}&=-F_x\\ M\ddot{y}&=-F_y\\ 0&=R-Mg\end{aligned}\right\}\quad\ldots\ldots\ldots\ldots\ldots(2),$$

and

$$\left.\begin{aligned} M.\frac{2a^2}{5}\frac{d\omega_x}{dt}&=-F_y.a\\ M.\frac{2a^2}{5}\frac{d\omega_y}{dt}&=F_x.a\\ M.\frac{2a^2}{5}\frac{d\omega_z}{dt}&=0\end{aligned}\right\}\quad\ldots\ldots\ldots\ldots\ldots(3).$$

The resultant friction must be opposite to the instantaneous motion of A and equal to μMg.

Hence
$$\frac{F_y}{F_x}=\frac{\dot{y}+a\omega_x}{\dot{x}-a\omega_y} \quad \text{.............................(4)},$$

and
$$F_x^2+F_y^2=\mu^2M^2g^2 \quad \text{.........................(5)}.$$

Equations (2) and (3) give
$$\frac{F_y}{F_x}=\frac{\ddot{y}}{\ddot{x}}=\frac{-\dot{\omega}_x}{\dot{\omega}_y}=\frac{\ddot{y}+a\dot{\omega}_x}{\ddot{x}-a\dot{\omega}_y}.$$

Hence (4) gives
$$\frac{\ddot{y}+a\dot{\omega}_x}{\dot{y}+a\omega_x}=\frac{\ddot{x}-a\dot{\omega}_y}{\dot{x}-a\omega_y}.$$

$$\therefore \quad \log(\dot{y}+a\omega_x)=\log(\dot{x}-a\omega_y)+\text{const.}$$

$$\therefore \quad \frac{\dot{y}+a\omega_x}{\dot{x}-a\omega_y}=\frac{v+a\Omega_x}{u-a\Omega_y}=0, \text{ by (1).}$$

Hence (4) and (5) give $F_y=0$ and $F_x=\mu Mg$.

From (2) it follows that the centre moves under the action of a constant force parallel to the axis of x and hence it describes an arc of a parabola, whose axis lies along the negative direction of the axis of x.

(1), (2), and (3) now give
$$\left.\begin{aligned}\dot{x}&=-\mu gt+u\\ \dot{y}&=\text{const.}=v\end{aligned}\right\} \quad \text{.............................(6)},$$

and
$$\left.\begin{aligned}a\omega_x&=\text{const.}=a\Omega_x\\ a\omega_y&=\frac{5}{2}\mu gt+a\Omega_y\end{aligned}\right\} \quad \text{.........................(7)}.$$

At time t the velocity of the point of contact parallel to Ox
$$=\dot{x}-a\omega_y=u-a\Omega_y-\frac{7}{2}\mu gt,$$

and parallel to Oy it $=\dot{y}+a\omega_x=v+a\Omega_x=0$, by (1).

The velocity of the point of contact vanishes and pure rolling begins when
$$t=\frac{2}{7\mu g}(u-a\Omega_y),$$

and then
$$\frac{\dot{y}}{\dot{x}}=\frac{v}{u-\mu gt}=\frac{7v}{5u+2a\Omega_y},$$

i.e. the direction of motion when pure rolling commences is inclined at $\tan^{-1}\dfrac{7v}{5u+2a\Omega_y}$ to the original direction of motion of the point of contact.

On integrating (6), it is easily seen that pure rolling commences at the point whose coordinates are

$$\frac{2(u-a\Omega_y)(6u+a\Omega_y)}{49\mu g} \text{ and } \frac{2v(u-a\Omega_y)}{7\mu g}.$$

It is easily seen that the motion continues to be one of pure rolling, and the motion of the centre is now in a straight line.

EXAMPLES

1. If a homogeneous sphere roll on a fixed rough plane under the action of any forces, whose resultant passes through the centre of the sphere, shew that the motion is the same as if the plane were smooth and the forces reduced to five-sevenths of their given value.

2. A sphere is projected obliquely up a perfectly rough plane; shew that the equation of the path of the point of contact of the sphere and plane is $y=x\tan\beta-\dfrac{5}{14}\dfrac{gx^2}{V^2}\dfrac{\sin\alpha}{\cos^2\beta}$, where α is the inclination of the plane to the horizon, and V is the initial velocity at an angle β to the horizontal line in the plane.

3. A homogeneous sphere is projected, so as to roll, in any direction along the surface of a rough plane inclined at α to the horizontal; shew that the coefficient of friction must be $>\frac{2}{7}\tan\alpha$.

4. A perfectly rough sphere, of mass M and radius a, is rotating with angular velocity Ω about an axis at right angles to the direction of motion of its centre. It impinges directly on another rough sphere of mass m which is at rest. Shew that after separation the component velocities of the two spheres at right angles to the original direction of motion of the first sphere are respectively $\dfrac{2}{7}\dfrac{m}{M+m}a\Omega$ and $\dfrac{2}{7}\dfrac{M}{M+m}a\Omega$.

5. A homogeneous sphere spinning about its vertical axis moves on a smooth horizontal table and impinges directly on a perfectly rough vertical cushion. Shew that the kinetic energy of the sphere is diminished by the impact in the ratio $2e^2(5+7\tan^2\theta):10+49e^2\tan^2\theta$, where e is the coefficient of restitution of the ball and θ is the angle of reflection.

6. A sphere, of radius a, rotating with angular velocity ω about an axis inclined at an angle β to the vertical, and moving in the vertical plane containing that axis with velocity u in a direction making an angle α with the horizon, strikes a perfectly rough horizontal plane. Find the resulting motion, and shew that the vertical plane containing the new direction of motion makes an angle $\tan^{-1}\left[\dfrac{2a\omega\sin\beta}{5u\cos\alpha}\right]$ with the original plane.

7. A ball, moving horizontally with velocity u and spinning about a vertical axis with angular velocity ω, impinges directly on an equal ball at rest. Shew that the maximum deviation of the first ball from its initial direction of motion produced by the impact is $\tan^{-1}\dfrac{\mu(1+e)}{1-e}$, where μ is the coefficient of friction and e of restitution between the balls, and shew that the least value of ω which will produce this deviation is $\dfrac{7\mu u}{2a}(1+e)$.

8. Shew that the loss of kinetic energy at the impact of two perfectly rough inelastic uniform spheres, of masses M and M', which are moving before impact with their centres in one plane, is $\dfrac{MM'}{14(M+M')}(2u^2+7v^2)$, where u and v are the relative velocities before impact of the points of contact tangentially, in the plane of motion of the centres, and normally.

CHAPTER XVII

ON THE PRINCIPLES OF THE CONSERVATION OF MOMENTUM AND CONSERVATION OF ENERGY

235. If x, y, and z be the coordinates of any point of a body at time t referred to fixed axes, its equations of motion are, by Art. 164,

$$\frac{d}{dt}\Sigma m\frac{dx}{dt}=\Sigma X \quad \text{.........................(1)},$$

$$\frac{d}{dt}\Sigma m\frac{dy}{dt}=\Sigma Y \quad \text{.........................(2)},$$

$$\frac{d}{dt}\Sigma m\frac{dz}{dt}=\Sigma Z \quad \text{.........................(3)},$$

$$\frac{d}{dt}\Sigma m\left(y\frac{dz}{dt}-z\frac{dy}{dt}\right)=\Sigma\,(yZ-zY) \quad \text{..................(4)},$$

$$\frac{d}{dt}\Sigma m\left(z\frac{dx}{dt}-x\frac{dz}{dt}\right)=\Sigma\,(zX-xZ) \quad \text{..................(5)},$$

and

$$\frac{d}{dt}\Sigma m\left(x\frac{dy}{dt}-y\frac{dx}{dt}\right)=\Sigma\,(xY-yX) \quad \text{..................(6)}.$$

Suppose the axis of x to be such that the sum of the resolved parts of the external forces parallel to it is zero throughout the motion, *i.e.* such that $\Sigma X=0$ always.

Equation (1) then gives

$$\frac{d}{dt}\Sigma m\frac{dx}{dt}=0,$$

i.e.

$$\Sigma m\frac{dx}{dt}=\text{constant} \quad \text{.......................(7)},$$

or

$$M\frac{d\bar{x}}{dt}=\text{constant},$$

where $\bar{x}$ is the x-coordinate of the centre of gravity.

Equation (7) states that in this case the total momentum of the body measured parallel to the axis of x remains constant throughout the motion.

This is the Principle of the **Conservation of Linear Momentum.**

Again suppose the external forces to be such that the sum of their moments about the axis of x is zero, *i.e.* such that $\Sigma(yZ-zY)=0$.

Then, by equation (4), we have

$$\frac{d}{dt}\Sigma m\left(y\frac{dz}{dt}-z\frac{dy}{dt}\right)=0,$$

and $$\therefore \quad \Sigma m\left(y\frac{dz}{dt}-z\frac{dy}{dt}\right)=\text{constant} \quad \ldots\ldots\ldots\ldots\ldots\ldots(8).$$

Now $y\frac{dz}{dt}-z\frac{dy}{dt}=$ the moment about the axis of x of the velocity of the mass m, and hence equation (8) states that the total moment of momentum of the system about the axis of x is constant.

Hence the Principle of the **Conservation of the Moment of Momentum** (or Angular Momentum), *viz.*

If the sum of the moments of the external forces, acting on a rigid body, about a given line be zero throughout the motion, the moment of momentum of the body about that line remains unaltered throughout the motion.

236. The same theorems are true in the case of impulsive forces. For if the duration of the impulse be a small time τ we have, as in Art. 166, on integrating equation (1),

$$\left[\Sigma m\frac{dx}{dt}\right]_0^\tau=\int_0^\tau \Sigma X dt=\Sigma X_1,$$

where X_1 is the impulse of the forces parallel to the axis of x, *i.e.* the change in the total momentum parallel to the axis of x is equal to the sum of the impulses of the forces in that direction.

If then the axis of x be such that the sum of the impulses parallel to it vanish, there is no change in the total momentum parallel to it,

i.e. the total momentum parallel to the axis of x before the action of the impulsive forces = the total momentum in that direction after their action.

Again, integrating equation (4), we have

$$\left\{\Sigma m\left(y\frac{dz}{dt}-z\frac{dy}{dt}\right)\right\}_0^\tau=\int_0^\tau \Sigma(yZ-zY)dt=\Sigma[yZ_1-zY_1],$$

i.e. the change in the angular momentum about the axis of x is equal to the sum of the moments of the impulses of the forces about that same direction.

If then the axis of x be such that the sum of the moments of the impulsive forces about it vanishes, there is no change in the angular momentum about it,

i.e. the angular momentum about it just before the action of the impulsive forces = the angular momentum about the same line just after their action.

237. *Ex.* 1. *A bead, of mass m, slides on a circular wire, of mass M and radius a, and the wire turns freely about a vertical diameter. If ω and ω' be the angular velocities of the wire when the bead is respectively at the ends of a horizontal and vertical diameter, shew that* $\frac{\omega'}{\omega}=\frac{M+2m}{M}$.

The moment of inertia of the wire about any diameter $=M\frac{a^2}{2}$. Wherever the bead may be on the wire, the action of it on the wire is equal and opposite to that of the wire on it.

Hence the only external forces acting on the system are (1) the action of the vertical axis AA', which has no moment about AA', and (2) the weights of the bead and wire, neither of which has any moment with respect to the vertical axis AA'.

Hence the moment of momentum of the system (wire and bead) about AA' is constant throughout the motion. Also the velocity of the bead along the wire has no moment about AA', since its direction intersects AA.

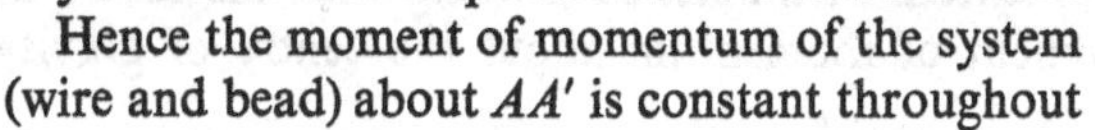

When the bead is at A, the moment of momentum about AA' is $M\frac{a^2}{2}.\omega'$; also, when it is at B, this moment is $M\frac{a^2}{2}\omega+ma^2\omega$. Equating these two, we have $\frac{\omega'}{\omega}=\frac{M+2m}{M}$.

Ex. 2. *A rod, of length $2a$, is moving on a smooth table with a velocity v perpendicular to its length and impinges on a small inelastic obstacle at a distance c from its centre. When the end leaves the obstacle, shew that the angular velocity of the rod is* $\frac{3cv}{4a^2}$.

Both at the impact, and throughout the subsequent motion whilst the rod is in contact with the obstacle, the only action on the rod is at the obstacle itself. Hence there is no change in the moment of momentum about the obstacle. But before the impact this moment was Mcv. Also, if ω be the angular velocity of the rod when its end is leaving the obstacle, its moment of momentum about the obstacle is, by Art. 191, $M\left(\frac{a^2}{3}+a^2\right)\omega$, *i.e.* $M.\frac{4a^2}{3}\omega$. Equating these two, we have $\omega=\frac{3cv}{4a^2}$.

If ω' were the angular velocity of the rod immediately after the impact, we have, similarly, $Mcv=M\left(\frac{a^2}{3}+c^2\right)\omega'$.

Ex. 3. *A uniform circular plate is turning in its own plane about a point A on its circumference with uniform angular velocity* ω; *suddenly A is released and another point B of the circumference is fixed; shew that the angular velocity about B is* $\frac{\omega}{3}(1+2\cos\alpha)$, *where* α *is the angle that AB subtends at the centre.*

In this case the only impulsive force acting on the plate is at B and its moment about B vanishes.

Hence, by Art. 235, the moment of momentum about B is the same after the fixing as before. If ω' be the required angular velocity, the moment of momentum after the fixing $=M(a^2+k^2)\omega'=M.\frac{3a^2}{2}\omega'$.

The moment of momentum before the fixing

= the moment of momentum of a mass M moving with the centre of gravity + the moment of momentum about the centre of gravity (Art. 191)

$=Ma\omega.a\cos\alpha+Mk^2\omega=M\omega a^2(\cos\alpha+\tfrac{1}{2})$,

since before the fixing the centre O was moving at right angles to AO with velocity $a\omega$.

Hence $M\frac{3a^2}{2}\omega'=M\omega a^2\left(\cos\alpha+\frac{1}{2}\right)$. $\therefore\ \omega'=\omega\frac{1+2\cos\alpha}{3}$.

It is clear that ω' is always less than ω, so that the energy, $\frac{1}{2}m(k^2+a^2)\omega'^2$, after the impact is always less than it was before. This is a simple case of the general principle that kinetic energy is always diminished whenever an impact, or anything in the nature of a jerk, takes place.

If $\alpha=120°$, *i.e.* if the arc AB is one-third of the circumference, the disc is brought to rest.

Ex. 4. *A uniform square lamina, of mass M and side 2a, is moving freely about a diagonal with uniform angular velocity* ω *when one of the corners not in that diagonal becomes fixed; shew that the new angular velocity is* $\frac{\omega}{7}$, *and that the impulse of the force on the fixed point is* $\frac{\sqrt{2}}{7}.Ma\omega$.

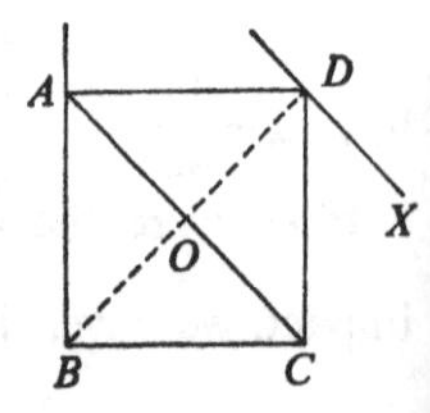

Let AC be the original axis of rotation.

As in Art. 149, the moment of inertia about it is $M.\frac{a^2}{3}$. Let the initial direction of rotation be such that B was moving upwards from the paper.

Let D be the point that becomes fixed and ω' the resulting angular velocity about DX, a line parallel to AC. Since the impulsive force at the fixing acts at D its moment about DX vanishes. Hence the moment of momentum about DX is unaltered by the fixing.

After the fixing it $= Mk^2\omega'$

$$= M\left[\frac{a^2}{3} + DO^2\right]\omega' = M\left(\frac{a^2}{3} + 2a^2\right)\omega' = M.\frac{7a^2}{3}\omega'.$$

Also before the fixing it, by Art. 191,

$=$ moment of momentum about $AC +$ the moment of momentum of a particle M at O and moving with it $= M.\frac{a^2}{3}\omega$.

Equating these two quantities, we have $\omega' = \frac{\omega}{7}$.

Similarly, the moment of momentum about DB after the fixing $=$ the moment of momentum before $=$ zero.

Hence after the fixing the square is moving about DX with angular velocity $\frac{\omega}{7}$.

Again, before the fixing the centre of gravity O was at rest, and after the fixing it is moving with velocity $DO.\omega'$, *i.e.* $\sqrt{2}a.\frac{\omega}{7}$, about D. The change in its momentum is therefore $M\frac{\sqrt{2}a\omega}{7}$, and this, by Art. 166, is equal to the impulse of the force required.

EXAMPLES

1. If the Earth, supposed to be a uniform sphere, had in a certain period contracted slightly so that its radius was less by $\frac{1}{n}$th than before, shew that the length of the day would have shortened by $\frac{48}{n}$ hours.

2. A heavy circular disc is revolving in a horizontal plane about its centre which is fixed. An insect, of mass $\frac{1}{n}$*th* that of the disc, walks from the centre along a radius and then flies away. Shew that the final angular velocity is $\frac{n}{n+2}$ times the original angular velocity of the disc.

3. A uniform circular board, of mass M and radius a, is placed on a perfectly smooth horizontal plane and is free to rotate about a vertical axis through its centre; a man, of mass M', walks round the edge of the board whose upper surface is rough enough to prevent his slipping; when he has walked completely round the board to his starting-point, shew that the board has turned through an angle $\frac{M'}{M+2M'} \cdot 4\pi$.

4. A circular ring, of mass M and radius a, lies on a smooth horizontal plane, and an insect, of mass m, resting on it starts and walks round it with uniform velocity v relative to the ring. Shew that the centre of the ring describes a circle with angular velocity

$$\frac{m}{M+2m} \cdot \frac{v}{a}.$$

5. If a merry-go-round be set in motion and left to itself, shew that in order that a man may (1) move with the greatest velocity, (2) be most likely to slip, he must place himself at a distance from the centre equal to (1) $k\sqrt{n}$, (2) $k\sqrt{\frac{n}{3}}$, k being the radius of gyration of the machine about its axis and n the ratio of its weight to that of the man.

6. A uniform circular wire, of radius a, lies on a smooth horizontal table and is movable about a fixed point O on its circumference. An insect, of mass equal to that of the wire, starts from the other end of the diameter through O and crawls along the wire with a uniform velocity v relative to the wire. Shew that at the end of time t the wire has turned through an angle $\frac{vt}{2a} - \frac{1}{\sqrt{3}} \tan^{-1}\left[\frac{1}{\sqrt{3}} \tan \frac{vt}{2a}\right]$.

[When the diameter OA has turned through an angle ϕ from its initial position, let the insect be at P so that $\angle ACP = \theta = \frac{vt}{a}$, where C is the centre of the wire. Since the moment of momentum about O is constant,

$$\therefore \quad m(k^2+a^2)\dot{\phi} + m\left[4a^2 \cos^2 \frac{\theta}{2}\dot{\phi} + v.2a \cos^2 \frac{\theta}{2}\right] = \text{constant} = 0.]$$

7. A small insect moves along a uniform bar, of mass equal to itself and of length $2a$, the ends of which are constrained to remain on the circumference of a fixed circle, whose radius is $\frac{2a}{\sqrt{3}}$. If the insect start from the middle point of the bar and move along the bar with relative velocity V, shew that the bar in time t will turn through an angle.

$$\frac{1}{\sqrt{3}} \tan^{-1} \frac{Vt}{a}.$$

8. A circular disc is moving with an angular velocity Ω about an axis through its centre perpendicular to its plane. An insect alights on its edge and crawls along a curve drawn on the disc in the form of a lemniscate with uniform relative angular velocity $\frac{1}{8}\Omega$, the curve touching the edge of the disc. The mass of the insect being $\frac{1}{18}$th of that of the disc, shew that the angle turned through by the disc when the insect gets to the centre is $\frac{24}{\sqrt{7}} \tan^{-1} \frac{\sqrt{7}}{3} - \frac{\pi}{4}$.

9. A rod OA can turn freely in a horizontal plane about the end O and lies at rest. An insect, whose mass is one-third that of the rod, alights on the end A and commences crawling along the rod with uniform velocity V; at the same instant the rod is set in rotation about O in such a way that the initial velocity of A is V; when the insect reaches O prove that the rod has rotated through a right angle, and that the angular velocity of the rod is then twice the initial angular velocity.

10. A particle, of mass m, moves within a rough circular tube, of mass M, lying on a smooth horizontal plane and initially the tube is at rest while the particle has an angular velocity round the tube. Shew that by the time the relative motion ceases the fraction $\frac{M}{M+2m}$ of the initial kinetic energy has been dissipated by friction.

[The linear momentum of the common centre of gravity, and the moment of momentum about it, are both constant throughout the motion.]

11. A rod, of length $2a$, is moving about one end with uniform angular velocity upon a smooth horizontal plane. Suddenly this end is loosed and a point, distant b from this end, is fixed; find the motion, considering the cases when $b < = > \frac{4a}{3}$.

12. A circular plate rotates about an axis through its centre perpendicular to its plane with angular velocity ω. This axis is set free and a point in the circumference of the plate fixed; shew that the resulting angular velocity is $\frac{\omega}{3}$.

13. Three equal particles are attached to the corners of an equilateral triangular area ABC, whose mass is negligible, and the system is rotating in its own plane about A. A is released and the middle point of AB is suddenly fixed. Shew that the angular velocity is unaltered.

14. A uniform square plate $ABCD$, of mass M and side $2a$, lies on a smooth horizontal plane; it is struck at A by a particle of mass M' moving with velocity V in the direction AB, the particle remaining attached to the plate. Determine the subsequent motion of the system, and shew that its angular velocity is $\frac{M'}{M+4M'} \cdot \frac{3V}{2a}$.

15. A lamina in the form of an ellipse is rotating in its own plane about one of its foci with angular velocity ω. This focus is set free and the other focus at the same instant is fixed; shew that the ellipse now rotates about it with angular velocity $\omega . \frac{2-5e^2}{2+3e^2}$.

16. An elliptic area, of eccentricity e, is rotating with angular velocity ω about one latus-rectum; suddenly this latus-rectum is loosed and the other fixed. Shew that the new angular velocity is

$$\omega . \frac{1-4e^2}{1+4e^2}.$$

17. A uniform circular disc is spinning with angular velocity ω about a diameter when a point P on its rim is suddenly fixed. If the radius vector to P make an angle α with this diameter, show that the angular velocities after the fixing about the tangent and normal at P are $\frac{1}{5}\omega \sin \alpha$ and $\omega \cos \alpha$.

18. A cube is rotating with angular velocity ω about a diagonal when suddenly the diagonal is let go, and one of the edges which does not meet this diagonal is fixed; shew that the resulting angular velocity about this edge is $\frac{\omega}{12} . \sqrt{3}$.

19. If an octant of an ellipsoid bounded by three principal planes be rotating with uniform angular velocity ω about the axis a, and if this axis be suddenly freed and the axis b fixed, shew that the new angular velocity is $\frac{2ab\omega}{\pi(a^2+c^2)}$.

CONSERVATION OF ENERGY

238. In many previous articles we have met with examples in which the change of kinetic energy of a particle, or system of particles, is equal to the work done on the particle, or system of particles.

The formal enunciation of the principle may be given as follows:

If a system move under the action of finite forces, and if the geometrical equations of the system do not contain the time explicitly, the change in the kinetic energy of the system in passing from one configuration to any other is equal to the corresponding work done by the forces.

By the principles of Art. 161, the forces $X-m\frac{d^2x}{dt^2}$, $Y-m\frac{d^2y}{dt^2}$, $Z-m\frac{d^2z}{dt^2}$ acting at the point (x, y, z) and similar forces acting on the other particles of the system are a system of forces in equilibrium.

Let δx, δy, δz be small virtual displacements of the particle m at (x, y, z) consistent with the geometrical conditions of the system at the time t.

Then the principle of Virtual Work states that

$$\Sigma\left[\left(X-m\frac{d^2x}{dt^2}\right)\delta x+\left(Y-m\frac{d^2y}{dt^2}\right)\delta y+\left(Z-m\frac{d^2z}{dt^2}\right)\delta z\right]=0.$$

If the geometrical relations of the system do not contain the time explicitly, then we may replace δx, δy, δz by the actual displacements $\frac{dx}{dt}\delta t$, $\frac{dy}{dt}\delta t$, and $\frac{dz}{dt}\delta t$.

Hence the above equation gives

$$\Sigma m\left[\frac{d^2x}{dt^2}\frac{dx}{dt}+\frac{d^2y}{dt^2}\frac{dy}{dt}+\frac{d^2z}{dt^2}\frac{dz}{dt}\right]=\Sigma\left(X\frac{dx}{dt}+Y\frac{dy}{dt}+Z\frac{dz}{dt}\right).$$

Integrating with respect to t, we have

$$\frac{1}{2}\Sigma m\left[\left(\frac{dx}{dt}\right)^2+\left(\frac{dy}{dt}\right)^2+\left(\frac{dz}{dt}\right)^2\right]_{t_1}^{t_2}=\Sigma\int(Xdx+Ydy+Zdz)\ \ldots(1),$$

i.e. the change in the kinetic energy of the system from time t_1 to time t_2 is equal to the work done by the external forces on the body from the configuration of the body at time t_1 to the configuration at time t_2.

239. When the forces are such that $\int(Xdx+Ydy+Zdz)$ is the complete differential of some quantity V, *i.e.* when the forces have a potential V, the quantity $\Sigma m\int(Xdx+Ydy+Zdz)$ is independent of the path pursued from the initial to the final position of the body, and depends only on the configuration of the body at the times t_1 and t_2. The forces are then said to be **conservative.**

Let the configurations of the body at times t_1 and t_2 be called A and B. The equation (1) of the previous article gives

$$\left.\begin{array}{r}\text{Kinetic Energy at the time } t_2\\ -\text{Kinetic Energy at the time } t_1\end{array}\right\}=\int_A^B \delta V = V_B - V_A \text{ ......(2).}$$

The potential energy of the body in any position is the work the forces do on it whilst it moves from that position to a standard position. Let its configuration in the latter position be called C.

Then the potential energy at the time t_1

$$=\int_A^C (Xdx+Ydy+Zdz)=\int_A^C \delta V = V_C - V_A \text{ ....(3).}$$

So the potential energy at the time t_2

$$=\int_B^C (Xdx+Ydy+Zdz)=\int_B^C \delta V = V_C - V_B.$$

Hence the equation (2) gives

$$\left.\begin{array}{r}\text{Kinetic Energy at the time } t_2\\ -\text{Kinetic Energy at the time } t_1\end{array}\right\}=\left\{\begin{array}{l}\text{Potential Energy at the time } t_1\\ -\text{Potential Energy at the time } t_2,\end{array}\right.$$

i.e. the sum of the kinetic and potential energies at the time t_2 = the sum of the same quantities at the time t_1.

Hence *when a body moves under the action of a system of Conservative Forces the sum of its Kinetic and Potential Energies is constant throughout the motion.*

240. As an illustration of the necessity that the geometrical equations must not contain the time explicitly, let the body be a particle moving on a smooth plane which revolves uniformly round a horizontal axis through which it passes, the particle starting from a position of rest relative to the plane.

Let OA be the plane at time t and P the position of the particle then; OB and Q' the corresponding positions at time $t+\delta t$.

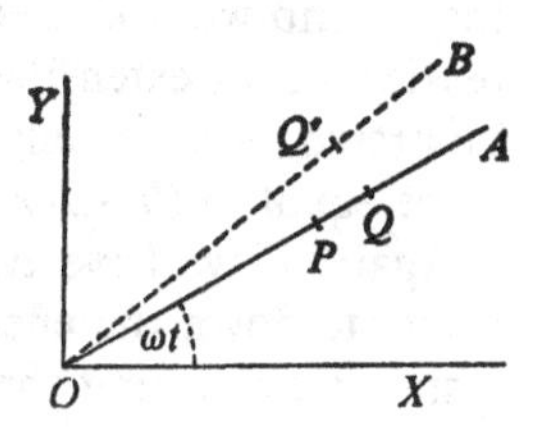

Then $\frac{dx}{dt}, \frac{dy}{dt}$ are the velocities at time t, so that $\frac{dx}{dt}\delta t, \frac{dy}{dt}\delta t$ are the corresponding distances parallel to the axes described in time δt; hence $\frac{dx}{dt}\delta t, \frac{dy}{dt}\delta t$ are the projections on the axes of PQ'.

Now δx, δy are by the Theory of Virtual Work the projections of a small displacement which is *consistent with the geometrical conditions at time t, i.e.* of a small displacement along the plane OA.

Hence δx and δy are the projections of some such displacement as PQ.

Hence in this case $\frac{dx}{dt}\delta t$ and $\frac{dy}{dt}\delta t$ cannot be replaced by δx and δy.

Also in this case the geometrical relation is $\frac{y}{x} = \tan AOX = \tan \omega t$, so that it contains the time explicitly.

The same argument holds for the general case where the geometrical relation is

$$\phi(x, y, z, t) = 0 \quad \text{.........................(1).}$$

For the latter at each time t gives a surface on which P must lie and also the virtual displacement PQ.

But $\frac{dx}{dt}\delta t$, $\frac{dy}{dt}\delta t$, $\frac{dz}{dt}\delta t$ are the projections on the axes of PQ', where Q' lies on the neighbouring surface

$$\phi(x, y, z, t+\delta t) = 0 \quad \text{.......................(2).}$$

Hence δx, δy, δz cannot be replaced by $\frac{dx}{dt}\delta t$, $\frac{dy}{dt}\delta t$, $\frac{dz}{dt}\delta t$ unless the surfaces (1) and (2) coincide, *i.e.* unless the geometrical conditions at time t coincide with those at time $t+dt$, and then the geometrical equation cannot contain the time explicitly.

241. In the result of Art. 238 all forces may be omitted which do not come into the equation of Virtual Work, *i.e.* all forces whose Virtual Work is zero.

Thus rolling friction may be omitted because the point of application of such a force of friction is instantaneously at rest; but sliding friction must not be omitted since the point of application is not at rest.

So the reactions of smooth fixed surfaces may be omitted, and generally all forces whose direction is perpendicular to the direction of motion of the point of application.

Similarly the tensions of an inextensible string may be left out; for they do no work since the length of such a string is constant; but the tension of an extensible string must be included; for the work done in stretching an extensible string from length a to length b is known to be equal to $(b-a) \times$ the mean of the initial and final tensions.

Again, if we have two rigid bodies which roll on one another and we write down the energy equation for the two bodies treated as one system we can omit the reaction between them.

242. *The kinetic energy of a rigid body, moving in any manner, is at any instant equal to the kinetic energy of the whole mass, supposed collected at its centre of inertia and moving with it, together with the kinetic energy of the whole mass relative to its centre of inertia.*

Let $(\bar{x}, \bar{y}, \bar{z})$ be the coordinates of the centre of inertia of the body at any time t referred to axes fixed in space, and let (x, y, z) be the coordinates then of any element m of the body; also let (x', y', z') be the coordinates of m relative to G at the same instant, so that $x=\bar{x}+x'$, $y=\bar{y}+y'$, and $z=\bar{z}+z'$.

Then the total kinetic energy of the body

$$=\frac{1}{2}\Sigma m\left[\left(\frac{dx}{dt}\right)^2+\left(\frac{dy}{dt}\right)^2+\left(\frac{dz}{dt}\right)^2\right]$$

$$=\frac{1}{2}\Sigma m\left[\left(\frac{d\bar{x}}{dt}+\frac{dx'}{dt}\right)^2+\left(\frac{d\bar{y}}{dt}+\frac{dy'}{dt}\right)^2+\left(\frac{d\bar{z}}{dt}+\frac{dz'}{dt}\right)^2\right]$$

$$=\frac{1}{2}\Sigma m\left[\left(\frac{d\bar{x}}{dt}\right)^2+\left(\frac{d\bar{y}}{dt}\right)^2+\left(\frac{d\bar{z}}{dt}\right)^2\right]$$

$$+\frac{1}{2}\Sigma m\left[\left(\frac{dx'}{dt}\right)^2+\left(\frac{dy'}{dt}\right)^2+\left(\frac{dz'}{dt}\right)^2\right]$$

$$+\Sigma m\frac{d\bar{x}}{dt}\frac{dx'}{dt}+\Sigma m\frac{d\bar{y}}{dt}\frac{dy'}{dt}+\Sigma m\frac{d\bar{z}}{dt}\frac{dz'}{dt} \quad \text{....................(1).}$$

Now, since (x', y', z') are the coordinates of m relative to the centre of inertia G,

$\therefore \dfrac{\Sigma mx'}{\Sigma m}=$ the x-coordinate of the centre of inertia G relative to G itself $=0$.

$\therefore \Sigma mx'=0$, for all values of t.

$$\therefore \Sigma m\frac{dx'}{dt}=\frac{d}{dt}\Sigma mx'=0.$$

$$\therefore \Sigma m\frac{d\bar{x}}{dt}\frac{dx'}{dt}=\frac{d\bar{x}}{dt}.\Sigma m\frac{dx'}{dt}=0.$$

Similarly when x is changed into y or z.

Also $$\Sigma m\left[\left(\frac{d\bar{x}}{dt}\right)^2+\left(\frac{d\bar{y}}{dt}\right)^2+\left(\frac{d\bar{z}}{dt}\right)^2\right]$$

$$=M\times\left[\left(\frac{d\bar{x}}{dt}\right)^2+\left(\frac{d\bar{y}}{dt}\right)^2+\left(\frac{d\bar{z}}{dt}\right)^2\right]$$

$=Mv^2$, where v is the velocity of the centre of inertia.

And $$\frac{1}{2}\Sigma m\left[\left(\frac{dx'}{dt}\right)^2+\left(\frac{dy'}{dt}\right)^2+\left(\frac{dz'}{dt}\right)^2\right]$$

$$=\frac{1}{2}\Sigma m\times\text{the square of the velocity of } m \text{ relative to } G$$

$$=\text{the kinetic energy of the body relative to } G.$$

Hence (1) gives

The total kinetic energy of the body = the kinetic energy of the mass M, if it be supposed collected into a particle at the centre of inertia G and to move with the velocity of G + the kinetic energy of the body relative to G.

243. *Kinetic energy relative to the centre of inertia in space of three dimensions.*

Let ω_x, ω_y, ω_z be the angular velocities of the body about lines through G parallel to the axes.

Then, as in Art. 227,

$$\frac{dx'}{dt}=z\omega_y-y\omega_z,\quad \frac{dy'}{dt}=x\omega_z-z\omega_x,\ \text{and}\ \frac{dz'}{dt}=y\omega_x-x\omega_y.$$

Therefore the kinetic energy relative to the centre of inertia

$$=\frac{1}{2}\Sigma m\left[\left(\frac{dx'}{dt}\right)^2+\left(\frac{dy'}{dt}\right)^2+\left(\frac{dz'}{dt}\right)^2\right]$$

$$=\frac{1}{2}\Sigma m[\omega_x^2(y^2+z^2)+\omega_y^2(z^2+x^2)+\omega_z^2(x^2+y^2)-2yz\omega_y\omega_z -2zx\omega_z\omega_x-2xy\omega_x\omega_y]$$

$$=\frac{1}{2}[A\omega_x^2+B\omega_y^2+C\omega_z^2-2D\omega_y\omega_z-2E\omega_z\omega_x-2F\omega_x\omega_y],$$

where A, B, C are the moments of inertia about axes through the centre of inertia G and D, E, F are the products of inertia about the same axes.

If these axes are the principal axes of the body at G, the kinetic energy becomes $\frac{1}{2}[A\omega_x^2+B\omega_y^2+C\omega_z^2]$.

244. *Ex.* 1. *An ordinary window-blind, of length l and mass m, attached to a horizontal roller, of mass M, and having a horizontal rod, of mass μ, fixed to its free end, is allowed to unroll under the action of gravity. Neglecting friction, shew that the length of the blind unrolled in time t is* $\frac{\mu l}{m}(\cosh \alpha t-1)$, *where* $\alpha=\sqrt{\frac{g}{l}\frac{2m}{M+2m+2\mu}}$, *the thickness of the blind being negligible.*

When the blind has unrolled a distance x, each point of it is moving with a velocity $\dot{x}$; the angular velocity of the roller is then $\frac{\dot{x}}{a}$, where a is its radius.

The total kinetic energy

$$=\tfrac{1}{2}m.\dot{x}^2+\tfrac{1}{2}\mu.\dot{x}^2+\tfrac{1}{2}Mk^2.\omega^2$$
$$=\tfrac{1}{2}m\dot{x}^2+\tfrac{1}{2}\mu\dot{x}^2+\tfrac{1}{4}M\dot{x}^2$$
$$=\frac{\dot{x}^2}{4}.\frac{2mg}{la^2}=\frac{1}{2}.\frac{mg}{la^2}.\dot{x}^2.$$

The principle of Energy and Work gives

$$\frac{1}{2}\frac{mg}{la^2}\dot{x}^2=\mu gx+\frac{mx}{l}g.\frac{x}{2},$$

i.e.

$$\dot{x}^2=a^2\left[x^2+\frac{2\mu}{m}xl\right].$$

$$\therefore\ t.a=\int\frac{dx}{\sqrt{\left(x+\frac{\mu l}{m}\right)^2-\frac{\mu^2l^2}{m^2}}}=\cosh^{-1}\frac{x+\frac{\mu l}{m}}{\frac{\mu l}{m}},$$

the constant vanishing since x and t are both zero together.

$$\therefore\ x=\frac{\mu l}{m}[\cosh at-1].$$

Ex. 2. *A uniform rod, of length* 2*a*, *hangs in a horizontal position being supported by two vertical strings, each of length l, attached to its ends, the other extremities being attached to fixed points. The rod is given an angular velocity* ω *about a vertical axis through its centre; find its angular velocity when it has turned through any angle, and shew that it will rise through a distance* $\frac{a^2\omega^2}{6g}$.

Prove also that the time of a small oscillation about the position of equilibrium is $2\pi\sqrt{\frac{l}{3g}}$. (Bifilar Suspension.)

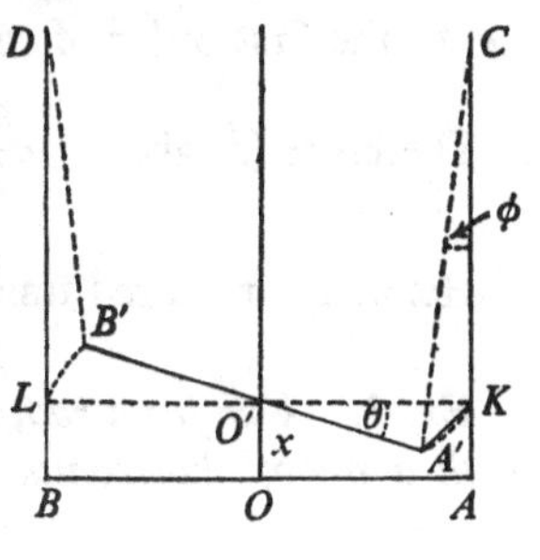

Let AB be the initial position of the rod with the strings CA and DB vertical, $A'B'$ its position when it has risen through a vertical distance x and turned through an angle θ. Let the horizontal plane through $A'B'$ cut CA and BD in K and L, and let $\angle A'CA=\phi$.

The equation of energy then gives

$$\tfrac{1}{2}m\dot{x}^2+\tfrac{1}{2}mk^2\dot{\theta}^2=\tfrac{1}{2}mk^2\omega^2-mgx \qquad \text{.................(1).}$$

Now, since the angle $A'KC$ is a right angle,

$$\therefore\ x=AC-CK=l-l\cos\phi \qquad \text{...................(2),}$$

where l is the length of a vertical string.

Also $$l\sin\phi=A'K=2a\sin\frac{\theta}{2} \qquad \text{.......................(3).}$$

$$\therefore\ \dot{x}=l\sin\phi\dot{\phi}=\tan\phi.a\cos\frac{\theta}{2}\dot{\theta}=\frac{a^2\sin\theta\dot{\theta}}{\sqrt{l^2-4a^2\sin^2\frac{\theta}{2}}}.$$

Hence equation (1) gives

$$\frac{1}{2}ma^2\dot{\theta}^2\left[\frac{a^2\sin^2\theta}{l^2-4a^2\sin^2\frac{\theta}{2}}+\frac{1}{3}\right]=\frac{1}{2}m.\frac{a^2}{3}.\omega^2-mgx \qquad \text{.........(4).}$$

This equation gives the angular velocity in any position. The rod comes to instantaneous rest when $\dot{\theta}=0$, *i.e.* when $x=\dfrac{a^2\omega^2}{6g}$.

For a small oscillation we have, on taking moments about O', if T be the tension of either string,

$$m\frac{a^2}{3}\ddot{\theta}=-2T\sin\phi\times\text{perpendicular from } O' \text{ on } A'K$$

$$=-T\sin\phi.2a\cos\frac{\theta}{2}=-\frac{2a^2}{l}.T\sin\theta \qquad \text{..............(5).}$$

Also $$2T\cos\phi-mg=m\ddot{x}=m\frac{d^2}{dt^2}\left(\frac{l}{2}\phi^2\right),\ \text{when } \phi \text{ is small,}$$

$$=\frac{ma^2}{2l}\frac{d^2}{dt^2}(\theta^2)=\frac{ma^2}{2l}[2\dot{\theta}^2+2\theta\ddot{\theta}] \qquad \text{.................(6),}$$

i.e. to the first order of small quantities, $2T-mg=0$.

Therefore (5) gives $m\dfrac{a^2}{3}\ddot{\theta}=-\dfrac{ma^2}{l}g.\theta$, *i.e.* $\ddot{\theta}=-\dfrac{3g}{l}\theta$.

Hence the required time $=2\pi\sqrt{\dfrac{l}{3g}}$.

Ex. 3. *A uniform rod, of length* 2a, *is placed with one end in contact with a smooth horizontal table and is then allowed to fall; if* α *be its initial inclination to the vertical, shew that its angular velocity when it is inclined at an angle θ is*

$$\left\{\frac{6g}{a}\cdot\frac{\cos\alpha-\cos\theta}{1+3\sin^2\theta}\right\}^{\frac{1}{2}}.$$

Find also the reaction of the table.

There is no horizontal force acting on the rod; hence its centre of inertia G has no horizontal velocity during the motion since it had none initially. Hence G describes a vertical straight line GO. When inclined at θ to the vertical its kinetic energy

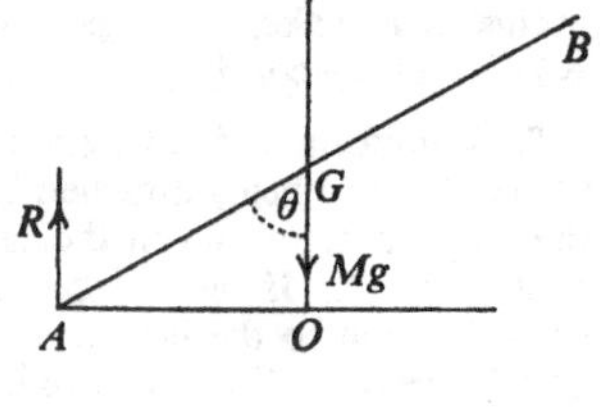

$$=\frac{1}{2}M\left\{\frac{d}{dt}(a\cos\theta)\right\}^2+\frac{1}{2}M.\frac{a^2}{3}\dot{\theta}^2$$

$$=\tfrac{1}{2}Ma^2\dot{\theta}^2\left(\sin^2\theta+\tfrac{1}{3}\right).$$

Equating this to the work done, *viz.* $Mga(\cos\alpha-\cos\theta)$, we get

$$\dot{\theta}^2=\frac{6g}{a}\frac{\cos\alpha-\cos\theta}{1+3\sin^2\theta} \quad\text{.........................(1).}$$

Differentiating, we have $\ddot{\theta}=\dfrac{3g\sin\theta}{a}\,\dfrac{4-6\cos\alpha\cos\theta+3\cos^2\theta}{(1+3\sin^2\theta)^2}$.

Also, for the vertical motion of G, we have

$$R-Mg=M\frac{d^2}{dt^2}(a\cos\theta)=M[-a\sin\theta\ddot{\theta}-a\cos\theta\dot{\theta}^2].$$

On substitution, we have

$$R=Mg\,\frac{4-6\cos\theta\cos\alpha+3\cos^2\theta}{(1+3\sin^2\theta)^2}.$$

EXAMPLES

1. A uniform rod, of given length and mass, is hinged at one end to a fixed point and has a string fastened to its other end which, after passing over a light pulley in the same horizontal line with the fixed point, is attached to a particle of given weight. The rod is initially horizontal and is allowed to fall; find how far the weight goes up.

2. A light elastic string of natural length $2a$ has one end, A, fixed and the other, B, attached to one end of a uniform rod BC of length $2a$ and mass m. This can turn freely in a vertical plane about its other end C, which is fixed at a distance $2a$ vertically below A. Initially the rod is vertical, and, on being slightly displaced, falls until it is horizontal, and then rises again. Shew that the modulus of elasticity is

$$mg(3+2\sqrt{2}).$$

3. A uniform rod moves in a vertical plane, its ends being in contact with the interior of a fixed smooth sphere; when it is inclined at an angle θ to the horizon, shew that the square of its angular velocity is $\frac{6cg}{3c^2+a^2}(\cos\theta-\cos\alpha)$, where α is the initial value of θ, $2a$ is the length of the rod, and c is the distance of its middle point from the centre of the sphere.

4. A hemisphere, of mass M and radius a, is placed with its plane base on a smooth table, and a heavy rod, of mass m, is constrained to move in a vertical line with one end P on the curved surface of the hemisphere; if at any time t the radius to P makes an angle θ with the vertical, shew that $a\dot{\theta}^2[M\cos^2\theta+m\sin^2\theta]=2mg(\cos\alpha-\cos\theta)$.

5. A uniform rod, of length $2a$, is held with one end on a smooth horizontal plane, this end being attached by a light inextensible string to a point in the plane; the string is tight and in the same vertical plane as the rod and makes with it an acute angle α. If the rod be now allowed to fall under the action of gravity, find its inclination to the horizon when the string ceases to be tight, and shew that its angular velocity Ω just before it becomes horizontal is given by the equation

$$6a\Omega^2=g\sin\alpha\,(8+\cos^2\alpha).$$

6. A uniform straight rod, of length $2a$, has two small rings at its ends which can respectively slide on thin smooth horizontal and vertical wires Ox and Oy. The rod starts at an angle α to the horizon with an angular velocity $\sqrt{\frac{3g}{2a}(1-\sin\alpha)}$, and moves downwards. Shew that it will strike the horizontal wire at the end of time

$$2\sqrt{\frac{a}{3g}}\log\left\{\cot\left(\frac{\pi}{8}-\frac{\alpha}{4}\right)\tan\frac{\pi}{8}\right\}.$$

7. A straight uniform rod, of mass m, is placed at right angles to a smooth plane of inclination α with one end in contact with it; the rod is then released. Shew that, when its inclination to the plane is ϕ, the reaction of the plane will be $mg\frac{3(1-\sin\phi)^2+1}{(3\cos^2\phi+1)^2}\cos\alpha$.

8. A hoop, of mass M, carrying a particle of mass m fixed to a point of its circumference, rolls down a rough inclined plane; find the motion.

9. Two like rods AB and BC, each of length $2a$, are freely jointed at B; AB can turn round the end A and C can move freely on a vertical straight line through A. Initially the rods are held in a horizontal line, C being in coincidence with A, and they are then released. Shew that when the rods are inclined at an angle θ to the horizontal, the angular velocity of either is $\sqrt{\frac{3g}{a}\cdot\frac{\sin\theta}{1+3\cos^2\theta}}$.

10. A sphere, of radius b, rolls without slipping down the cycloid

$$x=a(\theta+\sin\theta),\quad y=a(1-\cos\theta).$$

It starts from rest with its centre on the horizontal line $y=2a$. Shew that the velocity V of its centre when at its lowest point is given by

$$V^2=g\tfrac{10}{7}(2a-b).$$

11. A string, of length $2l$, is attached to two points in the same horizontal plane at a distance $2b$ and carries a particle m at its middle point; a uniform rod, of length $2a$ and mass M, has at each end a ring through which the string passes and is let fall from a symmetrical position in the straight line joining the ends of the string; shew that the rod will not reach the particle if

$$(l+b-2a).(M+2m)\,M<2(2a-b)m^2.$$

If $M=m$ and $b=a$, and the particle be given a small vertical displacement when it is in a position of equilibrium, shew that the time of a small oscillation is $\frac{2\pi}{3}\sqrt{\frac{2\sqrt{3}a}{g}}$.

12. Two equal perfectly rough spheres are placed in unstable equilibrium one on the top of the other, the lower sphere resting on a smooth table. If the equilibrium be disturbed, shew that the spheres will continue to touch at the same point, and that when the line joining their centres is inclined at an angle θ to the vertical its angular velocity ω is given by the equation $a^2\omega^2\,(5\sin^2\theta+7)=10ga\,(1-\cos\theta)$, where a is the radius of each sphere.

13. An inextensible uniform band, of small thickness τ, is wound round a thin fixed axis so as to form a coil of radius b. The coil is unrolled until a length a hangs freely and then begins to unroll freely under the action of gravity, starting from rest. Shew that, if the small horizontal motion be neglected, the time which will elapse before the hanging part is of length x is approximately

$$b\sqrt{\frac{\pi}{2g\tau}}\left[\log\frac{x+\sqrt{x^2-a^2}}{a}+\frac{\tau}{2\pi b^2}\sqrt{x^2-a^2}\right].$$

14. A roll of cloth, of small thickness ϵ, lying at rest on a perfectly rough horizontal table is propelled with initial angular velocity Ω so that the cloth unrolls. Apply the Principle of Energy to shew that the radius of the roll will diminish from a to r (so long as r is not small compared with a) in time $\frac{2\pi}{\epsilon}\sqrt{\frac{1}{3g}}\{\sqrt{c^3-r^3}-\sqrt{c^3-a^3}\}$, where $3\Omega^2a^4=4(c^3-a^3)g$. Is the application of the principle correct?

245. In many cases of motion the application of the principles of this Chapter will give two first integrals of the motion, and hence determine the motion.

Ex. A perfectly rough inelastic sphere, of radius a, is rolling with velocity v on a horizontal plane when it meets a fixed obstacle of height h. Find the condition that the sphere will surmount the obstacle and, if it does, shew that it will continue rolling on the plane with velocity $\left(1-\frac{5h}{7a}\right)^2 v$.

Let Ω be the angular velocity immediately after the impact about the point of contact, K, with the obstacle.

The velocity of the centre before the impact was v in a horizontal direction, and the angular velocity was $\frac{v}{a}$ about the centre.

Since the moment of momentum about K is unaltered, as the only impulsive force acts at K, we have

$$m(k^2+a^2)\Omega=mv(a-h)+mk^2\frac{v}{a}.$$

$$\therefore\ \Omega=\frac{v(7a-5h)}{7a^2}\ \text{..........................}(1).$$

Let ω be the angular velocity of the sphere about K when the radius to K is inclined at θ to the horizontal. The equation of Energy gives

$$\frac{1}{2}m.\frac{7a^2}{5}(\omega^2-\Omega^2)=-mg(h+a\sin\theta-a) \quad \text{...........(2).}$$

Also, if R be the normal reaction at this instant, we have, since the acceleration of the centre is $a\omega^2$ towards K,

$$ma\omega^2=mg\sin\theta-R \quad \text{.......................(3).}$$

(2) gives
$$\omega^2=\Omega^2-\frac{10}{7}\cdot\frac{g}{a^2}(h+a\sin\theta-a) \quad \text{..................(4),}$$

and (3) gives
$$\frac{R}{m}=\frac{g}{7a}[10h-10a+17a\sin\theta]-a\Omega^2 \quad \text{...............(5).}$$

In order that the sphere may surmount the obstacle without leaving it, (i) ω must not vanish before the sphere gets to its highest point, *i.e.* ω^2 must be positive when $\theta=90°$, and (ii) R must not be negative when it is least, *i.e.* when $\sin\theta=\frac{a-h}{a}$.

The first condition gives $\Omega^2>\frac{10gh}{7a^2}$, and the second gives $\Omega^2<\frac{g(a-h)}{a^2}$.

Hence, from (1),

$$v>\frac{a}{7a-5h}\sqrt{70gh}, \quad \text{and} \quad v<\frac{7a}{7a-5h}\sqrt{g(a-h)}.$$

For both these conditions to be true it is clear that $h \not> \frac{7a}{17}$.

If these conditions are satisfied so that the sphere surmounts, without leaving, the obstacle, its angular velocity when it hits the plane again is Ω. If its angular velocity immediately after hitting the plane be ω_1, we have, by the Principle of the Conservation of Momentum,

$$m\frac{7a^2}{5}\omega_1=m.a\Omega.(a-h)+m\frac{2a^2}{5}\Omega,$$

since just before the impact the centre was moving with velocity $a\Omega$ perpendicular to the radius to the obstacle.

$$\therefore \omega_1=\Omega\left(1-\frac{5h}{7a}\right)=\left(1-\frac{5h}{7a}\right)^2\cdot\frac{v}{a},$$

so that the sphere will continue to roll on the plane with velocity $v\left(1-\frac{5h}{7a}\right)^2$.

EXAMPLES

1. A smooth uniform rod is moving on a horizontal table about one end which is fixed; it impinges on an inelastic particle whose distance from the fixed end was $\frac{1}{n}$*th* of the length of the rod; find the ratio of the velocity of the particle when it leaves the rod to its initial velocity.

[For the impact we have $M.\frac{4a^2}{3}\omega = M\frac{4a^2}{3}\omega' + m\frac{4a^2}{n^2}\omega'$.

The Principles of Energy and Momentum then give

$$\frac{1}{2}M.\frac{4a^2}{3}\dot{\theta}^2 + \frac{1}{2}m(\dot{x}^2 + x^2\dot{\theta}^2) = \frac{1}{2}M.\frac{4a^2}{3}\omega'^2 + \frac{1}{2}m\frac{4a^2}{n^2}\omega'^2,$$

and
$$M.\frac{4a^2}{3}\dot{\theta} + mx^2\dot{\theta} = M\frac{4a^2}{3}\omega' + m\frac{4a^2}{n^2}\omega'.]$$

2. A uniform rod, of mass M, is moving on a smooth horizontal table about one end which is fixed; it drives before it a particle, of mass nM, which initially was at rest close to the fixed end of the rod; when the particle is at a distance $\frac{1}{x}$th of the length of the rod from the fixed end, shew that its direction of motion makes with the rod an angle

$$\cot^{-1}\sqrt{1+\frac{3n}{x^2}}.$$

3. A uniform rod, of length $2a$, lying on a smooth horizontal plane passes through a small ring on the plane which allows it to rotate freely. Initially the middle point of the rod is very near the ring, and an angular velocity ω is impressed on it; find the motion, and shew that when the rod leaves the ring the velocity of its centre is $\frac{\sqrt{5}}{4}a\omega$, and its angular velocity is $\frac{\omega}{4}$.

4. A piece of a smooth paraboloid, of mass M, cut off by a plane perpendicular to the axis rests on a smooth horizontal plane with its vertex upwards. A particle, of mass m, is placed at the highest point and slightly displaced; shew that when the particle has descended a distance x, the square of the velocity of the paraboloid is

$$\frac{2m^2gax}{(M+m)\{(M+m)x+Ma\}}.$$

[The horizontal momentum of the system is always zero and its kinetic energy is equal to the work done by gravity.]

5. A thin spherical shell, of mass M and radius R, is placed upon a smooth horizontal plane and a smooth sphere, of mass m and radius r, slides down its inner surface starting from a position in which the line of centres is horizontal. Shew that when the line of centres makes an angle ϕ with the horizontal the velocity of the shell M is given by

$$V^2 = 2g\frac{m^2}{(M+m)(M+m\cos^2\phi)}(R-r)\sin^3\phi.$$

[Compare with the example of Art. 202.]

6. A fine circular tube, of radius a and mass M, lies on a smooth horizontal plane; within it are two equal particles, each of mass m, connected by an elastic string in the tube, whose natural length is equal to the semi-circumference. The particles are in contact and fastened together, the string being stretched round the circumference. If the particles become separated, shew that the velocity of the tube when the string has regained its natural length is $\sqrt{\frac{2\pi\lambda ma}{M(M+2m)}}$, where λ is the modulus of elasticity.

If one of the particles be fixed in the tube and the tube be movable about the point at which it is fixed, shew that the corresponding velocity of the centre of the tube is $\frac{1}{\sqrt{2}}$ times its value in the first case.

7. A heavy pendulum can turn freely about a horizontal axis, and a bullet is fired into it at a depth p below the axis with a velocity which is horizontal and perpendicular to the axis; the pendulum is observed to swing through an angle θ before coming to rest; shew that the velocity of the bullet was $2\sin\frac{\theta}{2}\sqrt{\left(1+\frac{Mk^2}{mp^2}\right)\left(1+\frac{Mh}{mp}\right).gp}$, where M and m are the masses of the pendulum and bullet, and h and k are the depth of the centre of inertia below, and the radius of gyration about, the axis of the pendulum.

8. To the pendulum of the previous question is attached a rifle in a horizontal position at a depth p below the axis and from it is fired a bullet of mass m; shew that the velocity of the bullet is

$$2.\frac{M'}{m}.\frac{k'}{p}\sin\frac{\theta}{2}.\sqrt{gh'},$$

where M' is the mass of the pendulum and gun, and h' and k' are the depth of the centre of inertia below, and the radius of gyration about, the axis of M'.

9. A thin uniform circular wire, of radius a, can revolve freely about a vertical diameter, and a bead slides freely along the wire. If initially the wire was rotating with angular velocity Ω, and the bead was at rest relatively to the wire very close to its highest point, shew that when the bead is at its greatest distance from the axis of rotation the angular velocity of the wire is $\Omega.\frac{n}{n+2}$, and that the angular velocity of the bead relative to the wire is $\sqrt{\frac{n\Omega^2}{n+2}+\frac{2g}{a}}$, the mass of the wire being n times that of the bead.

10. Two uniform rods AB, BC are jointed at B and can rotate on a smooth horizontal plane about the end A which is fixed. From the principles of the Conservation of Energy and Momentum obtain equations to give their angular velocities in any position.

11. One end of a light inextensible string, of length a, is attached to a fixed point O of a smooth horizontal table and the other end to one extremity of a uniform rod of length $12a$. When the string and rod are at rest in one straight line a perpendicular blow is applied to the rod at its middle point. Apply the principles of Energy and Momentum to shew that when in the subsequent motion the string and rod are at right angles they have the same angular velocity.

12. AB, BC, and CD are three equal uniform rods lying in a straight line on a smooth table, and they are freely jointed at B and C. A blow is applied at the centre of BC in a direction perpendicular to BC. If ω be the initial angular velocity of AB or CD, and θ the angle they make with BC at any time, shew that the angular velocity then is $\frac{\omega}{\sqrt{1+\sin^2\theta}}$.

13. A uniform rod, moving perpendicularly to its length on a smooth horizontal plane, strikes a fixed circular disc of radius b at a point distant c from the centre of the rod. Find the magnitude of the impulse; and shew that, if there be no sliding between the rod and the disc, the centre of the rod will come into contact with the disc after a time

$$\frac{\sqrt{k^2+c^2}}{2bcv}\left\{c\sqrt{k^2+c^2}+k^2\log\frac{c+\sqrt{k^2+c^2}}{k}\right\},$$

where k is the radius of gyration of the rod about its centre, and v is the initial velocity of the rod.

14. A uniform rod, of length $2a$, is freely jointed at one end to a small ring, whose mass is equal to that of the rod. The ring is free to slide on a smooth horizontal wire and initially it is at rest and the rod is vertical and below the ring and rotating with angular velocity $\sqrt{\frac{3g}{a}}$ in a vertical plane passing through the wire. When the rod is inclined at an angle θ to the vertical, shew that its angular velocity is

$$\sqrt{\frac{3g}{a}\cdot\frac{1+4\cos\theta}{8-3\cos^2\theta}},$$

and find the velocity of the ring then.

[The horizontal momentum of the system is constant throughout the motion, and the change in the kinetic energy is equal to the work done against gravity.]

15. A uniform rod AB hangs from a smooth horizontal wire by means of a small ring at A; a blow is given to the end A which causes it to start off with velocity u along the wire; shew that the angular velocity ω of the rod when it is inclined at an angle θ to the horizontal is given by the equation $\omega^2(1+3\cos^2\theta)$ $=\frac{9u^2}{16a^2}-\frac{6g}{a}(1-\sin\theta)$.

16. A hoop, of radius a, rolling on a horizontal road with velocity v comes into collision with a rough inelastic kerb of height h, which is perpendicular to the plane of the hoop. Shew that, if the hoop is to clear the kerb without jumping, v must be

$$>\frac{2a}{2a-h}\sqrt{gh} \quad\text{and}\quad <\frac{2a}{2a-h}\sqrt{g(a-h)}.$$

17. An inelastic uniform sphere, of radius a, is moving without rotation on a smooth table when it impinges on a thin rough horizontal rod, at right angles to its direction of motion and at a height b from the plane; shew that it will just roll over the rod if its velocity be

$$\frac{a}{a-b}\sqrt{\frac{14gb}{5}},\ \text{and } b \text{ be } <\frac{7a}{17}.$$

18. A sphere, of radius a, rolling on a rough table comes to a slit, of breadth b, perpendicular to its path; if V be its velocity, shew that the condition it should cross the slit without jumping is

$$V^2>\frac{100}{7}ga(1-\cos\alpha)\sin^2\alpha\frac{14-10\sin^2\alpha}{(7-10\sin^2\alpha)^2},$$

where $\quad b=2a\sin\alpha, \quad \text{and} \quad 17ga\cos\alpha>7V^2+10ga.$

19. A sphere, of radius a, rests between two thin parallel perfectly rough rods A and B in the same horizontal plane at a distance apart equal to $2b$; the sphere is turned about A till its centre is very nearly vertically over A; it is then allowed to fall back; shew that it will rock between A and B if $10b^2 < 7a^2$, and that, after the nth impact, it will turn till the radius vector to the point of contact is inclined at an angle θ_n to the vertical given by the equation

$$\cos\theta_n = \frac{\sqrt{a^2-b^2}}{a} + \frac{a-\sqrt{a^2-b^2}}{a}\left(1-\frac{10b^2}{7a^2}\right)^{2n}.$$

Shew also that the successive maximum heights of the centre above its equilibrium position form a descending geometrical progression.

20. An inelastic cube slides down a plane inclined at α to the horizon and comes with velocity V against a small fixed nail. If it tumble over the nail and go on sliding down the plane, shew that the least value of V^2 is $\frac{16ga}{3}[\sqrt{2}-\cos\alpha-\sin\alpha]$.

21. A cube, of side $2a$, rests with one edge on a rough horizontal plane and the opposite edge vertically over the first; it falls over and hits the plane; shew that it will start rotating about another edge and that its centre of inertia will rise to a height $\frac{a}{16}(15+\sqrt{2})$.

22. A uniform cube, of side $2a$, rolls round four parallel edges in succession along a rough horizontal plane. Initially, with a face just in contact, Ω is the angular velocity round the edge which remains in contact till the first impact. Shew that the cube will continue to roll forward after the nth impact as long as $4^{2n-1} < a\Omega^2(\sqrt{2}+1) \div 3g$.

23. A rectangular parallelepiped, of mass $3m$, having a square base $ABCD$ rests on a horizontal plane and is movable about CD as a hinge. The height of the solid is $3a$ and the side of the base a. A particle m moving with a horizontal velocity v strikes directly the middle of that vertical face which stands on AB, and sticks there without penetrating. Shew that the solid will not upset unless $v^2 > \frac{53}{9}ga$.

24. A uniform cubical block stands on a railway truck, which is moving with velocity V, two of its faces being perpendicular to the direction of motion. If the lower edge of the front face of the block be hinged to the truck and the truck be suddenly stopped, shew that the block will turn over if V is greater than $\frac{4}{3}\sqrt{3ga(\sqrt{2}-1)}$, where $2a$ is the side of the block.

25. A string, of length b, with a particle of mass m attached to one end, is fastened to a point on the edge of a circular disc, of mass M and radius a, free to turn about its centre. The whole lies on a smooth table with the string along a radius produced, and the particle is set in motion. Shew that the string will never wrap round the disc if $aM < 4bm$.

26. A uniform rod, of length $2a$ and mass nm, has a string attached to it at one end, the other end of the string being attached to a particle, of mass m; the rod and string being placed in a straight line on a smooth table, the particle is projected with velocity V perpendicular to the string; shew that the greatest angle that the string makes with the rod is

$$2\sin^{-1}\sqrt{\frac{(n+1)a}{12b}},$$

and that the angular velocities of the rod and string then are each $\frac{V}{a+b}$, where b is the length of the string.

[The linear momentum of the centre of inertia of the system and the angular momentum about it are both constant; also the kinetic energy is constant.]

27. A smooth circular disc is fixed on a smooth horizontal table and a string, having masses M and m at its ends, passes round the smooth rim leaving free straight portions in the position of tangents; if m be projected perpendicularly to the tangent to it with velocity V, and the length at any instant of this tangent be η, shew that

$$(M+m)\eta^2\dot{\eta}^2 = V^2[(M+m)a^2 + m(\eta^2 - b^2)],$$

where a is the radius of the disc and b is the initial value of η.

[The total kinetic energy is constant, and also the moment of momentum about the centre of the disc.]

28. A homogeneous elliptic cylinder rests on a rough plane; shew that the least impulsive couple that will make it roll along the plane is

$$m\sqrt{\frac{g(a-b)(a^2+5b^2)}{2}},$$

where m is the mass and a, b the semi-axes.

29. Explain why a boy in a swing can increase the arc of his swing by crouching when at the highest point of his swing and standing erect when at the lowest point.

30. One end of a square door has a horizontal hinge in a vertical wall. To the middle point A of the opposite edge is fastened a cord which passes over a small pulley let into the wall at the point B reached by A when the door is closed, B being above the hinge. At the other end of the cord is a hanging weight. The door is slowly opened until it is horizontal and then let go. Find the angular velocity of the door in any position, and shew that its kinetic energy on slamming is less than the work done in opening it in the ratio $M:M+3m$, where M, m are the masses of the door and weight.

31. A sphere is free to rotate about a fixed diameter and a fly whose mass is $\frac{2}{5}$ths that of the sphere alights at a point on the surface and crawls with uniform velocity along a rhumb line (of angle α). Shew that, when the fly reaches the pole, the sphere will have turned through one or other of the angles whose sum is $\frac{1}{\sqrt{2}} \tan\alpha \log \frac{\sqrt{2}+1}{\sqrt{2}-1}$.

32. A horizontal wheel, with buckets on its circumference, revolves about a frictionless vertical axis. Water falls into the buckets at a uniform rate of mass m per unit of time. Treating the buckets as small compared with the wheel, find the angular velocity of the wheel after time t, if Ω be its initial value; and shew that, if I be the moment of inertia of the wheel and buckets about the vertical axis, and a be the radius of the circumference on which the buckets are placed, the angle turned through by the wheel in time t is $\frac{I\Omega}{ma^2} \log_e\left(1+\frac{ma^2t}{I}\right)$.

33. A man, of mass m, stands at A on a horizontal lamina which can rotate freely about a fixed vertical axis O, and originally both man and lamina are at rest. The man proceeds to walk on the lamina and ultimately describes (relative to the lamina) a closed circle having $OA(=a)$ as diameter, and returns to the point of starting on the lamina. Shew that the lamina has moved through an angle relative to the ground given by

$$\pi\left[1-\sqrt{\frac{I}{I+ma^2}}\right],$$

where I is the moment of inertia of the lamina about the axis.

CHAPTER XVIII

LAGRANGE'S EQUATIONS IN GENERALISED COORDINATES

246. In the previous chapter we have shewn how we can directly write down equations which do not involve reactions. In the present chapter we shall obtain equations which will often give us the whole motion of the system.

They will be obtained in terms of any coordinates that we may find convenient to use, the word coordinates being extended to mean any independent quantities which, when they are given, determine the positions of the body or bodies under consideration. The number of these coordinates must be equal to the number of independent motions that the system can have, *i.e.* they must be equal in number to that of the degrees of freedom of the system.

247. Lagrange's Equations.

Let (x, y, z) be the coordinates of any particle m of the system referred to rectangular axes, and let them be expressed in terms of a certain number of independent variables θ, ϕ, ψ ... so that, if t be the time, we have

$$x=f(t, \theta, \phi, ...) \quad \text{.........................(1)},$$

with similar expressions for y and z.

These equations are not to contain $\dot{\theta}, \dot{\phi}$ ... or any other differential coefficients with regard to the time.

As usual, let dots denote differential coefficients with regard to the time, and let $\frac{dx}{d\theta}, \frac{dx}{d\phi}$ ... denote partial differential coefficients.

Then, differentiating (1), we have

$$\dot{x}=\frac{dx}{dt}+\frac{dx}{d\theta}.\dot{\theta}+\frac{dx}{d\phi}.\dot{\phi}+ \quad \text{........................(2).}$$

On differentiating (2) partially with regard to $\dot{\theta}$, we have

$$\frac{d\dot{x}}{d\dot{\theta}}=\frac{dx}{d\theta} \quad \text{...............................(3).}$$

Again, differentiating (2) with regard to θ, we have

$$\frac{d\dot{x}}{d\theta}=\frac{d^2x}{d\theta dt}+\frac{d^2x}{d\theta^2}.\dot{\theta}+\frac{d^2x}{d\theta d\phi}.\dot{\phi}+...$$

$$=\frac{d}{dt}\left[\frac{dx}{d\theta}\right] \quad \text{...(4).}$$

If T be the kinetic energy of the system, then

$$T=\tfrac{1}{2}\Sigma m[\dot{x}^2+\dot{y}^2+\dot{z}^2] \quad \text{......................(5).}$$

Now the reversed effective forces and the impressed forces form a system of forces in equilibrium, so that their equation of virtual work vanishes; in other words the virtual work of the effective forces = the virtual work of the impressed forces.

The first of these, for a variation of θ only,

$$=\Sigma m\left[\ddot{x}\frac{dx}{d\theta}+\ddot{y}\frac{dy}{d\theta}+\ddot{z}\frac{dz}{d\theta}\right]\delta\theta$$

$$=\frac{d}{dt}\Sigma m\left[\dot{x}\frac{dx}{d\theta}+\ldots+\ldots\right]\delta\theta-\Sigma m\left[\dot{x}\frac{d}{dt}\left(\frac{dx}{d\theta}\right)+\ldots+\ldots\right]\delta\theta$$

$$=\frac{d}{dt}\Sigma m\left[\dot{x}\frac{d\dot{x}}{d\dot{\theta}}+\ldots+\ldots\right]\delta\theta-\Sigma m\left[\dot{x}\frac{d\dot{x}}{d\theta}+\ldots+\ldots\right]\delta\theta,$$

by equations (3) and (4),

$$=\frac{d}{dt}\cdot\frac{d}{d\dot{\theta}}\Sigma m\,.\,\tfrac{1}{2}(\dot{x}^2+\dot{y}^2+\dot{z}^2)\delta\theta-\frac{d}{d\theta}\Sigma m\,.\,\tfrac{1}{2}[\dot{x}^2+\dot{y}^2+\dot{z}^2]\delta\theta$$

$$=\left[\frac{d}{dt}\frac{dT}{d\dot{\theta}}-\frac{dT}{d\theta}\right]\delta\theta, \text{ by equation (5), ...(6).}$$

Again, if V be the Work function, we have the virtual work of the impressed forces, for a variation of θ alone,

$$=\Sigma m\left[X\frac{dx}{d\theta}+Y\frac{dy}{d\theta}+Z\frac{dz}{d\theta}\right]\delta\theta$$

$$=\left[\frac{dV}{dx}\frac{dx}{d\theta}+\frac{dV}{dy}\frac{dy}{d\theta}+\frac{dV}{dz}\frac{dz}{d\theta}\right]\delta\theta=\frac{dV}{d\theta}\,.\,\delta\theta \quad \text{...........(7).}$$

Equating (6) and (7), we have

$$\frac{d}{dt}\left(\frac{dT}{d\dot{\theta}}\right)-\frac{dT}{d\theta}=\frac{dV}{d\theta} \quad \text{........................(8).}$$

Similarly, we have the equations

$$\frac{d}{dt}\left(\frac{dT}{d\dot{\phi}}\right)-\frac{dT}{d\phi}=\frac{dV}{d\phi},$$

$$\frac{d}{dt}\left(\frac{dT}{d\dot{\psi}}\right)-\frac{dT}{d\psi}=\frac{dV}{d\psi},$$

and so on, there being one equation corresponding to each independent coordinate of the system.

These equations are known as Lagrange's equations in Generalised Coordinates.

COR. If K be the potential energy of the system, since V=a constant$-K$, equation (8) becomes

$$\frac{d}{dt}\left(\frac{dT}{d\dot{\theta}}\right)-\frac{dT}{d\theta}+\frac{dK}{d\theta}=0.$$

If we put $T-K=L$, so that L is equal to the difference between the kinetic and potential energies then, since V does not contain $\dot{\theta}$, $\dot{\phi}$, etc., this equation can be written in the form

$$\frac{d}{dt}\left[\frac{dL}{d\dot{\theta}}\right]-\frac{dL}{d\theta}=0.$$

L is called the Langrangian Function.

248. When a system is such that the coordinates of any particle of it can be expressed in terms of independent coordinates by equations which do not contain differential coefficients with regard to the time, the system is said to be holonomous.

249. *Ex.* 1. *A homogeneous rod OA, of mass m_1 and length 2a, is freely hinged at O to a fixed point; at its other end is freely attached another homogeneous rod AB, of mass m_2 and length 2b; the system moves under gravity; find equations to determine the motion.*

Let G_1 and G_2 be the centres of mass of the rods, and θ and ϕ their inclinations to the vertical at time t.

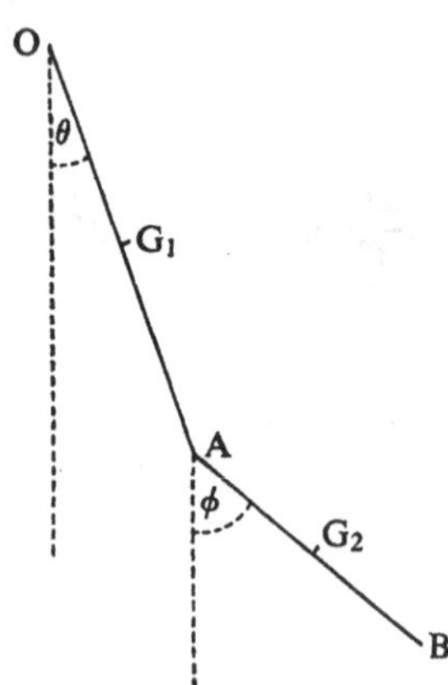

The kinetic energy of OA is

$$\frac{1}{2}m_1.\frac{4a^2}{3}.\dot{\theta}^2.$$

G_2 is turning round A with velocity $b\dot{\phi}$, whilst A is turning round O with velocity $2a\dot{\theta}$. Hence the square of the velocity of G_2

$$=(2a\dot{\theta}\cos\theta+b\dot{\phi}\cos\phi)^2+(2a\dot{\theta}\sin\theta+b\dot{\phi}\sin\phi)^2$$
$$=4a^2\dot{\theta}^2+b^2\dot{\phi}^2+4ab\dot{\theta}\dot{\phi}\cos(\theta-\phi).$$

Also the kinetic energy of the rod about G_2

$$=\frac{1}{2}m_2.\frac{b^2}{3}\dot{\phi}^2.$$

$$\therefore\ T=\frac{1}{2}m_1.\frac{4a^2}{3}\dot{\theta}^2+\frac{1}{2}m_2[4a^2\dot{\theta}^2+b^2\dot{\phi}^2+4ab\dot{\theta}\dot{\phi}\cos(\phi-\theta)]+\frac{1}{2}m_2.\frac{b^2}{3}\dot{\phi}^2$$

$$=\frac{1}{2}.\left(\frac{m_1}{3}+m_2\right).4a^2\dot{\theta}^2+\frac{1}{2}m_2.\frac{4b^2}{3}\dot{\phi}^2+\frac{1}{2}m_2.4ab\dot{\theta}\dot{\phi}\cos(\phi-\theta)\ldots(1).$$

Also the work function V

$$=m_1 ga \cos \theta + m_2 g(2a \cos \theta + b \cos \phi) + C \quad \text{.........(2).}$$

Lagrange's θ-equation then gives

$$\frac{d}{dt}\left[\left(\frac{m_1}{3}+m_2\right).4a^2\dot{\theta}+m_2.2ab\dot{\phi} \cos \overline{\phi-\theta}\right]-2m_2 ab\dot{\theta}\dot{\phi} \sin (\phi-\theta)$$

$$=\frac{d}{dt}\left(\frac{dT}{d\dot{\theta}}\right)-\frac{dT}{d\theta}=\frac{dV}{d\theta}$$

$$=-(m_1+2m_2)ga \sin \theta,$$

i.e. $$\left(\frac{m_1}{3}+m_2\right)4a\ddot{\theta}+2m_2 b[\ddot{\phi} \cos \overline{\phi-\theta}-\dot{\phi}^2 \sin \overline{\phi-\theta}]$$

$$=-g(m_1+2m_2) \sin \theta \text{......(3).}$$

So the ϕ-equation is

$$\frac{d}{dt}\left[m_2\left\{\frac{4b^2}{3}\dot{\phi}+2ab\dot{\theta} \cos (\phi-\theta)\right\}\right]+m_2.2ab\dot{\theta}\dot{\phi} \sin (\phi-\theta)$$

$$=-m_2 bg \sin \phi,$$

i.e. $$\frac{4b}{3}\ddot{\phi}+2a\ddot{\theta} \cos (\phi-\theta)+2a\dot{\theta}^2 \sin (\phi-\theta)=-g \sin \phi \text{ ......(4).}$$

Multiplying (3) by $a\dot{\theta}$, (4) by $m_2 b\dot{\phi}$, adding and integrating, we have

$$\frac{1}{2}\left(\frac{m_1}{3}+m_2\right)4a^2\dot{\theta}^2+\frac{1}{2}m_2.\frac{4b^2}{3}\dot{\phi}^2+\frac{1}{2}m_2.4ab\dot{\theta}\dot{\phi} \cos (\phi-\theta)$$

$$=(m_1+2m_2)ga \cos \theta+m_2 gb \cos \phi+C \quad \text{...........(5).}$$

This is the equation of Energy.

Again multiplying (3) by a, and (4) by $m_2 b$ we have, on adding,

$$\frac{d}{dt}\left[\left(\frac{m_1}{3}+m_2\right).4a^2\dot{\theta}+m_2\left\{\frac{4b^2}{3}\dot{\phi}+2ab(\dot{\theta}+\dot{\phi}) \cos (\phi-\theta)\right\}\right]$$

$$=-ag(m_1+2m_2) \sin \theta-m_2 bg \sin \phi.$$

This is the equation derived by taking moments about O for the system.

Ex. 2. A uniform rod, of length 2a, can turn freely about one end, which is fixed. Initially it is inclined at an acute angle α to the downward drawn vertical and it is set rotating about a vertical axis through its fixed end with angular velocity ω. Shew that, during the motion, the rod is always inclined to the vertical at an angle which is $\gtrless \alpha$, according as $\omega^2 \gtrless \frac{3g}{4a \cos \alpha}$, and that in each case its motion is included between the inclinations α and

$$\cos^{-1}[-n+\sqrt{1-2n \cos \alpha+n^2}],$$

where $$n=\frac{a\omega^2 \sin^2 \alpha}{3g}.$$

If it be slightly disturbed when revolving steadily at a constant inclination α, *shew that the time of a small oscillation is* $2\pi\sqrt{\dfrac{4a\cos\alpha}{3g(1+3\cos^2\alpha)}}$.

At any time t let the rod be inclined at θ to the vertical, and let the plane through it and the vertical have turned through an angle ϕ from its initial position.

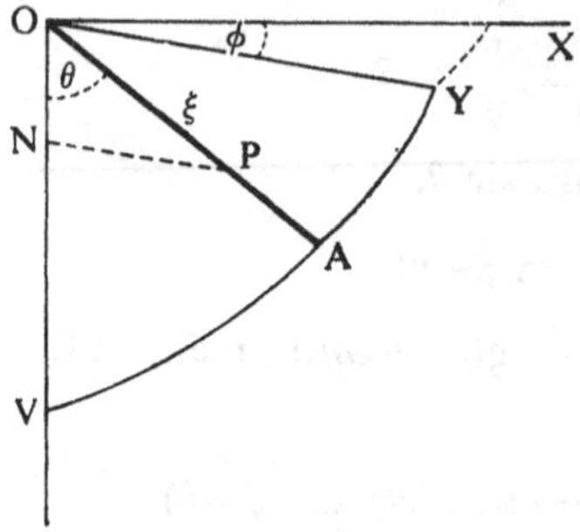

Consider any element $\dfrac{d\xi}{2a}\,.\,m$ of the rod at P, where $OP=\xi$.

If PN be drawn perpendicular to the vertical through the end O of the rod, then P is moving perpendicular to NP with velocity $\dot\phi\,.\,NP$, *i.e.*

$$\xi \sin\theta\dot\phi,$$

and it is moving perpendicular to OP in the plane VOA with velocity $\xi\dot\theta$.

Hence the kinetic energy of this element

$$=\frac{1}{2}\cdot\frac{d\xi}{2a}.m[\xi^2\sin^2\theta\dot\phi^2+\xi^2\dot\theta^2].$$

Therefore the whole kinetic energy T

$$=\frac{1}{2}\cdot\frac{m}{2a}(\sin^2\theta\dot\phi^2+\dot\theta^2)\int_0^{2a}\xi^2 d\xi=\frac{2ma^2}{3}(\dot\phi^2\sin^2\theta+\dot\theta^2).$$

Also the work function V

$$=m.g.a\cos\theta+C.$$

Hence Lagrange's equations give

$$\frac{d}{dt}\left[\frac{4ma^2}{3}\dot\theta\right]-\frac{2ma^2}{3}\dot\phi^2.2\sin\theta\cos\theta=-mga\sin\theta,$$

and

$$\frac{d}{dt}\left[\frac{4ma^2}{3}\dot\phi\sin^2\theta\right]=0,$$

i.e.

$$\ddot\theta-\dot\phi^2\sin\theta\cos\theta=-\frac{3g}{4a}\sin\theta \quad\text{..................(1)},$$

and

$$\dot\phi\sin^2\theta=\text{constant}=\omega\sin^2\alpha \quad\text{..................(2)}.$$

(1) and (2) give, on the elimination of $\dot\phi$,

$$\ddot\theta-\frac{\omega^2\sin^4\alpha}{\sin^3\theta}\cos\theta=\frac{-3g}{4a}\sin\theta \quad\text{..................(3)}.$$

Steady motion. The rod goes round at a constant inclination α to the vertical if $\ddot{\theta}$ be zero when $\theta=\alpha$, *i.e.* if

$$\omega^2=\frac{3g}{4a\cos\alpha} \quad \text{.........................(4).}$$

When ω has not this particular value, equation (3) gives on integration

$$\dot{\theta}^2+\frac{\omega^2\sin^4\alpha}{\sin^2\theta}=\frac{3g}{2a}\cos\theta+C=\omega^2\sin^2\alpha+\frac{3g}{2a}(\cos\theta-\cos\alpha)\text{...(5),}$$

from the initial conditions.

$$\therefore\ \dot{\theta}^2=\frac{3gn}{a}\left[1-\frac{\sin^2\alpha}{\sin^2\theta}\right]+\frac{3g}{2a}(\cos\theta-\cos\alpha)$$

$$=\frac{3g}{2a}\,\frac{\cos\alpha-\cos\theta}{\sin^2\theta}[\cos^2\theta+2n\cos\theta-1+2n\cos\alpha].$$

Hence $\dot{\theta}$ is zero when $\theta=\alpha$, *i.e.* initially, or when

$$\cos^2\theta+2n\cos\theta-1+2n\cos\alpha=0,$$

i.e. when $$\cos\theta=-n+\sqrt{1-2n\cos\alpha+n^2} \quad \text{.................(6).}$$

[The + sign must be taken; for the − sign would give a value of $\cos\theta$ numerically greater than unity.]

The motion is therefore included between the values $\theta=\alpha$ and $\theta=\theta_1$ where $\cos\theta_1$ is equal to the right hand of (6).

Now $\theta_1 \gtrless \alpha$, *i.e.* the rod will rise higher than or fall below its initial position, according as $\cos\theta_1 \lesseqgtr \cos\alpha$,

i.e. according as $\sqrt{1-2n\cos\alpha+n^2}\lesseqgtr\cos\alpha+n$,

i.e. according as $\sin^2\alpha\lesseqgtr 4n\cos\alpha$,

i.e. according as $\dfrac{\sin^2\alpha}{4\cos\alpha}\lesseqgtr\dfrac{a\omega^2\sin^2\alpha}{3g}$,

i.e. according as $\omega^2\gtrless\dfrac{3g}{4a\cos\alpha}$,

i.e. according as the initial angular velocity is greater or less than that for steady motion at the inclination α.

It is clear that equation (2) might have been obtained from the principle that the moment of momentum about the vertical OV is constant.

Also the Principle of the Conservation of Energy gives

$$\frac{2ma^2}{3}(\dot{\phi}^2\sin^2\theta+\dot{\theta}^2)=mga(\cos\theta-\cos\alpha)+\frac{2ma^2}{3}\omega^2\sin^2\alpha.$$

On substituting for $\dot{\phi}$ from (2) this gives equation (5).

Small oscillations about the steady motion.

By (4) the value of ω^2 is $\dfrac{3g}{4a\cos\alpha}$ for steady motion. If ω^2 has this value, then (3) gives

$$\ddot{\theta}=\frac{3g}{4a}\left[\frac{\sin^4\alpha}{\cos\alpha}\frac{\cos\theta}{\sin^3\theta}-\sin\theta\right] \quad \text{.....................(7).}$$

Here put $\theta=\alpha+\psi$, where ψ is small, and therefore

$$\sin\theta=\sin\alpha+\psi\cos\alpha,$$

and

$$\cos\theta=\cos\alpha-\psi\sin\alpha.$$

Hence (7) gives

$$\ddot{\psi}=\frac{3g\sin\alpha}{4a}[(1-\psi\tan\alpha)(1+\psi\cot\alpha)^{-3}-(1+\psi\cot\alpha)]$$

$$=-\frac{3g\sin\alpha}{4a}.\psi[4\cot\alpha+\tan\alpha]$$

$$=-\psi.\frac{3g(1+3\cos^2\alpha)}{4a\cos\alpha},$$

on neglecting squares of ψ.

Hence the required time

$$=2\pi\sqrt{\frac{4a\cos\alpha}{3g(1+3\cos^2\alpha)}}.$$

Ex. 3. *Four equal rods, each of length* 2*a*, *are hinged at their ends so as to form a rhombus ABCD. The angles B and D are connected by an elastic string and the lowest end A rests on a horizontal plane whilst the end C slides on a smooth vertical wire passing through A; in the position of equilibrium the string is stretched to twice its natural length and the angle BAD is* 2*α. Shew that the time of a small oscillation about this position is*

$$2\pi\left\{\frac{2a(1+3\sin^2\alpha)\cos\alpha}{3g\cos 2\alpha}\right\}^{\frac{1}{2}}.$$

When the rods are inclined at an angle θ to the vertical, the component velocities of the centre of either of the upper rods are

$$\frac{d}{dt}[3a\cos\theta] \text{ and } \frac{d}{dt}(a\sin\theta), \textit{ i.e. } -3a\sin\theta.\dot{\theta} \text{ and } a\cos\theta.\dot{\theta}.$$

Hence T, the total kinetic energy,

$$=2\times\frac{1}{2}m\left[\frac{4a^2}{3}\dot{\theta}^2+(-3a\sin\theta\dot{\theta})^2+(a\cos\theta\dot{\theta})^2+\frac{a^2}{3}\dot{\theta}^2\right]$$

$$=8ma^2\dot{\theta}^2[\tfrac{1}{3}+\sin^2\theta].$$

Also the work function V

$$= -mg.2.(a\cos\theta + 3a\cos\theta) - 2\lambda\int_c^{2a\sin\theta}\frac{x-c}{c}.dx$$

$$= -8mga\cos\theta - \frac{\lambda}{c}(2a\sin\theta - c)^2,$$

where $2c$ is the stretched length of the string and λ the modulus of elasticity.

Lagrange's equation is therefore

$$\frac{d}{dt}\left[16a^2m\dot{\theta}\left(\frac{1}{3}+\sin^2\theta\right)\right] - 16ma^2\dot{\theta}^2\sin\theta\cos\theta$$

$$= 8mga\sin\theta - \frac{4\lambda}{c}a\cos\theta(2a\sin\theta - c) \ldots\ldots(1).$$

Now we are given that $\dot{\theta}$ and $\ddot{\theta}$ are zero when $\theta = \alpha$, and that $c = a\sin\alpha$.

$$\therefore \lambda = \frac{2mgc}{a\cos\alpha}.$$

In (1) putting $\theta = \alpha + \psi$, where ψ is small, and neglecting squares and products of ψ and $\dot{\psi}$, we have

$$16a^2m\ddot{\psi}(\tfrac{1}{3}+\sin^2\alpha)$$

$$= 8mga(\sin\alpha + \psi\cos\alpha) - \frac{8mg}{\cos\alpha}[\cos\alpha - \psi\sin\alpha][a\sin\alpha + 2\psi a\cos\alpha]$$

$$= -8amg\psi(\cos\alpha - \sin\alpha\tan\alpha),$$

i.e.

$$\ddot{\psi} = -\frac{3g\cos 2\alpha}{2a\cos\alpha(1+3\sin^2\alpha)}.\psi.$$

Therefore the required time $= 2\pi\sqrt{\dfrac{2a\cos\alpha(1+3\sin^2\alpha)}{3g\cos 2\alpha}}$.

Ex. 4. *Small oscillations about the stable position of equilibrium for the case of Ex.* 1 *when the masses and lengths of the rods are equal.*

When $m_1 = m_2$ and $a = b$, the equations (3) and (4) of Ex. 1 become

$$\frac{16}{3}\ddot{\theta} + 2\ddot{\phi}\cos(\phi-\theta) - 2\dot{\phi}^2\sin(\phi-\theta) = -\frac{3g}{a}\sin\theta,$$

and

$$2\ddot{\theta}\cos(\phi-\theta) + \frac{4}{3}\ddot{\phi} + 2\dot{\theta}^2\sin(\phi-\theta) = -\frac{g}{a}\sin\phi.$$

The stable position is given by $\theta = \phi = 0$. Taking θ and ϕ to be small, and neglecting $\dot{\theta}^2$ and $\dot{\phi}^2$, and putting θ and ϕ for $\sin\theta$ and $\sin\phi$, these equations become

$$\left(\frac{16}{3}D^2+\frac{3g}{a}\right)\theta+2D^2\phi=0 \quad \text{.....................(1)},$$

and
$$2D^2.\theta+\left(\frac{4}{3}D^2+\frac{g}{a}\right)\phi=0 \quad \text{.......................(2)},$$

where D stands for $\frac{d}{dt}$.

Eliminating ϕ, we have

$$\left(\frac{4}{3}D^2+\frac{g}{a}\right)\left(\frac{16}{3}D^2+\frac{3g}{a}\right)\theta-4D^4\theta=0,$$

i.e.
$$\left(D^4+3\frac{g}{a}D^2+\frac{27}{28}\frac{g^2}{a^2}\right)\theta=0.$$

To solve this equation put $\theta=L_p\cos(pt+\alpha_p)$, and we have

$$p^4-\frac{3g}{a}p^2+\frac{27}{28}\frac{g^2}{a^2}=0,$$

giving
$$p_1^2=\frac{3g}{14a}(7+2\sqrt{7}), \text{ and } p_2^2=\frac{3g}{14a}(7-2\sqrt{7}).$$

$$\therefore\ \theta=L_1\cos(p_1t+\alpha_1)+L_2\cos(p_2t+\alpha_2),$$

so that the motion of θ is given as the compounding of two simple harmonic motions of periods $\frac{2\pi}{p_1}$ and $\frac{2\pi}{p_2}$.

Similarly, we obtain $\phi=M_1\cos(p_1t+\alpha_1)+M_2\cos(p_2t+\alpha_2)$.

The constants L_1, L_2, M_1, and M_2 are not independent. For if we substitute the values of θ and ϕ in equations (1) and (2) we obtain two relations between them.

These are found to be $\frac{L_1}{M_1}=-\frac{2\sqrt{7}-1}{9}$ and $\frac{L_2}{M_2}=\frac{2\sqrt{7}+1}{9}$.

The independent constants ultimately arrived at may be determined from the initial conditions of the motion.

Equations (1) and (2) may be otherwise solved as follows. Multiply (2) by λ and add to (1), and we have

$$D^2\left[\left(\frac{16}{3}+2\lambda\right)\theta+\left(2+\frac{4}{3}\lambda\right)\phi\right]+\frac{g}{a}(3\theta+\lambda\phi)=0 \quad \text{.........(3)}.$$

Choose λ so that $\frac{\frac{16}{3}+2\lambda}{3}=\frac{2+\frac{4}{3}\lambda}{\lambda}$, *i.e.* $\lambda=\frac{-1\pm\sqrt{28}}{3}$.

Putting these values in (3), we have, after some reduction,

$$D^2[9\theta-(2\sqrt{7}+1)\phi]=-\frac{3g}{14a}[7+2\sqrt{7}][9\theta-(2\sqrt{7}+1)\phi] \quad ...(4),$$

and $$D^2[9\theta+(2\sqrt{7}-1)\phi]=-\frac{3g}{14a}[7-2\sqrt{7}][9\theta+(2\sqrt{7}-1)\phi] \quad ...(5).$$

$$\therefore\ 9\theta-(2\sqrt{7}+1)\phi=A\cos(p_1t+\alpha_1),$$

and $$9\theta+(2\sqrt{7}-1)\phi=B\cos(p_2t+\alpha_2).$$

This method of solution has the advantage of only bringing in the four necessary arbitrary constants.

250. If in the last example we put

$$9\theta-(2\sqrt{7}+1)\phi=X,$$

and $$9\theta+(2\sqrt{7}-1)\phi=Y$$

the equations (4) and (5) become

$$\frac{d^2X}{dt^2}=-\lambda X, \text{ and } \frac{d^2Y}{dt^2}=-\mu Y,$$

where λ and μ are numerical quantities.

The quantities X and Y, which are such that the corresponding equations each contain only X or Y, are called **Principal Coordinates** or **Normal Coordinates.**

More generally let the Kinetic Energy T, and the Work Function V, be expressible in terms of θ, ϕ, ψ, $\dot{\theta}$, $\dot{\phi}$, and $\dot{\psi}$ in the case of a small oscillation about a position of equilibrium, so that

$$T=A_{11}\dot{\theta}^2+A_{22}\dot{\phi}^2+A_{33}\dot{\psi}^2+2A_{12}\dot{\theta}\dot{\phi}+2A_{13}\dot{\theta}\dot{\psi}+2A_{23}\dot{\phi}\dot{\psi} \quad ...(1),$$

and $$V=C+a_1\theta+a_2\phi+a_3\psi+a_{11}\theta^2+a_{22}\phi^2+a_{33}\psi^2+... \quad ......(2).$$

If θ, ϕ, ψ be expressed in terms of X, Y, Z by linear equations of the form

$$\theta=\lambda_1X+\lambda_2Y+\lambda_3Z,$$

$$\phi=\mu_1X+\mu_2Y+\mu_3Z,$$

and $$\psi=\nu_1X+\nu_2Y+\nu_3Z,$$

and $$\lambda_1,\ \lambda_2,\ \lambda_3,\quad \mu_1,\ \mu_2,\ \mu_3,\quad \nu_1,\ \nu_2,\ \nu_3$$

be so chosen that on substitution in (1) and (2) there are no terms in T containing $\dot{Y}\dot{Z}$, $\dot{Z}\dot{X}$, $\dot{X}\dot{Y}$, and none in V containing YZ, ZX, XY, then X, Y, Z are called Principal or Normal Coordinates. For then

$$T=A_{11}'\dot{X}^2+A_{22}'\dot{Y}^2+A_{33}'\dot{Z}^2,$$

and $$V=C_1+a_1'X+a_2'Y+a_3'Z+a_{11}'X^2+a_{22}'Y^2+a_{33}'Z^2,$$

and the typical Lagrange equation is then

$$2A_{11}'\ddot{X}=a_1'+2a_{11}'X,$$

i.e. an equation containing X only.

On solving this and the two similar equations for Y and Z, we have θ given by sum of three simple harmonic motions.

Similarly if the original equations contained more coordinates than three.

251. Lagrange's Equations for Blows.

Let $\dot{x}_0$ and $\dot{x}_1$ denote the values of $\dot{x}$ before and after the action of the blows. Since the virtual moment of the effective impulses $\Sigma m(\dot{x}_1-\dot{x}_0)$, etc. is equal to the virtual moment of the impressed blows, we have, for a variation in θ only,

$$\Sigma m\left[(\dot{x}_1-\dot{x}_0)\frac{dx}{d\theta}+(\dot{y}_1-\dot{y}_0)\frac{dy}{d\theta}+(\dot{z}_1-\dot{z}_0)\frac{dz}{d\theta}\right]\delta\theta$$

$$=\Sigma m\left[X_1\frac{dx}{d\theta}+Y_1\frac{dy}{d\theta}+Z_1\frac{dz}{d\theta}\right]\delta\theta \quad \text{.................(1).}$$

Let T_0 and T_1 be the values of T just before and just after the blows. Then, from equations (3) and (5) of Art. 247,

$$\left(\frac{dT}{d\dot{\theta}}\right)_0=\Sigma m\left[\dot{x}\frac{d\dot{x}}{d\dot{\theta}}+\dot{y}\frac{d\dot{y}}{d\dot{\theta}}+\dot{z}\frac{d\dot{z}}{d\dot{\theta}}\right]_0$$

$$=\Sigma m\left[\dot{x}\frac{dx}{d\theta}+\dot{y}\frac{dy}{d\theta}+\dot{z}\frac{dz}{d\theta}\right]_0,$$

and
$$\left(\frac{dT}{d\dot{\theta}}\right)_1=\Sigma m\left[\dot{x}\frac{dx}{d\theta}+\dot{y}\frac{dy}{d\theta}+\dot{z}\frac{dz}{d\theta}\right]_1.$$

Hence the left hand of (1) is

$$\left[\left(\frac{dT}{d\dot{\theta}}\right)_1-\left(\frac{dT}{d\dot{\theta}}\right)_0\right]\delta\theta.$$

Also the right hand of (1)

$$=\left[\frac{dV_1}{dx}\frac{dx}{d\theta}+\frac{dV_1}{dy}\frac{dy}{d\theta}+\frac{dV_1}{dz}\frac{dz}{d\theta}\right]\delta\theta=\frac{dV_1}{d\theta}\delta\theta,$$

where δV_1 is the virtual work of the blows.

Hence if δV_1 be expressed in the form

$$\delta V_1=P\delta\theta+Q\delta\phi+\ldots,$$

the equation (1) can be written in the form

$$\left(\frac{dT}{d\dot{\theta}}\right)_1-\left(\frac{dT}{d\dot{\theta}}\right)_0=P \quad \text{........................(2),}$$

and similarly for the other equations.

The equation (2) may be obtained by integrating equation (8) of

Art. 247 between the limits 0 and τ, where τ is the infinitesimal time during which the blows last.

The integral of $\frac{d}{dt}\left(\frac{dT}{d\theta}\right)$ is $\left[\frac{dT}{d\theta}\right]_0^\tau$, *i.e.* $\left[\frac{dT}{d\theta}\right]_1 - \left[\frac{dT}{d\theta}\right]_0$.

Since $\frac{dT}{d\theta}$ is finite, its integral during the small time τ is ultimately zero.

The integral of $\frac{dV}{d\theta}$ is $\frac{dV_1}{d\theta}$.

Hence equation (2).

252. We give two examples of the preceding Article. Many of the examples of pp. 262—263 and also Ex. 2 of p. 266 and Ex. 14 of p. 270 may well be solved by this method.

Ex. 1. *Three equal uniform rods AB, BC, CD are freely jointed at B and C and the ends A and D are fastened to smooth fixed pivots whose distance apart is equal to the length of either rod. The frame being at rest in the form of a square, a blow J is given perpendicularly to AB at its middle point and in the plane of the square. Shew that the energy set up is* $\frac{3J^2}{40m}$, *where m is the mass of each rod. Find also the blows at the joints B and C.*

When AB, or CD, has turned through an angle θ, the energy of either is $\frac{1}{2}m.\frac{4a^2}{3}\dot{\theta}^2$, and that of BC, which remains parallel to AD, is $\frac{1}{2}m(2a\dot{\theta})^2$.

$$\therefore\ T=2.\frac{1}{2}m.\frac{4a^2}{3}\dot{\theta}^2+\frac{1}{2}m4a^2\dot{\theta}^2=\frac{10ma^2}{3}.\dot{\theta}^2.$$

$$\therefore\ \left(\frac{dT}{d\dot{\theta}}\right)_1=\frac{20ma^2}{3}\dot{\theta},\text{ and }\left(\frac{dT}{d\dot{\theta}}\right)_0=0.$$

Also $$\delta V_1=J.a\delta\theta.$$

Hence we have $$\frac{20ma^2}{3}\dot{\theta}=J.a,\ i.e.\ \dot{\theta}=\frac{3J}{20ma}.$$

$$\therefore\text{ required energy}=\frac{10ma^2\dot{\theta}^2}{3}=\frac{3J^2}{40m}.$$

If Y and Y_1 be the blows at the joints B and C then, by taking moments about A and D for the rods AB and DC, we have

$$m.\frac{4a^2}{3}\theta = J.a - Y.2a, \quad \text{and} \quad m.\frac{4a^2}{3}\theta = Y_1.2a.$$

$$\therefore\ Y = \frac{2J}{5}, \quad \text{and} \quad Y_1 = \frac{J}{10}.$$

Ex. 2. *Solve by the same method Ex.* 12 *of page* 270.

Let m_1 be the mass of the rod struck, and m_2 that of an adjacent rod, so that

$$\frac{m_1}{a} = \frac{m_2}{b} = \frac{1}{2}\frac{M'}{a+b}.$$

Let u be the velocity and ω the angular velocity communicated to to rod that is struck.

Then $$T = \frac{1}{2}.2m_1\left(u^2 + \frac{a^2}{3}\omega^2\right) + \frac{1}{2}m_2[(u+a\omega)^2 + (u-a\omega)^2]$$

$$= \frac{1}{2}M'u^2 + \frac{1}{6}M'a^2\omega^2\frac{a+3b}{a+b} \quad \text{..............................(1).}$$

Also the blow $$X = M[V - u - c\omega] \quad \text{.......................(2),}$$

and $$\delta V_1 = M[V - u - c\omega][\delta x + c\delta\theta] \quad \text{.................(3),}$$

where $u = \dot{x}$ and $\omega = \dot{\theta}$.

Hence the equations of the last article give

$$M'u = \frac{dV_1}{dx} = M(V - u - c\omega) \quad \text{.....................(4),}$$

and $$\frac{M'}{3}a^2\omega.\frac{a+3b}{a+b} = \frac{dV_1}{d\theta} = Mc[V - u - c\omega] \quad \text{...............(5).}$$

If $$\lambda \equiv \frac{3c^2}{a^2}\frac{a+b}{a+3b},$$

these give $$u = \frac{MV}{M(1+\lambda)+M'} \text{ and } c\omega = \frac{MV\lambda}{M(1+\lambda)+M'} \quad \text{.........(6).}$$

Also, by Ex. 3, Art. 207, the total loss of kinetic energy

$$= \tfrac{1}{2}X[V + (u + c\omega)] - \tfrac{1}{2}X[u + c\omega]$$
$$= \tfrac{1}{2}X.V = \tfrac{1}{2}MV[V - (u + c\omega)] = \text{etc.}$$

EXAMPLES

1. A bead, of mass M, slides on a smooth fixed wire, whose inclination to the vertical is α, and has hinged to it a rod, of mass m and length $2l$, which can move freely in the vertical plane through the wire. If the system starts from rest with the rod hanging vertically, shew that

$$\{4M + m(1 + 3\cos^2\theta)\}\, l\dot{\theta}^2 = 6(M+m)g\sin\alpha(\sin\theta - \sin\alpha),$$

where θ is the angle between the rod and the lower part of the wire.

2. A solid uniform sphere has a light rod rigidly attached to it which passes through its centre. This rod is so jointed to a fixed vertical axis that the angle, θ, between the rod and the axis may alter but the rod must turn with the axis. If the vertical axis be forced to revolve constantly with uniform angular velocity, shew that the equation of motion is of the form $\dot{\theta}^2=n^2 (\cos\theta-\cos\beta)(\cos\alpha-\cos\theta)$. Shew also that the total energy imparted to the sphere as θ increases from θ_1 to θ_2 varies as $\cos^2\theta_1-\cos^2\theta_2$.

3. A uniform rod, of mass $3m$ and length $2l$, has its middle point fixed and a mass m attached at one extremity. The rod when in a horizontal position is set rotating about a vertical axis through its centre with an angular velocity equal to $\sqrt{\frac{2ng}{l}}$. Shew that the heavy end of the rod will fall till the inclination of the rod to the vertical is $\cos^{-1}[\sqrt{n^2+1}-n]$, and will then rise again.

4. A rod OA, whose weight may be neglected, is attached at O to a fixed vertical rod OB, so that OA can freely rotate round OB in a horizontal plane. A rod XY, of length $2a$, is attached by small smooth rings at X and Y to OA and OB respectively. Find an equation to give θ, the inclination of the rod XY to the vertical at time t, if the system be started initially with angular velocity Ω about OB. Shew that the motion will be steady with the rod XY inclined at α to the vertical, if $\Omega^2=\frac{3g}{4a}\sec\alpha$, and that, if the rod be slightly disturbed from its position of steady motion, the time of a small oscillation is $4\pi\sqrt{\frac{a\cos\alpha}{3g(1+3\cos^2\alpha)}}$.

5. If in the preceding question the rod OA be compelled to rotate with constant angular velocity ω, shew that, if $4\omega^2 a>3g$, the motion will be steady when $\cos\alpha=\frac{3g}{4\omega^2 a}$, and that the time of a small oscillation will be $\frac{8\pi a\omega}{\sqrt{16\omega^4a^2-9g^2}}$.

[Reduce the system to rest by putting on the " centrifugal force " for each element of the rod XY, and apply the principle of Energy.]

6. Three equal uniform rods AB, BC, CD, each of mass m and length $2a$, are at rest in a straight line smoothly jointed at B and C. A blow I is given to the middle rod at a distance c from its centre O in a direction perpendicular to it; shew that the initial velocity of O is $\frac{2I}{3m}$, and that the initial angular velocities of the rods are

$$\frac{(5a+9c)I}{10ma^2},\quad \frac{6cI}{5ma^2},\quad \text{and}\quad \frac{(5a-9c)I}{10ma^2}.$$

7. Six equal uniform rods form a regular hexagon, loosely jointed at the angular points, and rest on a smooth table; a blow is given perpendicularly to one of them at its middle point; find the resulting motion and shew that the opposite rod begins to move with one-tenth of the velocity of the rod that is struck.

8. A framework in the form of a regular hexagon $ABCDEF$ consists of uniform rods loosely jointed at the corners and rests on a smooth table; a string tied to the middle point of AB is jerked in the direction of AB. Find the resulting initial motion and shew that the velocities along AB and DE of their middle points are in opposite directions and in the ratio of 59:4.

[Let u_1 and v_1 be the resulting velocities of the middle point of AB along and perpendicular to AB and ω_1 its angular velocity; and let u_2, v_2, and ω_2 give the motion of BC similarly, and so on. From the motion of the corners A, B, C, etc., we obtain

$$v_1 = \frac{u_6 - u_2}{\sqrt{3}} \quad \text{and} \quad a\omega_1 = \frac{u_6 - u_1 + u_2}{\sqrt{3}}, \text{ etc.}$$

Hence

$$T = \frac{1}{2}\Sigma m\left[u_1^2 + v_1^2 + \frac{a^2}{3}\omega_1^2\right] = \frac{m}{18}\Sigma[9u_1^2 + 3(u_6 - u_2)^2 + (u_6 - u_1 + u_2)^2].$$

Also $\delta V_1 = J.\delta x_1$, where $u_1 = \dot{x}_1$.

The equations written down by Art. 251 then completely determine the motion.]

9. A perfectly rough sphere lying inside a hollow cylinder, which rests on a perfectly rough plane, is slightly displaced from its position of equilibrium. Shew that the time of a small oscillation is

$$2\pi\sqrt{\frac{a-b}{g}\,\frac{14M}{10M+7m}},$$

where a is the radius of the cylinder, b that of the sphere, and M, m are the masses of the cylinder and sphere.

10. A perfectly rough sphere, of mass m and radius b, rests at the lowest point of a fixed spherical cavity of radius a. To the highest point of the movable sphere is attached a particle of mass m' and the system is disturbed. Shew that the oscillations are the same as those of a simple pendulum of length $(a-b)\dfrac{4m' + \frac{7m}{5}}{m + m'\left(2 - \frac{a}{b}\right)}$.

11. A hollow cylindrical garden roller is fitted with a counterpoise which can turn on the axis of the cylinder; the system is placed on a rough horizontal plane and oscillates under gravity; if $\dfrac{2\pi}{p}$ be the time of a small oscillation, shew that p is given by

$$p^2[(2M+M')k^2 - M'h^2] = (2M+M')gh$$

where M and M' are the masses of the roller and counterpoise, k is the radius of gyration of M' about the axis of the cylinder, and h is the distance of its centre of mass from the axis.

12. A thin circular ring, of radius a and mass M, lies on a smooth horizontal plane and two tight elastic strings are attached to it at opposite ends of a diameter, the other ends of the strings being fastened to fixed points in the diameter produced. Shew that for small oscillations in the plane of the ring the periods are the values of $\dfrac{2\pi}{p}$ given by $\dfrac{Mlp^2}{2T} - 1 = 0$ or $\dfrac{b}{l-b}$ or $\dfrac{l}{a}$, where b is the natural length, l the equilibrium length, and T the equilibrium tension of each string.

13. A uniform rod AB, of length $2a$, can turn freely about a point distant c from its centre, and is at rest at an angle α to the horizon when a particle is hung by a light string of length l from one end. If the particle be displaced slightly in the vertical plane of the rod, shew that it will oscillate in the same time as a simple pendulum of length

$$l\frac{a^2 + 3ac\cos^2\alpha + 3c^2\sin^2\alpha}{a^2 + 3ac}.$$

14. A plank, of mass M, radius of gyration k and length $2b$, can swing like a see-saw across a perfectly rough fixed cylinder of radius a . At its ends hang two particles, each of mass m, by strings of length l. Shew that, as the system swings, the lengths of its equivalent pendula are l and $\frac{Mk^2+2mb^2}{(M+2m)a}$.

15. At the lowest point of a smooth circular tube, of mass M and radius a, is placed a particle of mass M'; the tube hangs in a vertical plane from its highest point, which is fixed, and can turn freely in its own plane about this point. If the system be slightly displaced, shew that the periods of the two independent oscillations of the system are $2\pi\sqrt{\frac{2a}{g}}$ and $2\pi\sqrt{\frac{M}{M+M'}\cdot\frac{a}{g}}$.

16. A string AC is tied to a fixed point at A and has a particle attached at C, and another equal one at B the middle point of AC. The system makes small oscillations under gravity; if at zero time ABC is vertical and the angular velocities of AB, BC are ω and ω', shew that at time t the inclinations θ and ϕ to the vertical of AB, BC are given by the equations $\phi+\sqrt{2}\theta=\frac{\omega'+\sqrt{2}\omega}{n}\sin nt$ and $\phi-\sqrt{2}\theta=\frac{\omega'-\sqrt{2}\omega}{n'}\sin n't$, where

$$AB=BC=a,\ n^2=\frac{g}{a}(2-\sqrt{2}) \text{ and } n'^2=\frac{g}{a}(2+\sqrt{2}).$$

17. A uniform straight rod, of length $2a$, is freely movable about its centre and a particle of mass one-third that of the rod is attached by a light inextensible string, of length a, to one end of the rod; shew that one period of principal oscillation is $(\sqrt{5}+1)\pi\sqrt{\frac{a}{g}}$.

18. A uniform rod, of length $2a$, which has one end attached to a fixed point by a light inextensible string, of length $\frac{5a}{12}$, is performing small oscillations in a vertical plane about its position of equilibrium. Find its position at any time, and shew that the periods of its principal oscillations are $2\pi\sqrt{\frac{5a}{3g}}$ and $\pi\sqrt{\frac{a}{3g}}$.

19. A uniform rod, of mass $5m$ and length $2a$, turns freely about one end which is fixed; to its other extremity is attached one end of a light string, of length $2a$, which carries at its other end a particle of mass m; shew that the periods of the small oscillations in a vertical plane are the same as those of simple pendulums of lengths $\frac{2a}{3}$ and $\frac{20a}{7}$.

20. A rough plank, $2a$ feet long, is placed symmetrically across a light cylinder, of radius a, which rests and is free to roll on a perfectly rough horizontal plane. A heavy particle whose mass is n times that of the plank is embedded in the cylinder at its lowest point. If the system is slightly displaced, shew that its periods of oscillation are the values of $\frac{2\pi}{p}\sqrt{\frac{a}{g}}$ given by the equation $4p^4-(n+12)p^2+3(n-1)=0$.

21. To a point of a solid homogeneous sphere, of mass M, is freely hinged one end of a homogeneous rod, of mass nM, and the other end is freely hinged to a fixed point. If the system make small oscillations under gravity about the position of equilibrium, the centre of the sphere and the rod being always in a vertical plane passing through the fixed point, shew that the periods of the principal oscillations are the values of $\frac{2\pi}{p}$ given by the equation

$$2ab(6+7n)\,p^4 - p^2 g\,\{10a\,(3+n)+21b\,(2+n)\} + 15g^2\,(2+n)=0,$$

where a is the length of the rod and b is the radius of the sphere.

CHAPTER XIX

SMALL OSCILLATIONS. INITIAL MOTIONS. TENDENCY TO BREAK

253. In the preceding chapters we have had several examples of small oscillations, and in the last chapter we considered the application of Lagrange's equations to some problems of this class.

When the oscillation is that of a single body and the motion is in one plane, it is often convenient to make use of the properties of the instantaneous centre. By Art. 214 we know that, if the motion be a small oscillation, we may take moments about the instantaneous centre I as if it were a fixed point, and the equation of motion becomes

$$Mk^2\frac{d^2\theta}{dt^2} = \text{the moment of the impressed forces about } I.$$

Since the motion is a small oscillation the right-hand member must be small, and therefore $\ddot{\theta}$ must be small. Hence any terms in Mk^2 which contain θ may be neglected since we are leaving out all quantities of the second order, *i.e.* we may in calculating Mk^2 take the body in its undisturbed position. In the right-hand member we have no small quantity as multiplier; hence in finding it we must take the disturbed position.

The student will best understand the theory by a careful study of an example.

254. *Ex. One-half of a thin uniform hollow cylinder, cut off by a plane through its axis, is performing small oscillations on a horizontal floor. Shew that if a be the radius of the cylinder, the time of a small oscillation is* $2\pi\sqrt{\frac{(\pi-2)a}{g}}$ *or* $2\pi\sqrt{\frac{(\pi^2-4)a}{2\pi g}}$, *according as the plane is rough enough to prevent any sliding, or smooth.*

Let C be the centre of the flat base of the cylinder, and G its centre of inertia, so that $CG=\frac{2a}{\pi}$; let N be the point of contact with the floor in the vertical plane through CG, and

$$\theta = \angle NCG.$$

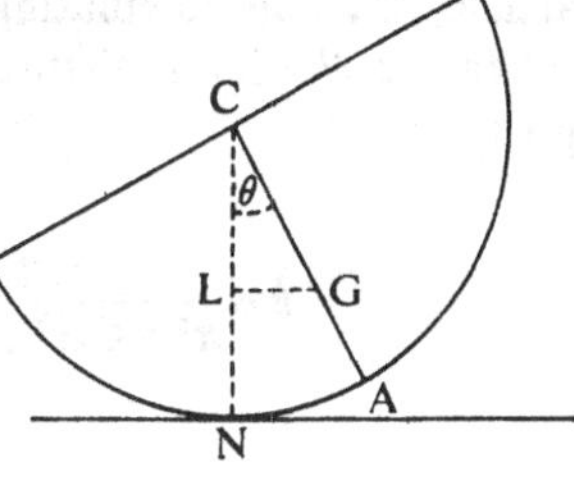

If the floor be perfectly rough, N is the instantaneous centre of rotation;

hence, taking moments about it we have, if k is the radius of gyration about G,

$$M[k^2+NG^2]\ddot{\theta}=-Mg.CG.\sin\theta \quad\text{.................(1).}$$

Now $$NG^2=a^2+CG^2-2a.CG.\cos\theta,$$

and $$M(k^2+CG^2)=\text{moment of inertia about } C=Ma^2.$$

Hence (1) gives $$\ddot{\theta}=-g.\frac{2a}{\pi}\frac{\sin\theta}{2a^2-2a.CG.\cos\theta}$$

$$=-\frac{g}{\pi}\frac{\theta}{a-CG}, \text{ since } \theta \text{ is very small,}$$

$$=-\frac{g\theta}{a(\pi-2)} \quad\text{.......................................(2).}$$

Hence the required time is $2\pi\sqrt{\dfrac{(\pi-2)a}{g}}$.

Next, let the plane be perfectly smooth, and draw GL perpendicular to CN; then L is the instantaneous centre. For, since there are no horizontal forces acting on the body, G moves in a vertical straight line, and hence the instantaneous centre is in GL. So, since N moves horizontally, the centre must be in NC; hence it is at L. Taking moments about L, we have

$$M[k^2+LG^2]\ddot{\theta}=-M.g.CG\sin\theta \quad\text{...........(3),}$$

i.e. $$M[k^2+CG^2.\sin^2\theta]\ddot{\theta}=-M.g.CG.\sin\theta.$$

Hence, when θ is very small,

$$\ddot{\theta}=-g.\frac{2a}{\pi}\frac{\theta}{a^2-CG^2+CG^2.\theta^2}$$

$$=-\frac{2ag}{\pi}\frac{\theta}{a^2-CG^2}=-\frac{2g}{a}\frac{\pi}{\pi^2-4}\theta \quad\text{..............(4).}$$

Hence the required time $=2\pi\sqrt{\dfrac{a(\pi^2-4)}{2\pi g}}$.

If we had applied the principle enunciated at the end of the previous article, then in calculating the left-hand side of (1) we should have taken for NG its undisturbed value, *viz.* AG, *i.e.* $a-CG$, and then (1) gives

$$\ddot{\theta}=\frac{-g.CG.\theta}{k^2+CG^2+a^2-2a.CG}=\frac{-g.\dfrac{2a}{\pi}}{2a^2-2a.\dfrac{2a}{\pi}}\theta, \text{ etc.}$$

In calculating the left hand of (3), for LG we take its value in the undisturbed position which is zero.

(3) then gives $$\ddot{\theta}=\frac{-g.\frac{2a}{\pi}}{a^2-CG^2}.\theta, \text{ etc.}$$

EXAMPLES

1. A thin rod, whose centre of mass divides it into portions of lengths b and c, rests in a vertical plane inside a smooth bowl of radius a; if it be slightly displaced, shew that its time of oscillation is the same as that of a simple pendulum of length $\frac{k^2+a^2-bc}{\sqrt{a^2-bc}}$, where k is the radius of gyration of the rod about its centre of mass.

2. Two rings, of masses m and m', are connected by a light rigid rod and are free to slide on a smooth vertical circular wire of radius a . If the system be slightly displaced from its position of equilibrium, shew that the length of the simple equivalent pendulum is $\frac{(m+m')a}{\sqrt{m^2+m'^2+2mm'\cos\alpha}}$, where α is the angle subtended by the rod at the centre of the wire.

3. Two uniform rods, of the same mass and of the same length $2a$ and freely jointed at a common extremity, rest upon two smooth pegs which are in the same horizontal plane so that each rod is inclined at the same angle α to the vertical; shew that the time of a small oscillation, when the joint moves in a vertical straight line through the centre of the line joining the pegs is $2\pi\sqrt{\frac{a}{9g}\frac{1+3\cos^2\alpha}{\cos\alpha}}$.

4. Two heavy uniform rods, AB and AC, each of mass M and length $2a$, are hinged at A and placed symmetrically over a smooth cylinder, of radius c, whose axis is horizontal. If they are slightly and symmetrically displaced from the position of equilibrium, shew that the time of a small oscillation is $2\pi\sqrt{\frac{a\sin\alpha}{3g}\cdot\frac{1+3\sin^2\alpha}{1+2\sin^2\alpha}}$,

where $$a\cos^3\alpha=c\sin\alpha.$$

5. A solid elliptic cylinder rests in stable equilibrium on a rough horizontal plane. Shew that the time of a small oscillation is

$$\pi\sqrt{\frac{b}{g}\frac{a^2+5b^2}{a^2-b^2}}.$$

6. A homogeneous hemisphere rests on an inclined plane, rough enough to prevent any sliding, which is inclined at α to the horizon. If it be slightly displaced, shew that its time of oscillation is the same as that of a pendulum of length $\frac{2a}{5}\left[\frac{28-40\sin^2\alpha}{\sqrt{9-64\sin^2\alpha}}-5\cos\alpha\right]$, where a is the radius of the hemisphere.

7. A sphere, whose centre of gravity G is at a distance c from its centre C is placed upon a perfectly smooth horizontal table; shew that the time of a small oscillation of its centre of gravity about its geometrical centre is $2\pi\sqrt{\frac{k^2}{cg}}\left[1+\frac{4c^2+k^2}{4k^2}\sin^2\frac{\alpha}{2}\right]$, where k is its radius of gyration about G, and α is the initial small angle which CG makes with the vertical.

8. A uniform rod is movable about its middle point, and its ends are connected by elastic strings to a fixed point; shew that the period of the rod's oscillations about a position of equilibrium is $\frac{2\pi b}{c}\sqrt{\frac{mb}{6\lambda}}$, where m is the mass of the rod, λ the modulus of elasticity of either string, b its length in the position of equilibrium, and c the distance of the fixed point from the middle point of the rod.

9. A uniform beam rests with one end on a smooth horizontal plane, and the other end is supported by a string of length l which is attached to a fixed point; shew that the time of a small oscillation in a vertical plane is $2\pi\sqrt{\frac{2l}{g}}$.

10. A uniform heavy rod OA swings from a hinge at O, and an elastic string is attached to a point C in the rod, the other end of the string being fastened to a point B vertically below O. In the position of equilibrium the string is at its natural length and the coefficient of elasticity is n times the weight of the rod. If the rod be held in a horizontal position and then set free, prove that, if ω be the angular velocity when it is vertical, then $\frac{2}{3}a^2\omega^2 = ga + ng\left[\frac{hc}{h-c} - \sqrt{h^2+c^2} + h - c\right]$, where $2a$ = length of the rod, $OC = c$, and $OB = h$.

Find also the time of a small oscillation and prove that it is not affected by the elastic string.

11. A uniform rod AB can turn freely in a vertical plane about the end A which is fixed. B is connected by a light elastic string, of natural length l, to a fixed point which is vertically above A and at a distance h from it. If the rod is in equilibrium when inclined at an angle α to the vertical, and the length of this string then is k, shew that the time of a small oscillation about this position is the same as that of a simple pendulum of length $\frac{2}{3}\frac{k^2(k-l)}{hl\sin^2\alpha}$.

12. A rhombus, formed of four equal rods freely jointed, is placed over a fixed smooth sphere in a vertical plane so that only the upper pair are in contact with the sphere. Shew that the time of a symmetrical oscillation in the vertical plane is $2\pi\sqrt{\frac{2a\cos\alpha}{3g(1+2\cos^2\alpha)}}$, where $2a$ is the length of each rod and α is the angle it makes with the vertical in a position of equilibrium.

13. A circular arc, of radius a, is fixed in a vertical plane and a uniform circular disc, of mass M and radius $\frac{a}{4}$, is placed inside so as to roll on the arc. When the disc is in a position of equilibrium, a particle of mass $\frac{M}{3}$ is fixed to it in the vertical diameter through the centre and at a distance $\frac{a}{6}$ from the centre. Shew that the time of a small oscillation about the position of equilibrium is $\frac{\pi}{6}\sqrt{\frac{83a}{g}}$.

14. A uniform rod rests in equilibrium in contact with a rough sphere, under the influence of the attraction of the sphere only. Shew that if displaced it will always oscillate, and that the period of a small oscillation is $2\pi\frac{l}{a}\frac{(a^2+l^2)^{\frac{3}{4}}}{(3\gamma m)^{\frac{1}{2}}}$, where γ is the constant of gravitation, m the mass and a the radius of the sphere, and $2l$ the length of the rod.

15. Two centres of force $\left[=\frac{\mu}{(\text{distance})^2}\right]$ are situated at two points S and H, where $SH=2b$. At the middle point of SH is fixed the centre of a uniform rod, of mass M and length $2a$; shew that the time of a small oscillation about the position of equilibrium is $2\pi(b^2-a^2)\div\sqrt{6\mu b}$.

16. A shop-sign consists of a rectangle $ABCD$ which can turn freely about its side AB which is horizontal. The wind blows horizontally with a steady velocity v and the sign is at rest inclined at an angle α to the vertical; assuming the wind-thrust on each element of the sign to be k times the relative normal velocity, find the value of α and shew that the time of a small oscillation about the position of equilibrium is

$$2\pi\sqrt{\frac{a}{g}\,\frac{4v^2\cos^2\alpha}{3v^2\cos\alpha-ga\sin^2\alpha}},\text{ where } BC=2a.$$

17. A heavy ring A, of mass nm, is free to move on a smooth horizontal wire; a string has one end attached to the ring and, after passing through another small fixed ring O at a depth h below the wire has its other end attached to a particle of mass m. Shew that the inclination θ of the string OA to the vertical is given by the equation

$$h(n+\sin^2\theta)\dot{\theta}^2=2g\cos^4\theta(\sec\alpha-\sec\theta),$$

where α is the initial value of θ.

Hence shew that the time of a small oscillation about the position of equilibrium is the same as that of a simple pendulum of length nh.

18. A straight rod AB, of mass m, hangs vertically, being supported at its upper end A by an inextensible string of length a . A string attached to B passes through a small fixed ring at a depth b below B and supports a mass M at its extremity. Shew that the rod, is displaced to a neighbouring vertical position, will remain vertical during the subsequent oscillation if $\frac{M+m}{M}=\frac{a}{b}$; and that the equivalent pendulum is of length $\frac{a-b}{2}$.

19. A uniform heavy rod AB is in motion in a vertical plane with its upper end A sliding without friction on a fixed straight horizontal bar. If the inclination of the rod to the vertical is always very small, shew that the time of a small oscillation is half that in the arc of a similar motion in which A is fixed.

20. A uniform rod, of length $2l$, rests in a horizontal position on a fixed horizontal cylinder of radius a; it is displaced in a vertical plane and rocks without any slipping; if ω be its angular velocity when inclined at an angle θ to the horizontal, shew that

$$\left(\frac{l^2}{3}+a^2\theta^2\right)\omega^2+2ga\,(\cos\theta+\theta\sin\theta)\text{ is constant.}$$

If the oscillation be small, shew that the time is $2\pi\sqrt{\frac{l^2}{3ga}}$.

21. A smooth circular wire, of radius a, rotates with constant angular velocity ω about a vertical diameter, and a uniform rod, of length $2b$, can slide with its ends on the wire. Shew that the position of equilibrium in which the rod is horizontal and below the centre of the wire is stable if $\omega^2<\frac{3gc}{(3a^2-4b^2)}$, where $c=\sqrt{a^2-b^2}$, and that then the time of a small oscillation about the stable position is $2\pi\sqrt{\frac{3a^2-2b^2}{3gc-\omega^2(3a^2-4b^2)}}$.

INITIAL MOTIONS

255. We sometimes have problems in which initial accelerations, initial reactions, and initial radii of curvature are required. We write down the equations of motion and the geometrical equations in the usual manner, differentiate the latter and simplify the results thus obtained by inserting for the variables their initial values, and by neglecting the initial velocities and angular velocities.

We then have equations to give us the second differentials of the coordinates for small values of the time t, and hence obtain approximate values of the coordinates in terms of t.

The initial values of the radius of curvature of the path of any point P is often easily obtained by finding the initial direction of its motion. This initial direction being taken as the axis of y, and y and x being its initial displacements expressed in terms of the time t, the value of the radius of curvature $= Lt \frac{y^2}{2x}$.

Some easy examples are given in the next article.

256. *Ex.* 1. *A uniform rod AB, of mass m and length 2a, has attached to it at its ends two strings, each of length l, and their other ends are attached to two fixed points, O and O′, in the same horizontal line; the rod rests in a horizontal position and the strings are inclined at an angle α to the vertical. The string O′B is now cut; find the change in the tension of the string OA and the instantaneous angular accelerations of the string and rod.*

When the string is cut let the string turn through a small angle θ whilst the rod turns through the small angle ϕ, and let T be the tension then. Let x and y be the horizontal and vertical coordinates of the centre of the rod at this instant, so that

$$x = l\sin(\alpha - \theta) + a\cos\phi = l(\sin\alpha - \theta\cos\alpha) + a \quad \text{.........(1).}$$

$$y = l\cos(\alpha - \theta) + a\sin\phi = l(\cos\alpha + \theta\sin\alpha) + a\phi \quad \text{......(2),}$$

squares of θ and ϕ being neglected.

The equations of initial motion are then

$$l\sin\alpha \,.\, \ddot{\theta} + a\ddot{\phi} = \ddot{y} = g - \frac{T}{m}\cos(\alpha - \theta) = g - \frac{T}{m}\cos\alpha \quad \text{......(3),}$$

$$-l\cos\alpha\ddot{\theta} = \ddot{x} = -\frac{T\sin(\alpha - \theta)}{m} = -\frac{T}{m}\sin\alpha \quad \text{.........(4),}$$

and $$\frac{a^2}{3}\ddot{\phi} = \frac{T}{m}a \,.\, \sin[90 - \phi - (\alpha - \theta)] = \frac{T}{m}\,.\, a\cos(\alpha + \phi - \theta)$$

$$= \frac{T}{m}\,.\, a\cos\alpha \text{...(5).}$$

Solving (3), (4), and (5) we have

$$T=\frac{mg\cos\alpha}{1+3\cos^2\alpha},\quad \ddot{\theta}=\frac{g}{l}\frac{\sin\alpha}{1+3\cos^2\alpha},\quad\text{and}\quad \ddot{\phi}=\frac{3g}{a}\frac{\cos^2\alpha}{1+3\cos^2\alpha}.$$

Ex. 2. *Two uniform rods, OA and AB, of masses m_1 and m_2 and lengths 2a and 2b, are freely jointed at A and move about the end O which is fixed. If the rods start from a horizontal position, find the initial radius of curvature and the initial path of the end B.*

By writing down Lagrange's equations for the initial state only, we have

$$T=\frac{1}{2}\left[m_1.\frac{4a^2}{3}\dot{\theta}^2+m_2\frac{b^2}{3}\dot{\phi}^2+m_2(2a\dot{\theta}+b\dot{\phi})^2\right],$$

and
$$V=m_1g.a\theta+m_2g(2a\theta+b\phi).$$

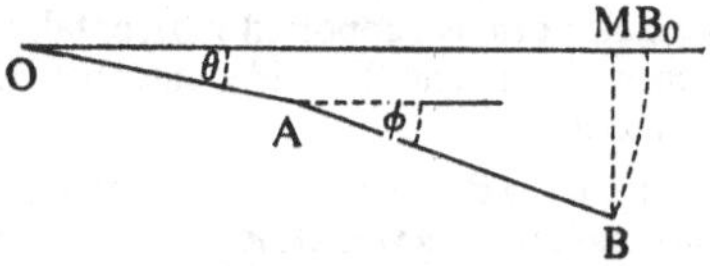

MB₀
O θ
A φ
B

Hence Lagrange's equations give

$$\left(\frac{m_1}{3}+m_2\right).4a\ddot{\theta}+2m_2b\ddot{\phi}=g(m_1+2m_2),$$

and
$$\frac{4b}{3}\ddot{\phi}+2a\ddot{\theta}=g.$$

Hence $$\ddot{\theta}=\frac{3g}{a}\frac{2m_1+m_2}{8m_1+6m_2}\quad\text{and}\quad \ddot{\phi}=-\frac{3g}{b}\frac{m_1}{8m_1+6m_2}.$$

Hence, for small values of t, we have

$$\theta=\frac{3gt^2}{2a}\frac{2m_1+m_2}{8m_1+6m_2}\quad\text{and}\quad \phi=-\frac{3gt^2}{2b}\frac{m_1}{8m_1+6m_2}\quad\text{......(1)},$$

since θ, ϕ, $\dot{\theta}$, and $\dot{\phi}$ are zero initially.

If x and y be the coordinates at this small time t of the end B of he rod, we have

$$x=2a\cos\theta+2b\cos\phi=2a+2b-a\theta^2-b\phi^2,$$

and
$$y=2a\sin\theta+2b\sin\phi=2a\theta+2b\phi.$$

$$\therefore\ \rho=\text{Lt}\frac{1}{2}\frac{BM^2}{MB_0}=\text{Lt}\frac{1}{2}\frac{y^2}{(2a+2b-x)}=\text{Lt}\frac{2(a\theta+b\phi)^2}{a\theta^2+b\phi^2}$$

$$=\frac{2ab(m_1+m_2)^2}{(2m_1+m_2)^2b+m_1{}^2a},\ \text{on substitution from (1).}$$

Also the initial path of B is easily seen to be the parabola

$$\frac{y^2}{2a+2b-x}=\frac{4(a\theta+b\phi)^2}{a\theta^2+b\phi^2}=\frac{4ab(m_1+m_2)^2}{am_1{}^2+b(2m_1+m_2)^2}.$$

EXAMPLES

1. Two strings of equal length have each an end tied to a weight C, and their other ends tied to two points A and B in the same horizontal line. If one be cut, shew that the tension of the other will be instantaneously altered in the ratio $1:2\cos^2\frac{C}{2}$.

2. A uniform beam is supported in a horizontal position by two props placed at its ends; if one prop be removed, shew that the reaction of the other suddenly changes to one-quarter of the weight of the beam.

3. The ends of a heavy beam are attached by cords of equal length to two fixed points in a horizontal line, the cords making an angle of 30° with the beam. If one of the cords be cut, shew that the initial tension of the other is two-sevenths of the weight of the beam.

4. An equilateral triangular disc is supported horizontally by three equal vertical threads attached to its corners. If one thread be cut, shew that the tension of each of the others is at once halved.

5. A uniform square lamina $ABCD$ is suspended by vertical strings attached to A and B so that AB is horizontal. Shew that, if one of the strings is cut, the tension of the other is instantaneously altered in the ratio 5:4.

6. A uniform circular disc is supported in a vertical plane by two threads attached to the ends of a horizontal diameter, each of which makes an angle α with the horizontal. If one of the threads be cut, shew that the tension of the other is suddenly altered in the ratio

$$2\sin^2\alpha : 1+2\sin^2\alpha.$$

7. A particle is suspended by three equal strings, of length a, from three points forming an equilateral triangle, of side $2b$, in a horizontal plane. If one string be cut, the tension of each of the others will be instantaneously changed in the ratio $\frac{3a^2-4b^2}{2(a^2-b^2)}$.

8. A circular disc, of radius a and weight W, is supported, with its plane horizontal, by three equal strings tied to three symmetrical points of its rim, their other ends being tied to a point at a height h above the centre of the disc. One of the strings is cut; shew that the tension of each of the others immediately becomes $W\times\frac{2h\sqrt{h^2+a^2}}{8h^2+a^2}$.

9. An equilateral triangle is suspended from a point by three strings, each equal in length to a side of the triangle, attached to its angular points; if one of the strings be cut, shew that the tensions of the other two are diminished in the ratio of 12:25.

10. A uniform hemispherical shell, of weight W, is held with its base against a smooth vertical wall and its lowest point on a smooth floor. The shell is then suddenly released. Shew that the initial thrusts on the wall and floor are respectively $\frac{3W}{10}$ and $\frac{17W}{20}$.

11. A circular half-cylinder supports two rods symmetrically placed on its flat surface, the rods being parallel to the axis of the cylinder, and rests with its curved surface on a perfectly smooth horizontal plane. If one of the rods be removed, determine the initial acceleration of the remaining rod.

12. A straight uniform rod, of mass m, passes through a smooth fixed ring and has a particle, of mass M, attached to one of its ends. Initially the rod is at rest with its middle point at the ring and inclined at an angle α to the horizontal. Shew that the initial acceleration of the particle makes with the rod an angle $\tan^{-1}\left(\dfrac{M\cot\alpha}{M+\dfrac{m}{3}}\right)$.

13. A uniform rod, of mass M and length $2a$, is movable about one end and is held in a horizontal position; to a point of the rod distant b from the fixed end is attached a heavy particle, of mass m, by means of a string. The rod is suddenly released; shew that the tension of the string at once changes to $\dfrac{Mmga(4a-3b)}{4Ma^2+3mb^2}$.

14. A horizontal rod, of mass m and length $2a$, hangs by two parallel strings, of length $2a$, attached to its ends; an angular velocity ω being suddenly communicated to it about a vertical axis through its centre, shew that the tension of each string is instantaneously increased by $\dfrac{ma\omega^2}{4}$.

15. A uniform rod, movable about one extremity, has attached to the other end a heavy particle by means of a string, the rod and string being initially in one horizontal straight line at rest; prove that the radius of curvature of the initial path of the particle is $\dfrac{4ab}{a+9b}$, where a and b are the lengths of the rod and string.

16. n rods, of lengths $a_1, a_2, \ldots a_n$, are jointed at their ends and lie in one straight line; a blow is given to one of them so that their initial angular accelerations are $\omega_1, \omega_2, \ldots \omega_n$. If one end of the rods be fixed, shew that the initial radius of curvature of the other end is

$$\frac{(a_1\omega_1+a_2\omega_2+\ldots+a_n\omega_n)^2}{a_1\omega_1^2+a_2\omega_2^2+\ldots+a_n\omega_n^2}.$$

17. A rod ABC, of length l, is constrained to pass through a fixed point B. A is attached to another rod OA, of length a, which can turn about a fixed point O situated at a distance d from B. The system is arranged so that A, O, B, C are all in a straight line in the order given. If A be given a small displacement, shew that the initial radius of curvature of the locus of C is $\dfrac{a(l-a-d)^2}{(a+d)^2-la}$.

18. A uniform smooth circular lamina, of radius a and mass M, movable about a horizontal diameter is initially horizontal, and on it is placed, at a distance c from the axis, a particle of mass m; shew that the initial radius of curvature of the path of m is equal to $12\dfrac{mc^3}{Ma^2}$.

[The distance of the particle from the axis being r when the inclination of the disc is a small angle θ, the equations of motion are

$$\ddot{r}-r\dot{\theta}^2=g\sin\theta=g\theta \quad \text{.............................(1)},$$

and

$$\frac{d}{dt}[Mk^2\dot{\theta}+mr^2\dot{\theta}]=mrg\cos\theta,$$

i.e.

$$\left(M\frac{a^2}{4}+mr^2\right)\ddot{\theta}+2mr\dot{r}\dot{\theta}=mrg \quad \text{.........................(2)}.$$

Now $\ddot{\theta}$, $\dot{\theta}$, and θ are respectively of the order 0, 1, 2 in t, and hence, from (1), $\ddot{r}$ is of the order 2, and therefore $\dot{r}$ and $r-c$ of the order 3 and 4 in t.

Hence, from (2), on neglecting powers of t, $\ddot{\theta}=\frac{4mcg}{Ma^2+4mc^2}=Ag$, say.

$$\therefore\ \dot{\theta}=Agt, \quad \text{and} \quad \theta=\tfrac{1}{2}Agt^2.$$

Therefore (1) gives $\ddot{r}=c.A^2g^2t^2+\frac{1}{2}Ag^2t^2=Ag^2(\frac{1}{2}+Ac)t^2$.

$$\therefore\ r-c=Ag^2\left(\frac{1}{2}+Ac\right)\frac{t^4}{12}.$$

Hence $$2\rho=\text{Lt}\frac{(r\sin\theta)^2}{r\cos\theta-c}=\text{Lt}\frac{c^2.\theta^2}{r-c-r\frac{\theta^2}{2}}$$

$$=\text{Lt}\frac{c^2.\frac{1}{4}A^2g^2t^4}{Ag^2\left(\frac{1}{2}+Ac\right)\frac{t^4}{12}-\frac{c}{2}.\frac{1}{4}A^2g^2t^4}\ \text{etc.}]$$

19. A uniform rod, of length $2a$ and mass M, can freely rotate about one end which is fixed; it is held in a horizontal position and on it is placed a particle, of mass m, at a distance b from the fixed end and it is then let go. Shew that the initial radius of curvature of the path of the particle is $\frac{9b^2}{4a-3b}\left(1+\frac{mb}{Ma}\right)$.

Find also the initial reaction between the rod and the particle.

20. A homogeneous rod $ACDB$, of length $2a$, is supported by two smooth pegs, C and D, each distant $\frac{a}{2}$ from the end of the rod, and the peg D is suddenly destroyed; shew that the initial radius of curvature of the path of the end B is $\frac{81a}{28}$, and that the reaction of the peg C is instantaneously increased in the ratio of 7 : 8.

21. In the previous question, if E be the middle point of AB, and the single rod AB be replaced by two iniform rods AE, EB freely jointed at E and each of the same density, shew that the same results are true.

22. A solid cylinder, of mass m, is placed on the top of another solid cylinder, of mass M, on a horizontal plane and, being slightly displaced, starts moving from rest. Shew that the initial radius of curvature of the path of its centre is $\left\{\frac{3M+2m}{3(M+m)}\right\}^2 c$, where c is the distance between its centres and all the surfaces are rough enough to prevent any sliding.

Tendency to Break

257. If we have a rod AB, of small section, which is in equilibrium under the action of any given forces, and if we consider separately the equilibrium of a portion PB, it is clear that the action of AP on PB at the section at P must balance the external forces acting on PB.

Now we know, from Statics, that the action at the section at P consists of a tension T along the tangent at P, a shear S perpendicular to T, and a couple G called the stress-couple. The external forces acting on PB being known we therefore obtain T, S, and G by the ordinary processes of resolving and taking moments.

If the rod be in motion we must, by D'Alembert's Principle, amongst the external forces include the reversed effective forces acting on the different elements of PB.

Now we know that in the case of a rod it is the couple G which breaks it, and we shall therefore take it as the measure of the tendency of the rod to break.

Hence the measure of the tendency to break at P is *the moment about P of all the forces, external and reversed impressed, on one side of P.*

The rod may be straight, or curved, but is supposed to be in one plane; it is also supposed to be of very small section; otherwise the problem is more complicated.

If we had a string instead of a rod the couple G would vanish, and in this case it is the tension T which causes it to break.

The following two examples will shew the method to be adopted in any particular case.

258. *Ex.* 1. *A uniform rod, of length $2a$, is moving in a vertical plane about one end O which is fixed; find the actions across the section of the rod at a point P, distant x from O.*

Consider any element dy of the rod at a point Q distant y from P.

Its weight is $\frac{dy}{2a}.mg$.

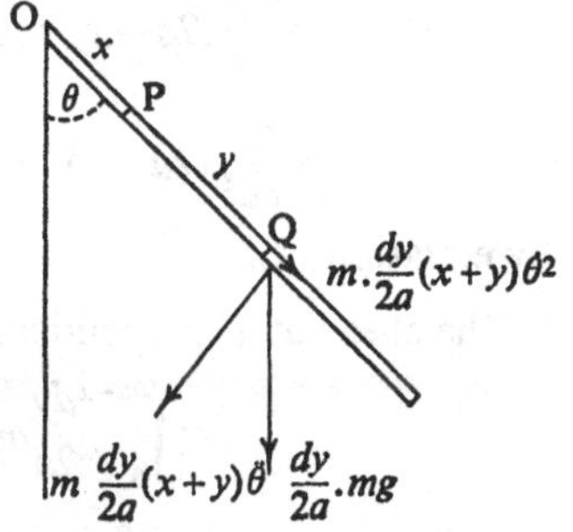

The *reversed* effective forces are

$$\frac{mdy}{2a}(x+y)\ddot{\theta} \quad \text{and} \quad \frac{mdy}{2a}(x+y)\dot{\theta}^2,$$

as marked.

These three forces, together with similar forces on all the other elements of the body, and the external forces give a system of forces in equilibrium.

The actions at P along and perpendicular to the rod and the stress couple at P, together with all such forces acting on the part PA, will be in equilibrium.

Hence the stress-couple at P in the direction ↻

$$=\int_0^{2a-x}\frac{dy}{2a}mg.y\sin\theta+\int_0^{2a-x}\frac{mdy}{2a}(x+y)\ddot{\theta}.y$$

$$=\frac{mg\sin\theta}{4a}(2a-x)^2+\frac{m}{2a}\ddot{\theta}\left[\frac{x}{2}(2a-x)^2+\frac{(2a-x)^3}{3}\right].$$

But, by taking moments about the end O of the rod, we have

$$\ddot{\theta} = -\frac{3g}{4a}\sin\theta \quad \text{.............................(1)},$$

and therefore

$$\dot{\theta}^2 = \frac{3g}{2a}(\cos\theta - \cos\alpha) \quad \text{.......................(2)},$$

where α was the initial inclination of the rod to the vertical.

Hence the stress-couple

$$= \frac{mg\sin\theta}{4a}(2a-x)^2 - \frac{3mg}{8a^2}\sin\theta(2a-x)^2\left[\frac{x}{2} + \frac{2a-x}{3}\right]$$

$$= -\frac{mg\sin\theta}{16a^2}x(2a-x)^2.$$

This is easily seen to be a maximum when $x = \frac{2a}{3}$; hence the rod will break, if it does, at a distance from O equal to one-third of the length of the rod.

The tension at P of the rod in the direction PO = the sum of the forces acting on PA in the direction PA

$$= \int_0^{2a-x} \frac{dy}{2a} mg\cos\theta + \int_0^{2a-x} \frac{m\,dy}{2a}(x+y)\dot{\theta}^2$$

$$= \frac{mg}{2a}(2a-x)\cos\theta + \frac{m\dot{\theta}^2}{4a}(2a-x)(2a+x)$$

$$= \frac{mg}{8a^2}(2a-x)\left[4a\cos\theta + 3(2a+x)(\cos\theta - \cos\alpha)\right] \quad \text{...(3)}$$

by equation (2).

The shear at P perpendicular to OP and upwards

$$= \int_0^{2a-x} \frac{dy}{2a} mg\sin\theta + \int_0^{2a-x} \frac{m\,dy}{2a}(x+y)\ddot{\theta}$$

$$= \frac{mg}{2a}(2a-x)\sin\theta + \frac{m\ddot{\theta}}{4a}(2a-x)(2a+x)$$

$$= \frac{mg\sin\theta}{16a^2}(2a-x)(2a-3x).$$

Ex. 2. *One end of a thin straight rod is held at rest, and the other is struck against an inelastic table until the rod breaks; shew that the point of fracture is at a distance from the fixed end equal to* $\frac{\sqrt{3}}{3}$ *times the total length of the rod.*

When the rod strikes the table let it be inclined at an angle α to the horizon; let ω be the angular velocity just before the impact and B the blow. Taking moments about the fixed end, we have

$$m.\frac{4a^2}{3}\omega = B.2a\cos\alpha \quad \text{.......................(1)},$$

where m is the mass and $2a$ the length of the rod.

Let us obtain the stress-couple at P.

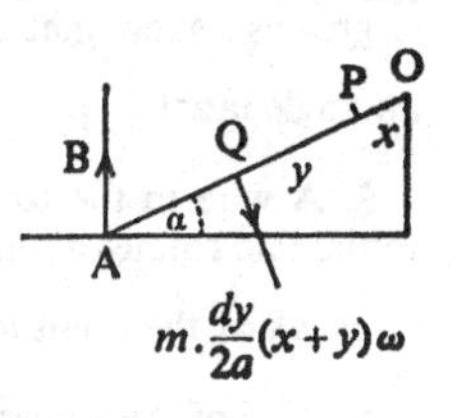

The effective impulse on an element $\frac{dy}{2a}.m$ at Q, where $PQ=y$, is $\frac{mdy}{2a}.(x+y).\omega$ upwards. Hence the reversed effective impulse at Q is in the direction marked.

Taking moments about P, the measure of the tendency to break

$$= B(2a-x)\cos\alpha - \int_0^{2a-x}\frac{dy}{2a}m(x+y)\omega.y$$

$$= B.(2a-x)\cos\alpha - \frac{m\omega}{12a}(2a-x)^2(4a+x)$$

$$= B\cos\alpha\left[(2a-x) - \frac{(2a-x)^2(4a+x)}{8a^2}\right] = \frac{B\cos\alpha}{8a^2}x(4a^2-x^2).$$

This is a maximum when $x=\frac{2a}{\sqrt{3}}$, and when B is big enough the rod will break here.

EXAMPLES

1. A thin straight rod, of length $2a$, can turn about one end which is fixed and is struck by a blow of given impulse at a distance b from the fixed end; if $b>\frac{4a}{3}$, shew that it will be most likely to snap at a distance from the fixed end equal to $2a\sqrt{\frac{3b-4a}{3b}}$.

If $b<\frac{4a}{3}$, prove that it will be most likely to snap at the point of impact.

2. A thin circular wire is cracked at a point A and is placed with the diameter AB through A vertical; B is fixed and the wire is made to rotate with angular velocity ω about AB. Find the tendency to break at any point P.

If it revolve with constant angular velocity in a horizontal plane about its centre, shew that the tendency to break at a point whose angular distance from the crack is α varies as $\sin^3\frac{\alpha}{2}$.

3. A semi-circular wire, of radius a, lying on a smooth horizontal table, turns round one extremity A with a constant angular velocity ω. If ϕ be the angle that any arc AP subtends at the centre, shew that the tendency to break at P is a maximum when $\tan \phi = \pi - \phi$.

If A be suddenly let go and the other end of the diameter through A fixed, the tendency to break due to the fixing is greatest at P where $\tan \frac{\phi}{2} = \phi$.

4. A cracked hoop rolls uniformly in a straight line on a perfectly rough horizontal plane. When the tendency to break at the point of the hoop opposite to the crack is greatest, shew that the diameter through the crack is inclined to the horizon at an angle $\tan^{-1}\left(\frac{2}{\pi}\right)$.

5. A wire in the form of the portion of the curve $r = a(1+\cos\theta)$ cut off by the initial line rotates about the origin with angular velocity ω. Shew that the tendency to break at the point $\theta = \frac{\pi}{2}$ is measured by $\frac{12\sqrt{2}}{5} m\omega^2 a^2$.

6. Two of the angles of a heavy square lamina, a side of which is a, are connected with two points equally distant from the centre of a rod of length $2a$, so that the square can rotate with the rod. The weight of the square is equal to that of the rod, and the rod when supported by its ends in a horizontal position is on the point of breaking. The rod is then held by its extremities in a vertical position and an angular velocity ω given to the square. Shew that the rod will break if $a\omega^2 > 3g$.

CHAPTER XX

MOTION OF A TOP

259. *A top, two of whose principal moments about the centre of inertia are equal, moves under the action of gravity about a fixed point O in the axis of unequal moment; find the motion if the top be initially set spinning about its axis which was initially at rest.*

Let OGC be the axis of the top, G the centre of inertia, OZ the vertical, ZOX the plane in which the axis OC was at zero time, OX and OY horizontal and at right angles.

At time t let OC be inclined at θ to the vertical, and let the plane ZOC have turned through an angle ψ from its initial position ZOX.

Let OA, OB be two perpendicular lines, each perpendicular to OC. Let A be the moment of inertia about OA or OB, and C that about OC.

At time t let ω_1, ω_2, and ω_3 be the angular velocities of the top about OA, OB, and OC.

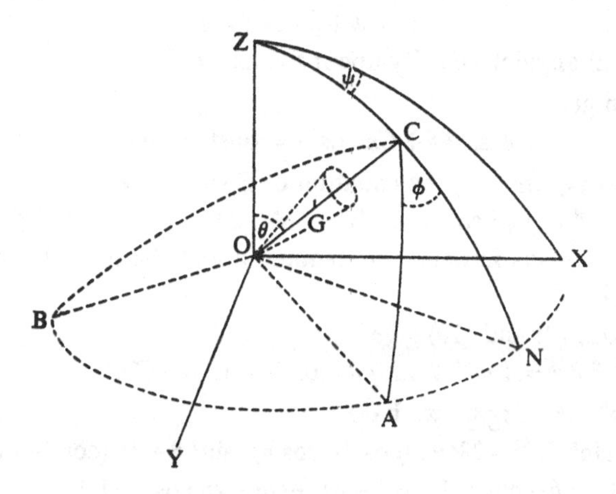

To obtain the relations between ω_1, ω_2, ω_3 and θ, ϕ, ψ consider the motions of A and C. If OC be unity, we have

$$\dot{\theta} = \text{velocity of } C \text{ along the arc } ZC = \omega_1 \sin\phi + \omega_2 \cos\phi \quad\ldots(1),$$

$$\dot{\psi}.\sin\theta = \dot{\psi} \times \text{perpendicular from } C \text{ on } OZ$$

$$= \text{velocity of } C \text{ perpendicular to the plane } ZOC$$

$$= -\omega_1 \cos\phi + \omega_2 \sin\phi \quad\ldots\ldots(2).$$

Also ω_3 = velocity of A along AB

$$= \dot{\phi} + \dot{\psi} \times \text{perpendicular from } N \text{ on } OZ$$

$$= \dot{\phi} + \dot{\psi} \sin(90° - \theta) = \dot{\phi} + \dot{\psi} \cos\theta \quad \text{.................(3).}$$

By Art. 229, the kinetic energy

$$T = \tfrac{1}{2}[A\omega_1^2 + A\omega_2^2 + C\omega_3^2]$$

$$= \tfrac{1}{2}A(\dot{\theta}^2 + \dot{\psi}^2 \sin^2\theta) + \tfrac{1}{2}C(\dot{\phi} + \dot{\psi}\cos\theta)^2 \quad \text{...........(4),}$$

by equations (1), (2), and (3).

Also $$V = Mg(h\cos i - h\cos\theta) \quad \text{...................(5),}$$

where $h = OG$ and i was the initial value of θ.

Hence Lagrange's equations give

$$\frac{d}{dt}[A\dot{\theta}] - A\dot{\psi}^2 \sin\theta\cos\theta + C(\dot{\phi} + \dot{\psi}\cos\theta)\,\dot{\psi}\sin\theta = Mgh\sin\theta \text{...(6),}$$

$$\frac{d}{dt}[C(\dot{\phi} + \dot{\psi}\cos\theta)] = 0 \text{........................(7),}$$

and $$\frac{d}{dt}[A\dot{\psi}\sin^2\theta + C\cos\theta\,(\dot{\phi} + \dot{\psi}\cos\theta)] = 0 \quad \text{...........(8).}$$

Equation (7) gives $\dot{\phi} + \dot{\psi}\cos\theta = \text{constant}$,

i.e. $$\omega_3 = \dot{\phi} + \dot{\psi}\cos\theta = n,$$

the original angular velocity about the axis OC.

(8) then gives

$$A\dot{\psi}\sin^2\theta + Cn\cos\theta = \text{const.} = Cn\cos i \quad \text{...........(9).}$$

Also, by (4) and (5), the equation of Energy gives

$$A(\dot{\theta}^2 + \dot{\psi}^2\sin^2\theta) + Cn^2 = Cn^2 + 2Mgh(\cos i - \cos\theta) \text{ ...(10),}$$

since the top was initially set spinning about the axis OC which was initially at rest.

Equations (9) and (10) give

$$A^2\sin^2\theta\dot{\theta}^2 = A\sin^2\theta \, . \, 2Mgh(\cos i - \cos\theta) - C^2n^2(\cos i - \cos\theta)^2,$$

i.e. if $C^2n^2 = A \, . \, 4Mghp$, we have

$$A\sin^2\theta \, . \, \dot{\theta}^2 = 2Mgh(\cos i - \cos\theta)\,[\sin^2\theta - 2p(\cos i - \cos\theta)]$$

$$= 2Mgh(\cos\theta - \cos i)\,[(\cos\theta - p)^2 - (p^2 - 2p\cos i + 1)]$$

$$= 2Mgh(\cos\theta - \cos i)\,[\cos\theta - p + \sqrt{p^2 - 2p\cos i + 1}]$$

$$[\cos\theta - p - \sqrt{p^2 - 2p\cos i + 1}] \quad \text{......(11).}$$

Hence $\dot{\theta}$ vanishes when $\theta = i$ or θ_1 or θ_2, where

$$\cos\theta_1 = p - \sqrt{p^2 - 2p\cos i + 1},$$

and $$\cos\theta_2 = p + \sqrt{p^2 - 2p\cos i + 1}.$$

[Clearly $\cos\theta_2 >$ unity and therefore θ_2 is imaginary.]

Also $\theta_1 > i$ since it is easily seen that $\cos\theta_1 < \cos i$, since

$$p - \cos i < \sqrt{p^2 - 2p\cos i + 1}.$$

Again, from (10), $\dot\theta^2$ is negative if $\theta < i$, *i.e.* if $\cos\theta > \cos i$; or again, from (11), if $\theta > \theta_1$, *i.e.* if

$$\cos\theta < p - \sqrt{p^2 - 2p\cos i + 1}.$$

Hence the top is never at a less inclination than i or at a greater inclination than θ_1, *i.e.* its motion is included between these limits.

Now (9) gives $A\dot\psi\sin^2\theta = Cn(\cos i - \cos\theta)$, so that $\dot\psi$ and n have the same sign.

Hence, so long as the centre of inertia G is above the point O, $\dot\psi$ and the angular velocity n of the top about its axis have the same sign. This is often expressed by saying that, if the centre of inertia is above the fixed point, the processional motion and the angular velocity are both direct or both retrograde.

[If G is below O, it is found that $\dot\psi$ and n have opposite signs.]

It is clear from equations (9) and (11) that both $\dot\theta$ and $\dot\psi$ vanish when $\theta = i$.

Also

$$\frac{d}{d\theta}\left[\frac{A}{Cn}\dot\psi\right] = \frac{d}{d\theta}\left[\frac{\cos i - \cos\theta}{\sin^2\theta}\right] = \frac{1 - 2\cos\theta\cos i + \cos^2\theta}{\sin^2\theta},$$

which is always positive when $\theta > i$.

Hence $\dot\psi$ continually increases, as θ increases, for values of θ between i and θ_1, and has its maximum value, which is easily seen to be $\frac{2Mgh}{Cn}$, when $\theta = \theta_1$.

The motion of the top may therefore be summed up thus; its angular velocity about its axis of figure remains constant throughout the motion and equal to the initial value n; the axis drops from the vertical until it reaches a position defined by $\theta = \theta_1$, and at the same time this axis revolves round the vertical with a varying angular velocity which is zero when $\theta = i$ and is a maximum when $\theta = \theta_1$.

The motion of the axis due to a change in θ only is called its "nutation."

Ex. 1. If the top be started when its axis makes an angle of 60° with the upward-drawn vertical, so that the initial spin about its axis is $\frac{A}{C}\sqrt{\frac{3Mgh}{A}}$, and the angular velocity of its axis in azimuth is $2\sqrt{\frac{Mgh}{3A}}$, its angular velocity in the meridian plane being initially zero, shew that the inclination θ of its axis to the vertical at any time t is given by the equation

$$\sec\theta = 1 + \operatorname{sech}\left\{\sqrt{\frac{Mgh}{A}}t\right\},$$

so that the axis continually approaches to the vertical without ever reaching it.

Ex. 2. Shew that the vertical pressure of the top on the point of support is equal to its weight when the inclination of its axis to the vertical is given by the least root of the equation

$$3AMgh\cos^2\theta - \cos\theta\,[C^2n^2 + 2AMbgh] + C^2n^2a - AMgh = 0,$$

where a and b are constants depending on the initial circumstances of the motion.

260. It can easily be seen from first principles that the axis of the top must have a precessional motion.

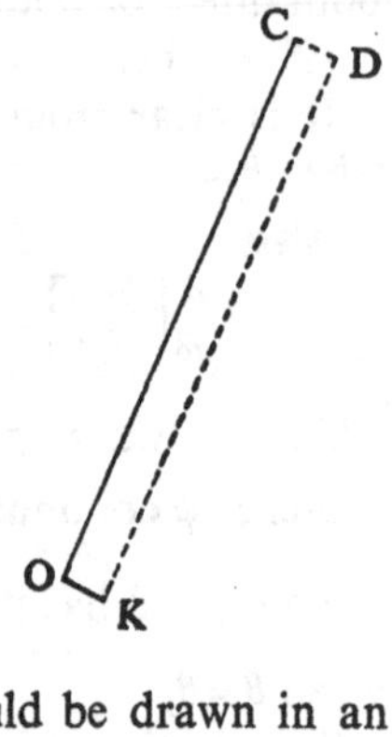

Let OC be a length measured along the axis of the top to represent the angular velocity n at time t. In time dt the weight of the cone, if G be above O, would tend to create an angular velocity which, with the usual convention as to sense, would be represented by a very small horizontal straight line OK perpendicular to OC.

The resultant of the two angular velocities represented by OK and OC is represented by OD, and the motion of the axis is thus a direct precession.

If the centre of inertia G be under O, OK would be drawn in an opposite direction and the motion would be retrograde.

261. *Two particular cases.*

If, as is generally the case, n is very large, so that p is very large also, then

$$\cos\theta_1 = p\left[1 - \left(1 - \frac{2}{p}\cos i + \frac{1}{p^2}\right)^{\frac{1}{2}}\right] = \cos i - \frac{\sin^2 i}{2p},$$

on neglecting squares of $\frac{1}{p}$.

Hence the motion is included between

$$\theta = i \text{ and } \theta = i + \frac{\sin i}{2p},$$

i.e. between i and $i + \frac{2AMgh\sin i}{C^2n^2}$.

Again if $i=0$, then $\cos\theta_1=1$, so that θ_1 is zero also and the axis remains vertical throughout the motion; but, if the axis is slightly displaced, the motion of the top is not necessarily stable.

262. *Steady motion of the top.* In this case the axis of the top describes a cone round the vertical with constant rate of rotation. Hence all through the motion

$$\theta=\alpha,\ \dot{\theta}=0,\ \ddot{\theta}=0 \text{ and } \dot{\psi}=\text{const.}=\omega.$$

The equation (6) of Art. 259 then gives

$$A\omega^2\cos\alpha-Cn\omega+Mgh=0 \quad\text{.....................(1).}$$

This equation gives two possible values of ω; in order that they may be real we must have

$$C^2n^2>4AMgh\cos\alpha.$$

We can shew that in either case the motion is stable. For, supposing that the disturbance is such that $\dot{\theta}$ is given a small value whilst $\dot{\psi}$ is unaltered initially, the equations (6), (7), and (8) of Art. 259 give

$$A\ddot{\theta}-A\dot{\psi}^2\sin\theta\cos\theta+Cn\dot{\psi}\sin\theta=Mgh\sin\theta,$$

and $$A\dot{\psi}\sin^2\theta+Cn\cos\theta=\text{const.}=A\omega\sin^2\alpha+Cn\cos\alpha.$$

Eliminating $\dot{\psi}$, we have

$$A^2\ddot{\theta}-\frac{\cos\theta}{\sin^3\theta}[A\omega\sin^2\alpha+Cn\cos\alpha-Cn\cos\theta]^2$$
$$+\frac{Cn}{\sin\theta}[A\omega\sin^2\alpha+Cn\cos\alpha-Cn\cos\theta]=AMgh\sin\theta.$$

Putting $\theta=\alpha+\theta_1$, where θ_1 is small, we have, after some reduction and using equation (1) above,

$$\ddot{\theta}_1=-\theta_1.\frac{A^2\omega^4-2AMgh\omega^2\cos\alpha+M^2g^2h^2}{A^2\omega^2}.$$

Now the numerator of the right hand is clearly always positive, so that the motion is stable for both values of ω given by (1).

Also the time of a small oscillation

$$=2\pi A\omega\div\sqrt{A^2\omega^4-2AMgh\omega^2\cos\alpha+M^2g^2h^2} \quad\text{.........(2).}$$

If the top be set in motion in the usual manner, then n is very great. Solving (1), we have

$$\omega=\frac{Cn\pm\sqrt{C^2n^2-4AMgh\cos\alpha}}{2A\cos\alpha}$$
$$=\frac{Cn}{2A\cos\alpha}\left[1\pm\left(1-\frac{2AMgh\cos\alpha}{C^2n^2}+\ldots\right)\right]$$
$$=\frac{Cn}{A\cos\alpha}\text{ or }\frac{Mgh}{Cn},\text{ very nearly.}$$

In the first of these cases the precession ω is very large and in the second case it is very small.

Also, when ω is very small, the time given by (2)

$$=\frac{2\pi A\omega}{Mgh}, \text{ nearly,}$$

$$=\frac{2\pi A}{Cn}.$$

This will be shewn independently in the next article.

263. *A top is set spinning with very great angular velocity, and initially its axis was at rest; to find the mean precessional motion and the corresponding period of nutation.*

From Art. 259 the equation for θ is

$$A\sin^2\theta\,.\,\dot\theta^2=2Mgh(\cos i-\cos\theta)\,[\sin^2\theta-2p(\cos i-\cos\theta)]\ldots(1).$$

If n, and therefore p, be great the second factor on the right-hand side cannot be positive unless $\cos i-\cos\theta$ be very small, *i.e.* unless θ be very nearly equal to i, *i.e.* unless the top go round inclined at very nearly the same angle to the vertical, and then, from (9), $\dot\psi$ is nearly constant and the motion nearly steady.

Put $\theta=i+x$, where x is very small, so that

$$\frac{\cos i-\cos\theta}{\sin\theta}=x \text{ approx.}$$

Then (1) becomes

$$A\dot x^2=2Mghx\,(\sin\theta-2px)$$

$$=2Mghx\,[\sin i-(2p-\cos i)x]$$

$$=2Mghx\,[\sin i-2px], \text{ since } p \text{ is very large.}$$

$$\therefore\ \dot x^2=\frac{4Mghp}{A}\left[\frac{x}{2p}\sin i-x^2\right]=\frac{C^2n^2}{A^2}[2qx-x^2],$$

where
$$q=\frac{\sin i}{4p}=\frac{AMgh\sin i}{C^2n^2}.$$

$$\therefore\frac{Cnt}{A}=\int\frac{dx}{\sqrt{2qx-x^2}}=\cos^{-1}\frac{q-x}{q}.$$

$$\therefore\ \theta=i+x=i+q\left[1-\cos\frac{Cnt}{A}\right].$$

Hence the period of the nutation

$$=2\pi\div\frac{Cn}{A}=\frac{2\pi A}{Cn}.$$

Again, $$\dot\psi=\frac{Cn}{A}\cdot\frac{\cos i-\cos\theta}{\sin^2\theta},$$

from equation (9) of Art. 259.

$$\therefore\ \dot\psi=\frac{Cn}{A}\cdot\frac{x}{\sin\theta}=\frac{Cn}{A}\cdot\frac{x}{\sin i}\ \text{approximately}$$

$$=\frac{Cnq}{A\sin i}\left(1-\cos\frac{Cn}{A}t\right)=\frac{Mgh}{Cn}\left(1-\cos\frac{Cnt}{A}\right).$$

$$\therefore\ \psi=\frac{Mgh}{Cn}t-\frac{AMgh}{C^2n^2}\sin\frac{Cnt}{A}.$$

The first term increases uniformly with the time, and the second is periodic and smaller, containing $\frac{1}{n^2}$.

Hence, to a first approximation, ψ increases at a mean rate of $\frac{Mgh}{Cn}$ per unit of time.

Thus, if a top be spun with very great angular velocity n, then, to start with, the axis makes small nutations of period $\frac{2\pi A}{Cn}$, and it precesses with a mean angular velocity approximately equal to $\frac{Mgh}{Cn}$. At first these oscillations are hardly noticeable; as n diminishes through the resistance of the air and friction they are most apparent, and finally we come to the case of Art. 259.

264. *A top is spinning with an angular velocity n about its axis which is vertical; find the condition of stability, if the axis be given a slight nutation.*

The work of Art. 262 will not apply here because in it we assumed that $\sin\alpha$ was not small.

We shall want the value of $\ddot\theta$ when θ is small; equation (6) of Art. 259 gives

$$A\ddot\theta=A\dot\psi^2\sin\theta\cos\theta-Cn\dot\psi\sin\theta+Mgh\sin\theta\ \ldots\ldots\ldots(1).$$

Also equation (9) gives

$$A\dot\psi\sin^2\theta=Cn(\cos i-\cos\theta)=Cn(1-\cos\theta)\ \ldots\ldots\ldots(2),$$

since the top was initially vertical.

θ being small, (2) gives

$$\dot\psi=\frac{Cn}{A}\,\frac{1}{1+\cos\theta}=\frac{Cn}{2A}+\text{terms involving }\theta^2\text{ etc.}$$

(1) then gives

$$A\ddot{\theta}=\frac{C^2n^2}{4A}.\theta-\frac{C^2n^2}{2A}.\theta+Mgh\theta+\text{terms involving }\theta^2\text{ etc.}$$

$$=-\left[\frac{C^2n^2}{4A}-Mgh\right]\theta.$$

Hence, if the top be given a small displacement from the vertical, the motion is stable, if

$$\frac{C^2n^2}{4A}>Mgh, \textit{ i.e.} \text{ if } n>\sqrt{\frac{4AMgh}{C^2}}.$$

Also the time of a nutation

$$=2\pi\sqrt{\frac{4A^2}{C^2n^2-4AMgh}}.$$

COR. If the body, instead of being a top, be a uniform sphere of radius a spinning about a vertical axis, and supported at its lowest point, then

$$h=a,\ A=M.\frac{7a^2}{5},\text{ and } C=M\frac{2a^2}{5}.$$

Therefore n must be greater than $\sqrt{\frac{35g}{a}}$.

If a=one foot, the least number of rotations per second in order that the motion may be stable

$$=\frac{n}{2\pi}=\frac{\sqrt{35.32}}{2\pi}=\text{about } 5\tfrac{1}{4}.$$

Ex. A circular disc, of radius a, has a thin rod pushed through its centre perpendicular to its plane, the length of the rod being equal to the radius of the disc; shew that the system cannot spin with the rod vertical unless the angular velocity is greater than $\sqrt{\frac{20g}{a}}$.

EXAMPLES

1. If a top is made by running a thin pin through the centre of a circular disc, of radius 3 inches, so that the length of the pin below the disc is 2 inches, prove that, for steady motion in which the rim does not touch the ground, the number of revolutions per second about the axis must exceed

$$\frac{80}{3\pi\sqrt[4]{13}}\ (=4{\cdot}5\text{ approx.}).$$

2. A top, of mass 8 lbs., spins with its point on a rough horizontal plane; its moment of inertia about its axis is $\frac{1}{8}$ lb.-ft.² and that about a perpendicular to its axis through its point is $2\frac{1}{8}$ lb.-ft.², and the distance of its centre of gravity from the point is 6 inches; shew that steady motion is possible with the axis inclined at 30° to the vertical, provided that the spin amounts to about 157 radians per second. With this limiting spin, shew that the precession is about 8·2 radians per second.

3. A top is formed of a thin disc, of radius $2a$, the axis being a needle of negligible weight passing through its centre C at right angles to its plane. The point of the needle is at a distance a from the disc. The top is set in motion with the end O of the needle in contact with a horizontal plane on which it does not slip, so that initially OC is inclined at an angle α to the vertical, and the resultant angular velocity is ω about a line bisecting the angle between OC and the vertical. Shew that the axis of the top will pass through the vertical after a time

$$\int_{\omega_0}^{\alpha} \frac{d\theta}{\sqrt{1-\cos^2\frac{\alpha}{2}\sec^2\frac{\theta}{2}+k(\cos\alpha-\cos\theta)}},$$

where $$k=\frac{g}{a\omega^2} \text{ and } k<\tfrac{1}{2}.$$

4. If initially the axis of a top is horizontal and it is set spinning with angular velocity ω in a horizontal plane, prove that the axis will start to rise if $Cn\omega>Mgh$, and that, when $Cn\omega=2Mgh$, the axis will rise to an angular distance $\cos^{-1}\left(\frac{A\omega}{Cn}\right)$, provided that $A\omega<Cn$, and will there be at instantaneous rest. Draw a rough sketch to indicate the general character of the path of the axis.

[A, C, and n have their usual meanings.]

5. A symmetrical top is set in motion on a rough horizontal plane with an angular motion about its axis of figure, the axis being inclined at an angle i to the vertical. Shew that between the greatest approach to and recess from the vertical, the centre of gravity describes an arc

$$h\tan^{-1}\left(\frac{\sin i}{p-\cos i}\right),$$

where p and h have the usual meanings.

6. A top, of vertex O, is set in motion so that the angular momentum Cn is equal to that about the vertical drawn upward through O. The inclination of the axis to the vertical was initially i, and every point on the axis was moving horizontally. Determine if the axis will become vertical in a finite time, and, if it does, shew that the velocity of the centre of gyration, P, at that instant is equal to

$$2\sqrt{gl}\tan\frac{i}{2}\sqrt{p-\cos^2\frac{i}{2}},$$

where $4A^2gp=C^2n^2l$, and $l=OP$.

What is the nature of the motion if $p<\cos^2\frac{i}{2}$?

MISCELLANEOUS EXAMPLES. I

1. A train, of mass 300 tons, is originally at rest upon a level track. It is acted on by a horizontal force which increases uniformly with the time in such a way that $F=0$ when $t=0$, and $F=5$ when $t=15$, F being measured in tons' weight and t in seconds. When in motion the train may be assumed to be acted upon by a constant frictional force equal to 3 tons' weight. Find the instant of starting, and shew that, when $t=15$, the speed of the train is 0·64 foot per second, whilst the H.P. required at this instant is about 13.

2. In starting a train the pull of the engine on the rails is at first constant and equal to P; and after the speed attains a certain value u the engine works at a constant rate R $(=Pu)$. When the engine has attained a speed v greater than u, shew that the time t and space x from the start are given by $t=\frac{M}{2R}(v^2+u^2)$ and $x=\frac{M}{3R}(v^3+\frac{1}{2}u^3)$, where M is the combined mass of the engine and train.

Calculate the time occupied, and space described, in attaining a speed of 45 miles per hour when the total mass is 300 tons, if the engine has 420 H.P. and can exert a pull equal to 12 tons' wt.

3. A unit particle is attracted by two centres of force, A and B, each of which attracts it with a force $\frac{\mu}{r^3}$ at distance r Shew that, if the particle is initially at rest at a point in AB produced distant $\sqrt{3}.a$ from the middle point of AB, it will arrive at B after a time

$$\frac{a^2}{\sqrt{\mu}}\left[1-\frac{1}{\sqrt{6}}\log_e(\sqrt{3}+\sqrt{2})\right],$$

where $2a$ is the distance AB.

4. A heavy particle, of mass m, is fastened at the middle point of an elastic string, of natural length $2a$, and the string is stretched between two points, $2l$ apart, in the same vertical line. If the particle starts from rest at a point midway between the two points, find the time of oscillation if the modulus of elasticity $\lambda \nless \frac{mga}{l-a}$.

What happens if $\lambda<\frac{mga}{l-a}$?

5. A plank, of length $2a$ and mass m, is placed with one end against a smooth vertical wall and the other end upon a smooth horizontal plane, its inclination to the horizontal being α. The plank is initially at rest and a monkey, of mass m', runs down it in such a way that the plank always remains at rest; shew that the square of his velocity when he has gone a distance x is $\frac{gx}{2\sin\alpha}\left[\frac{2(m+2m')}{m'}-\frac{x}{a}\right]$, and that the time he takes to get to the bottom of the plank is $\sqrt{\frac{2a\sin\alpha}{g}}\cos^{-1}\frac{m}{m+2m'}$.

6. A plank, of mass m, is placed on a rough plane inclined to the horizon at an angle α. A man of mass M runs down it. If the plank is not to slip, shew that the acceleration of the man must not be less than $\frac{M+m}{M}(\sin \alpha - \mu \cos \alpha)g$ nor greater than $\frac{M+m}{M}(\sin \alpha + \mu \cos \alpha)g$.

7. A chain, of length l, is placed along a line of greatest slope of a smooth plane whose inclination to the horizontal is α. If initially an end of the chain just hangs over the lower edge of the plane, prove that the chain will finally leave the plane in time $\sqrt{\frac{l}{g(1-\sin \alpha)}}.\log \cot \frac{\alpha}{2}$.

8. Referred to fixed axes the path of a particle is given by the equations $x = a \cos \omega t$, $y = b \sin \omega t$. Shew that, relatively to axes rotating with angular velocity ω, the path of the particle is a circle.

9. The greatest and least velocities of a planet in its orbit round the sun, which may be regarded as fixed, are 30 and 29·2 kilometres per second respectively. Shew that the eccentricity of the orbit is $\frac{1}{74}$.

10. A particle describes an ellipse with an acceleration which is always directed towards its centre; shew that the average value of its kinetic energy, taken with regard to the time, is equal to half the sum of its greatest and least kinetic energies.

11. A particle, of mass m, is held on a smooth table. A string attached to this particle passes through a hole in the table and supports a particle of mass $3m$. Motion is started by the particle on the table being projected with velocity V at right angles to the string. If a is the original length of the string on the table, shew that when the hanging weight has descended a distance $\frac{a}{2}$ (assuming this to be possible) its velocity will be

$$\frac{\sqrt{3}}{2}\sqrt{ga - V^2}.$$

12. A straight smooth tube is at rest in a horizontal position and contains a particle at A. The tube is rigidly attached to a point O vertically above A, and is made to rotate about O with constant angular velocity ω, so as to move in a vertical plane. If $OA = a$, shew that the distance of the particle from A at time t is $a \sinh \omega t + \frac{g}{2\omega^2}(\sinh \omega t - \sin \omega t)$.

13. A particle is projected vertically upwards with a velocity which would carry it to a height of 400 feet if there were no resistance; if the resistance varies as the square of the velocity, and the terminal velocity is 300 feet per second, shew that the height to which it actually rises is 352 feet, that its velocity on reaching the ground again is 141·2 feet per second, and that the total time of its flight is 9·4 seconds approximately.

14. A chain rests upon a smooth circular cylinder, whose radius is a and whose axis is horizontal; the length of the chain is equal to the semi-circumference of the cylinder. If the chain be slightly displaced, shew that its acceleration when a length x has slipped off the cylinder is

$$\frac{g}{\pi a}\left[x + a \sin \frac{x}{a}\right].$$

15. Two particles, of masses m and m', are joined by an elastic string of natural length a and of modulus λ; they are at rest with the string just tight when a force F begins and continues to act on the particle m in the direction away from m'.

Shew that at time t the distance between the particles is $a+\frac{2F}{mp^2}\sin^2\frac{pt}{2}$, where $p^2=\lambda\frac{m+m'}{amm'}$.

Find also the displacement of m at this time.

16. A safety device for lifts consists of an extension of the lift shaft below ground level; the floor of the lift is made to fit this well closely so that a pneumatic buffer is thus provided. A lift, weighing 3000 lbs., falls from a height of 30 feet above ground level into such a safety pit, 10 feet deep, the base of the lift being 8 feet by 5 feet. Shew that the distance x through which the lift will descend before it is stopped is given by the equation $10\log_e\frac{10-x}{10}+\frac{75}{72}+\frac{149}{144}x=0$, and hence that $x=4{\cdot}1$ feet nearly. Neglect air leakage, and assume that the pressure of the air varies inversely as its volume, and that atmospheric pressure is 15 lbs. per sq. in.

17. A heavy uniform string, of length l and mass $3m$, passes over a smooth horizontal peg and supports at one end a mass m and at the other end a mass $2m$. When there is equilibrium the mass m is pulled slowly downwards through a space $\frac{l}{9}$, and the system is then left to itself. Prove that, until the mass $2m$ reaches the peg, the space passed over by any point of the system at the end of time t is $\frac{l}{9}\left\{\cosh\sqrt{\frac{g}{l}}t-1\right\}$ and find the time in which the mass $2m$ will reach the peg and its velocity then.

18. A four-wheeled carriage is propelled by a force acting horizontally at a height h above the centre of gravity; the back and front axles are respectively at distances d_1 behind and d_2 in front of the centre of gravity. Neglecting the inertia of the wheels, shew that the greatest possible acceleration of the carriage is $\frac{gd_2}{h}$, and that the greatest retardation is $\frac{gd_1}{h}$; whilst, if the forces act at a depth h below the centre of gravity, the greatest acceleration is $\frac{gd_1}{h}$ and the greatest retardation is $\frac{gd_2}{h}$.

19. A hydrometer floats in a liquid with a volume V immersed; if the area of the cross-section of its stem is A, shew that the time of its oscillation about its position of equilibrium is $2\pi\sqrt{\frac{V}{Ag}}$.

20. A horizontal shelf is given a horizontal simple harmonic motion. The amplitude of the motion is a feet and n complete oscillations are performed per sec. A particle, of mass m lbs., is placed on the shelf at the instant when it is at the extremity of its motion. Shew that, if μ is less than $\frac{4\pi^2n^2a}{g}$, slipping between the particle and shelf will occur for a period t given by the equation $\frac{\sin 2\pi nt}{2\pi nt}=\frac{\mu g}{4\pi^2n^2a}$.

Shew that, if for a particular case this value of t is $\frac{1}{6n}$, the distance through which the particle moves relatively to the shelf in this time is

$$\frac{a}{2}\left(1-\frac{\pi\sqrt{3}}{6}\right).$$

21. In sinking a caisson in a muddy river bed, the resistance is found to increase in direct proportion to the depth in the mud.

A caisson, weighing 6 tons, sinks 4 feet under its own weight before coming to rest. Shew that, if a load of 8 tons is then suddenly added, it will sink 16 inches farther.

22. A uniform iron rod, of mass M, length a and specific gravity σ, hangs vertically just immersed in water from a light inextensible string which passes over a smooth peg and carries a counterpoise that maintains equilibrium.

A mass μM is gently added to the counterpoise; shew that, if μ exceeds a certain value, the rod will emerge from the water after a time

$$\sqrt{\frac{4a}{g}\{(\mu+2)\sigma-1\}}\,.\sin^{-1}\sqrt{\frac{1}{2\mu\sigma}}.$$

Discuss in general terms the subsequent motion.

[The counterpoise is quite clear of the water and the motion of the water is neglected.]

23. A weightless string AB consists of two portions AC, CB of unequal lengths and elasticities. The composite string is stretched and held in a vertical position with the ends A and B secured. A particle is attached to C and the steady displacement of C is found to be δ. Shew that a further small vertical displacement of C will cause the particle to execute a simple harmonic motion, and that the length of the simple equivalent pendulum is δ.

24. A particle moves under forces whose components parallel to a pair of fixed rectangular axes Ox, Oy are $-2k^2x+k\dot{y}$ and $-2k^2y-k\dot{x}$ per unit of mass. Interpret the equations giving the motion.

Shew that the path, relative to a second pair of rectangular axes rotating about the same origin with constant angular velocity k or $-2k$, is a circle.

25. A particle moves along a plane curve; v is its velocity when its distance from the origin is r, and ρ is the corresponding radius of curvature of its path; shew that the velocity of the foot of the perpendicular drawn from the origin upon the tangent to its path is $\frac{r}{\rho}v$.

26. A particle moves under a central attractive force which varies as the distance, and there is also a resisting force proportional to the velocity. Shew that the path may be an equiangular spiral.

27. A particle moves with a central acceleration $\mu u^2+\nu u^3$; find the orbit. If ν be small, shew that the path may approximately be represented by an ellipse whose axis revolves round the focus with a small angular velocity.

28. A straight tube, without mass, which moves on a horizontal table and contains a particle of mass m, is started with an angular velocity ω; find the position of the particle at the end of time t, and shew that, if θ be the angle turned through in that time, then $\tan\theta=\omega t$.

29. The angular displacement of a pendulum is given by $\theta=\theta_0 e^{-kt}\sin nt$. Shew that the successive maximum values of θ form a series in geometrical progression.

If the time of a complete oscillation is one second, and if the ratio of the first and fifth angular displacements on the same side is 4:1, shew that the time in swinging out from the position of equilibrium to an extreme displacement is 0·241 sec.

30. The horse-power required to propel a steamer of M tons displacement at its maximum speed of V feet per second is H. The resistance is proportional to the square of the speed, and the engine exerts a constant propeller thrust at all speeds. In time t from rest the steamer describes s feet, and acquires a velocity of v feet per second. Shew that

$$t=\frac{112}{55}\frac{MV^2}{Hg}\log_e\frac{V+v}{V-v}, \qquad s=\frac{112}{55}\frac{MV^3}{Hg}\log_e\frac{V^2}{V^2-v^2},$$

and

$$s=\frac{224}{55}\frac{MV^3}{Hg}\log\cosh\left(\frac{55}{224}\frac{Hgt}{MV^2}\right).$$

31. A loaded motor-car of 50 H.P. weighs 5000 lbs. and its full speed is 75 miles per hour; it is driven by a constant force at all speeds and the air resistance varies as the square of the velocity; shew that it acquires a speed of 45 miles per hour from rest in $47\frac{3}{4}$ secs., and that it has then described a distance of $1687\frac{1}{2}$ feet.

32. The horse-power required to propel a steamer of 10,000 tons displacement at a steady speed of 20 knots is 15,000. If the resistance is proportional to the square of the speed, and the engines exert a constant propeller thrust at all speeds, find the acceleration when the speed is 15 knots.

Shew that the time taken from rest to acquire a speed of 15 knots is about $1\frac{1}{2}$ minutes, given that $\log_e 7=1{\cdot}946$ and that one knot$=100$ feet per minute.

33. A train of total mass M is drawn by an engine which exerts a constant pull P at all speeds and the total resistances to the motion of the train are equal to $\mu\times(\text{velocity})^2$ per unit of its mass.

If $M=300$ tons, if the maximum speed on the level is 60 miles per hour, and if the horse-power then developed is 1500, shew that when climbing a slope of 1 in 100 the maximum speed is nearly 32 miles per hour.

34. The constant propelling force of the engines upon a ship, of M tons, is equal to P tons weight; the resistance to the motion varies as the square of the velocity and the limiting velocity is k. If, when the ship is going at full speed, the engines are reversed, shew that the ship is brought to rest in time $\frac{\pi}{4}\frac{Mk}{Pg}$ secs. after describing a distance $\frac{Mk^2}{2Pg}\log_e 2$.

35. An engine draws a total mass of M tons on the level and works at constant horse-power, overcoming a resistance to motion which varies as the square of the velocity. When the speed is u miles per hour, the tractive force is P lbs.-wt. and the limiting speed is v miles per hour; shew that it reaches a speed of V miles per hour ($V<v$) from the speed of u miles per hour in a distance

$$\frac{77}{8100}\frac{Mv^3}{Pu}\log_e\frac{v^3-u^3}{v^3-V^3}\text{ miles.}$$

If $M=264$, $P=20{,}000$, $u=15$, $v=60$, and $V=45$, shew that the distance is about 5080 feet.

36. A ship, of 1680 tons and of 230 feet in length, is travelling at full speed ahead 18 knots; the effective horse-power is then 2500. Shew that, if the engines are reversed, the ship can be stopped in about 7 lengths, assuming that the resistance is proportional to the square of the speed, and that the effective propeller thrust developed by the engines reversed is one-third of that at full speed ahead.

[1 knot$=6080$ feet per hour; $\log_e 4=1{\cdot}386$.]

37. The resistance to the motion of a train for speeds between 20 and 30 miles per hour may be taken to be $\frac{V^2}{400}+9$ in lbs.-wt. per ton, where V is the velocity in miles per hour. Steam is shut off when the speed is 30 miles per hour, and the train slows down under the given resistance. In what time will the speed fall to 20 miles per hour, and what distance will the train have described in that time?

38. The effective horse-power required to drive a ship of 15,000 tons at a steady speed of 20 knots is 25,000. Assuming the resistance to consist of two parts, one constant and one proportional to the square of the speed, these parts being equal at 20 knots, and that the propeller thrust is the same at all speeds, find the initial acceleration when starting from rest, and the acceleration when a speed of 10 knots is obtained.

Shew that this speed is attained from rest in about 93 secs. and the distance traversed is about 271 yards.

[One knot=100 feet per minute; $\log_e 4 = 1{\cdot}3863$ and $\log_e 3 = 1{\cdot}0986$.]

39. A spherical rain-drop falls through a cloud consisting of minute drops of water floating in air and occupying $\frac{1}{n}$*th* of the whole volume of the cloud; it is assumed that the rain-drop starts from rest, its radius being c, and that as it falls it picks up all the drops of water with which it comes into contact, its shape remaining spherical throughout. If, when it has fallen through a distance x, its radius is a and its velocity is v, shew that

$$x = 4n(a-c) \quad \text{and} \quad v^2 = \frac{8}{7}ng\left(a - \frac{c^7}{a^6}\right).$$

40. A uniform chain lies in a coil upon a smooth table, and a force equal to the weight of a length a of the chain is applied to one end. Shew that the length uncoiled in time t is $t\sqrt{ga}$. Shew also that the kinetic energy of the moving part of the chain at any time is equal to half the work done by the force.

41. A particle is projected horizontally with velocity $\sqrt{2ga}$ along the smooth surface of a sphere, of radius a, at the level of the centre; prove that the motion is confined between two horizontal planes at a distance $\frac{1}{2}(\sqrt{5}-1)a$ apart.

42. A particle moves under gravity on the surface of a smooth sphere of radius one metre; if the horizontal circles between which its motion is confined are at depths 40 and 50 centimetres below the centre of the sphere, shew that the velocity of the particle ranges between 404 and 428 centimetres per second.

43. A particle is projected horizontally under gravity with velocity V from a point on the inner surface of a smooth sphere at an angular distance α from the lowest point. Shew that, whatever be the value of V, this angular distance of the particle will not exceed $\pi-\alpha$ in the subsequent motion, and that the particle will not leave the surface if $3\sin\alpha > 1$.

Prove that in the subsequent motion the particle will leave the surface if

$$3\sin\alpha < 1 \text{ and } \frac{2V^2}{ag} - 7\cos\alpha \text{ lies between } \pm 3\sqrt{1-9\sin^2\alpha}.$$

44. The bob of a simple pendulum of length a is projected in a horizontal direction at right angles to the string with velocity $2\sqrt{ga}$ when the string is inclined at an angle α to the downward vertical. Shew that, if

$$4\sin^2\frac{\alpha}{2} + 6\sin\frac{\alpha}{2} - 1$$

is positive, the string will not become slack during the ensuing motion.

45. A particle is free to move within a smooth circular tube whose radius is a, which is compelled to rotate with constant angular velocity ω about a vertical axis in its own plane, whose distance is $b(>a)$ from its centre. Shew that the period of a small oscillation about the position of relative equilibrium is

$$\frac{2\pi}{\omega}\sqrt{\frac{a\sin\alpha}{b+a\sin^3\alpha}},$$

where α is the angle between the vertical and the radius to the particle when it is in equilibrium.

46. A simple pendulum, of length b, is initially at rest when the point of support is suddenly made to describe a vertical circle, of radius a, with uniform angular velocity ω, starting at the lowest point of the circle. Form the differential equation to give the inclination of the string to the vertical. Integrate it in the case when $\frac{a}{b}$ is small, and shew that in this case the inclination of the string will never exceed

$$\frac{a\omega}{b(n\sim\omega)}, \text{ where } n^2b=g.$$

47. A railway carriage, of mass M, impinges with velocity v on a carriage of mass M' at rest. The force necessary to compress a buffer through the full extent l is equal to the weight of a mass m. Assuming that the compression is proportional to the force, shew that the buffers will not be completely compressed if

$$v^2<2mgl\left(\frac{1}{M}+\frac{1}{M'}\right).$$

If v exceeds this limit, and the backing against which the buffers are driven is inelastic, the ratio of the final velocities of the carriages is

$$Mv-\left\{2mM'gl\left(1+\frac{M'}{M}\right)\right\}^{\frac{1}{2}} : Mv+\left\{2mMgl\left(1+\frac{M}{M'}\right)\right\}^{\frac{1}{2}}.$$

48. A motor car is driven and braked by the back wheels. The centre of gravity is at a height h above the ground and the back and front axles are respectively at horizontal distances d_1 behind and d_2 in front of the centre of gravity. Shew that, however great the horse-power, the maximum possible acceleration is $\frac{\mu g d_2}{d_1+d_2-\mu h}$, and the maximum retardation that can be produced by the brake is $\frac{\mu g d_2}{d_1+d_2+\mu h}$, where μ is the coefficient of friction.

If the car is driven and braked by the front wheels, shew that these quantities are respectively

$$\frac{\mu g d_1}{d_1+d_2+\mu h} \quad\text{and}\quad \frac{\mu g d_1}{d_1+d_2-\mu h}.$$

[The inertia of the wheels and driving gear is neglected.]

49. Two particles, of masses M and $2M$, are connected by an inextensible string passing over a smooth peg. From the particle of mass M another equal particle hangs by an elastic string, of natural length a and modulus λ equal to Mg. The system is initially supported with the strings vertical, the first being taut and the second at its natural length. The system is released from rest in this position. Shew that, provided the first string be sufficiently long, the motion will be simple harmonic with period $\pi\sqrt{\frac{3a}{g}}$.

Shew also that the extension of the second string after time t is

$$a\left[1-\cos\left(2t\sqrt{\frac{g}{3a}}\right)\right].$$

[Treat the strings as weightless.]

50. Two particles, of masses m_1 and m_2, are connected by a fine elastic string whose modulus of elasticity is λ and whose natural length is l. They are placed on a smooth table at a distance l apart, and equal impulses I in opposite directions in the line of the string act simultaneously on them, so that the string extends. Shew that in the ensuing motion the greatest extension is $I\sqrt{\dfrac{(m_1+m_2)l}{m_1m_2\lambda}}$, and that this value is attained in time $\dfrac{\pi}{2}\sqrt{\dfrac{m_1m_2l}{(m_1+m_2)\lambda}}$.

51. A circular disc, of mass M, lies on a smooth horizontal table; if a particle, of mass m, resting on the disc is attached to the centre by a spring which exerts a force μx when extended a length x, prove that the period of oscillations when the spring is extended and then set free is

$$2\pi\sqrt{\frac{Mm}{(M+m)\mu}}.$$

52. The component accelerations of a particle referred to axes, revolving with constant angular velocity ω, are $-4\omega v$ and $4\omega u$, where u and v are the component velocities parallel to these axes. Initially the particle is at the point $(0, -4b)$, and is at rest relative to the moving axes. Shew that its path relative to the moving axes is a four-cusped hypocycloid and that is path in space is a circle.

53. A particle is moving in a circle of radius a under the action of a force to the centre varying inversely as the fourth power of the distance; prove that, if slightly disturbed, it will ultimately be found on one of the curves

$$\frac{r}{a}=\frac{\cosh\theta+1}{\cosh\theta-2} \quad \text{or} \quad \frac{r}{a}=\frac{\cosh\theta-1}{\cosh\theta+2}.$$

If the force vary as the fifth power of the distance, shew that the corresponding curves are

$$\frac{r}{a}=\coth\frac{\theta}{\sqrt{2}} \quad \text{and} \quad \frac{r}{a}=\tanh\frac{\theta}{\sqrt{2}}.$$

54. A particle is projected towards the origin from infinity with any velocity and is acted upon by a force μu^3 at right angles to the radius vector; shew that it will describe a curve of the family $u=a\theta^{\frac{1}{2}}J_{\frac{1}{2}}(\theta)$, where $J_n(x)$ is the Bessel's function of the nth order, and find the velocity of projection in order that a particular curve may be described.

55. A particle is attached to a fixed point by a slightly elastic string and is projected at right angles to the string; shew that the polar equation of the path is approximately

$$r=c+c'\sin^2\left[\theta\sqrt{\frac{c}{2c'}}\right],$$

where c is the natural length of the string, which is supposed to be unstretched when the motion begins, and $c+c'$ is the greatest length it attains during the motion.

56. A fine straight tube, of length l, whose inner surface is smooth, is made to rotate in a vertical plane with uniform angular velocity ω about its middle point. At an instant when the tube is vertical a particle is dropped into it with negligible vertical velocity; prove that the particle will leave the tube by the end at which it enters, or the opposite end, according as l is greater, or less, than $\dfrac{g}{\omega^2}$.

Discuss the motion of the particle when l is equal to $\dfrac{g}{\omega^2}$.

57. One end of a light string, of length $\pi a+b$, is tied to a point of the circumference of a circle which is fixed to a horizontal table. The string is wrapped round the semi-circumference of the circle, and a length b of the string is straight and tangential to the circle. At the end of the straight portion is attached a particle of mass m which is projected with velocity V in a direction perpendicular to the straight portion. Shew that the string is completely unwound at the end of time $\dfrac{\pi^2 a+2\pi b}{2V}$, and that the tension of the string during the unwinding at time t from the commencement of the motion is $\dfrac{mV^2}{\sqrt{b^2+2\,Vat}}$.

58. A smooth circular wire, of radius a, is constrained to rotate about a vertical diameter with constant angular velocity ω, and a small bead rests on the wire at the lowest point. Shew that, if $a\omega^2>g$, the relative equilibrium is unstable and that, if the bead is slightly displaced, it will rise to a point whose vertical depth below the highest point of the wire is $\dfrac{2g}{\omega^2}$. Shew further that the work done by the constraining couple during the time occupied by the rise is twice the work done against gravity.

59. In the case of a nearly flat trajectory, with initial velocity V and a resistance equal to μ (velocity)^2, shew that the path of the projectile is approximately

$$y=x\left(\tan\alpha+\frac{g}{2\mu V^2}\right)+\frac{g}{4\mu^2V^2}(1-e^{2\mu x})$$

$$=x\tan\alpha-\frac{gx^2}{2V^2}-\frac{\mu g}{3V^2}x^3-\ldots\ldots,$$

where α is the (small) inclination to the horizontal of the path initially.

60. A golf ball owing to undercut is acted on at each point of its path by a force producing an acceleration $\mu vg\sin\alpha$ along the upward drawn normal and a retardation $\mu vg\cos\alpha$ along the tangent, where v is the velocity at the point. Shew that, at time t, the horizontal and vertical components of the velocity are

$$V\cos\beta-\mu g(x\cos\alpha+y\sin\alpha),$$

and

$$V\sin\beta-gt+\mu g(x\sin\alpha-y\cos\alpha),$$

where x and y are the horizontal and vertical coordinates, the motion being in two dimensions; and express these coordinates in terms of the time.

61. A particle is moving in a straight line under the action of a force towards a fixed point C in the line and proportional to the distance from C, in a medium whose resistance is proportional to the velocity. It makes damped oscillations with three consecutive positions of rest at distances a, b, c from a given point O on the line; shew that the distances from O of C and of the next position of rest are respectively

$$\frac{ac-b^2}{a-2b+c}\quad\text{and}\quad\frac{ac+bc-b^2-c^2}{a-b}.$$

62. A particle moving in a straight line is subject to a resistance which produces the retardation kv^3, where v is the velocity and k is a constant. Shew that v and the time t are given in terms of s, the distance described by the equations

$$v=\frac{u}{1+ksu},\quad\text{and}\quad t=\frac{s}{u}+\tfrac{1}{2}ks^2,$$

where u is the initial velocity.

A bullet left the rifle with a velocity 2400 ft. per sec., and had its velocity reduced to 2350 ft. per sec. when it had described a distance of 100 yards. Assuming that the air resistance varied as v^3, find the time taken in traversing 1000 yds., gravity being neglected.

63. An insect, of mass m, alights perpendicularly on one end of a flexible string, of mass M and length l, which is laid in a straight line on a smooth horizontal table, and proceeds to crawl with uniform velocity along the string. When it reaches the other end of the string, shew that that end will have moved through a distance $\frac{ml}{M}\log\left(1+\frac{M}{m}\right)$.

64. A weightless string, passing over a smooth peg, connects a weight P with a uniform string of weight $2P$ hanging vertically with its lower end just in contact with a horizontal table. When motion is allowed to take place, prove that the weight P ascends with uniform acceleration $\frac{g}{3}$, until the whole chain is coiled up on the table.

65. A driving belt, which weighs m lbs. per foot run, is moving at a uniform speed. Shew that the form assumed by the belt is a catenary whose shape does not depend on the particular speed of the belt. If the speed is altered from v_1 to v_2 feet per second, shew that the tension of the belt is everywhere increased by an amount equal to $m.\frac{v_2^2-v_1^2}{g}$ lbs. weight.

66. Shew that a uniform chain, of density m per unit of length, which is subject to no external forces, can run with constant velocity v in the form of any given curve provided that its tension is equal to mv^2.

67. A smooth surface has the form of a prolate spheroid of major axis (which is vertical) $2a$ and eccentricity e. A particle is describing on the inside of the spheroid a horizontal circle, whose plane is at a distance $a\cos\alpha$ below the centre of the spheroid; prove that the time of a small oscillation about the steady motion is $2\pi\sqrt{\frac{a\cos\alpha(1-e^2\cos^2\alpha)}{g(1+3\cos^2\alpha)}}$.

68. Two particles are connected by an elastic spring. If they vibrate freely in a straight line their period is $\frac{2\pi}{n}$. If they are set to rotate about one another with angular velocity ω, shew that the period of a small oscillation is $\frac{2\pi}{\sqrt{n^2+3\omega^2}}$.

69. The motion of a system depends on a single coordinate x; its energy at any instant is $\frac{1}{2}m\dot{x}^2+\frac{1}{2}ex^2$, and the time-rate of frictional damping of its energy is $\frac{1}{2}k\dot{x}^2$. Shew that the period of τ of its free oscillation is

$$2\pi\left(\frac{e}{m}-\frac{1}{16}\frac{k^2}{m^2}\right)^{-\frac{1}{2}}.$$

Shew that the forced oscillation sustained by a force of type $A\cos pt$ is at its maximum when $p^2=\frac{e}{m}-\frac{k^2}{8m^2}$, that the amplitude of this oscillation is then $\frac{A\tau}{\pi k}$, and that its phase lags behind that of the force by the amount $\tan^{-1}\frac{4mp}{k}$.

70. A particle, of mass m', is attached by a light inextensible string of length l to a ring of mass m which is free to slide on a smooth horizontal rod. Initially the masses are held with the string taut along the rod, and they are then set free. Prove that the greatest angular velocity of the string is $\{2g(m+m')/lm\}^{\frac{1}{2}}$.

Also shew that the time of a small oscillation about the vertical is

$$2\pi\{lm/g\,(m+m')\}^{\frac{1}{2}}.$$

71. A mass m is attached to a fixed point by a light spring and its time of oscillation vertically is $\frac{2\pi}{p_1}$. If a mass m' is suspended from m by a second spring and the period of m' when m is held fixed is $\frac{2\pi}{p_2}$, shew that, when both masses are free, the periods $\frac{2\pi}{n}$ of the normal modes of vertical vibrations of the system are given by the equation

$$n^4-\left\{p_1^2+\left(1+\frac{m'}{m}\right)p_2^2\right\}n^2+p_1^2p_2^2=0.$$

72. A particle, of mass m, moves in a resisting medium under a central attraction $m.P$; shew that the equation to the orbit is

$$\frac{d^2u}{d\theta^2}+u=\frac{P}{h^2u^2},$$

where $h=h_0e^{-\int\frac{R}{v}dt}$, and R is the resistance of the medium per unit of mass.

73. In a long railway journey performed with average velocity V, if the actual velocity be $v\equiv V+U\sin nt$ and if the resistances vary as the square of the velocity, shew that the average H.P. required is increased by $\frac{3}{2}\frac{U^2}{V^2}$ of what is required for uniform velocity V.

74. A particle moves from rest at a distance a towards a centre of force whose acceleration is μ times the distance; if the resistance to the motion is equal to kv^4, where v is the velocity, shew that, if squares of k are neglected, the time of falling to the centre of force is greater by $\frac{k\sqrt{\mu a^3}}{5}$ than it would be if there were no resisting medium, and that the amplitude of the swing is diminished by $\frac{16k\mu a^4}{5}$.

75. A particle moves in a straight line under a retardation kv^{m+1}, where v is the velocity at time t. Shew that, if u be the starting velocity, then

$$kt=\frac{1}{m}\left(\frac{1}{v^m}-\frac{1}{u^m}\right) \text{ and } ks=\frac{1}{m-1}\left(\frac{1}{v^{m-1}}-\frac{1}{u^{m-1}}\right).$$

A bullet fired with a horizontal velocity of 2500 ft. per sec. is travelling with a velocity of 1600 ft. per sec. at the end of one second. Assuming that $m=\frac{1}{2}$, find the value of k, and shew that the space described in one second is 2000 feet, neglecting the effect of gravity.

76. A particle moves on the surface of a rough circular cone under the action of no forces. It is projected with velocity V at right angles to a generator at a distance d from the vertex. Shew that, when it has moved through a distance s, its velocity v is given by

$$\log\frac{V}{v}=\frac{\mu s\cot\alpha}{\sqrt{s^2+d^2}},$$

where μ is the coefficient of friction and α is the half angle of the cone.

77. A particle moves on the surface of a sphere being acted upon by attractive forces to the ends of the polar axis each equal at distance r to $\frac{\mu m}{r^3}$; if it be projected with moment of momentum about that axis equal to $m\sqrt{\mu}$, its latitude increases uniformly with the time.

78. A smooth cup is formed by the revolution of the parabola $z^2=4ax$ about the axis of z, which is vertical. A particle is projected horizontally on the inner surface at a height z_0 with velocity $\sqrt{2kgz_0}$. Prove that, if $k=\frac{1}{4}$, the particle will describe a horizontal circle; but that, if $k=\frac{1}{30}$, its path will lie between the two planes $z=z_0$ and $z=\frac{1}{2}z_0$.

79. A train in the Northern hemisphere is travelling southward along a meridian of the Earth with velocity V; shew that, in latitude λ, it presses on the western rail with a force equal to $\dfrac{2V\omega}{g}\sin\lambda$ times its own weight, where ω is the angular velocity of the Earth about its axis.

80. A smooth cone, of vertical angle 2α, has its axis vertical and vertex upwards. A heavy particle moving on the outer surface is projected horizontally from a point at a distance R from the vertex with velocity $\sqrt{2gh}$. Shew that the particle goes to infinity, and that, for contact to be preserved, $h \not> \frac{1}{2}R\sin\alpha\,.\tan\alpha$.

81. A particle is projected along the surface of a smooth right circular cone, whose axis is vertical and vertex upwards, with a velocity due to the depth below the vertex. Shew that the equation to the path on the cone, when developed into a plane, is of the form $r^{\frac{3}{2}}\cos\dfrac{3\theta}{2}=a^{\frac{3}{2}}$

82. In latitude 45° N. a gun is fired due north at an object distant 20 kilometres, this being the maximum range of the gun. Shew that if the Earth's rotation has not been allowed for in aiming, the shell should fall about 44 metres east of the mark. Shew also that, if the shell is fired due south under similar conditions, the deviation will be twice as great and towards the west. [Air resistance is neglected.]

MISCELLANEOUS EXAMPLES. II

1. A fly-wheel turns on a horizontal axle of radius a in frictionless bearings. A fine cord wound round the axle carries a mass m at its ends; shew that, when the mass is released, the wheel turns with angular acceleration $\frac{mga}{I+ma^2}$, where I is the moment of inertia of the fly-wheel and axle.

The mass of the fly-wheel is 25 lb. and it may be treated as a uniform disc 8 inches in diameter; the hanging mass is 1 lb. and the diameter of the axle is 2 inches. Shew that after the mass has fallen 3 feet, the angular velocity of the wheel is about 112 revolutions per minute.

2. The lock of a carriage door will only engage if the angular velocity of the closing door exceeds ω. The door swings about vertical hinges, and has a radius of gyration k about the vertical axis through the hinges, whilst the centre of gravity of the door is at a distance a from the line of hinges. Shew that if the door be initially at rest and at right angles to the side of the train, which then commences to move with uniform acceleration f, the door will not close unassisted unless $f > \frac{1}{2}\frac{\omega^2 k^2}{a}$.

3. A door of uniform thickness, swinging about its hinges, is brought to rest by striking a stop on the ground at the point farthest from the door-post. Shew that the impulsive stress on the upper hinge is to that on the lower as $3h-a:h+a$, where $2h$ is the height of the door and a is the distance of either hinge from the nearest horizontal edge.

4. A wheel, 30 inches in diameter, which can rotate in a vertical plane about a horizontal axis through its centre O, carries a mass of $\frac{1}{4}$ lb. concentrated at a point P on its rim. The wheel is held with OP inclined at 30° above the horizontal and then released. Owing to a friction couple of constant magnitude L at the bearing, the first swing carries OP to a position 45° beyond the vertical. Determine the value of L, and prove that, in the next swing, OP will come to rest before reaching the vertical.

5. A uniform rod falls without rotation on to a smooth horizontal table; prove that its angular velocity after first striking the table will be a maximum when the rod makes before impact an angle $\cos^{-1}\frac{1}{\sqrt{3}}$ with the horizon.

6. A uniform rod AB is falling in a vertical plane and the end A is suddenly held fixed at an instant when the rod is horizontal, and the vertical components of the velocities of A and B are v_1 downwards and v_2 upwards. Prove that the end B will begin to rise round the end A if $v_1 < 2v_2$.

7. An inelastic uniform square lamina is held in a vertical plane, the diagonal through its lowest point making an angle α with the vertical. It is allowed to fall through a height h on to a horizontal plane, which is rough enough to prevent any slipping. Shew that the lamina will leave the plane immediately after the impact if h is greater than $\frac{a}{9}\frac{(1+3\cos^2\alpha)}{\sin^2\alpha\cos\alpha}$, where a is the length of a diagonal of the square.

8. If a body can only turn about a smooth horizontal axle, and when the body is at rest the axle is given an instantaneous horizontal velocity v in a direction perpendicular to its length, shew that the centre of mass will start off with a velocity $\frac{k^2-h^2}{k^2}v$, and that the initial angular velocity will be $\frac{vh}{k^2}$, where h is the distance of the centre of gravity from the axle and k is the radius of gyration about the axle.

9. A circular plate, of mass M and radius a, has a mass m fixed in it at a distance b from the centre. An axis through the centre of the plate and perpendicular to it can slide without friction horizontally while the plate revolves. If the plate is just disturbed from rest when m is in its highest position, find the angular velocities when the disc has made one-quarter and one-half a turn.

Determine the pressure on the axis in each case.

10. A uniform solid cylinder, of mass M and radius a, rolls on a rough inclined plane with its axis perpendicular to the line of greatest slope. As it rolls the cylinder winds up a light string which passes over a fixed light pulley and supports a freely hanging mass m, the part of the string between the pulley and cylinder being parallel to the lines of greatest slope. Find the motion of the cylinder, and prove that the tension of the string is

$$\frac{(3+4\sin\alpha)Mmg}{3M+8m},$$

where α is the inclination of the plane to the horizontal.

11. Four uniform rods, each of length $2a$ and mass m, are smoothly jointed together and lie in the form of a square on a smooth horizontal table. A horizontal blow of impulse J is applied at one corner in the direction of the diagonal there. Shew that the initial angular velocity of each rod is $\frac{3\sqrt{2}J}{16am}$, and that the kinetic energy generated is $\frac{5J^2}{16m}$.

[Use page 268, Ex. 3.]

12. A lamina is moving in its own plane with uniform angular velocity ω, and a given point of it is made to move in a straight line with uniform acceleration f. Prove that the locus in the lamina of the instantaneous centre of rotation is a spiral of the form $r\omega^2=f\theta$, and that the locus in space is a parabola of latus-rectum $\frac{2f}{\omega^2}$.

13. P is a fixed point on a circle of radius a which rolls with angular velocity ω on the outside of an equal circle whose centre is O; shew that the angular velocity of OP is $\frac{3\omega}{4}\left(1-\frac{a^2}{OP^2}\right)$.

14. A flywheel has a horizontal shaft of radius r; the moment of inertia about the axis of revolution is K. A string of negligible thickness is wound round the shaft and supports a mass M hanging vertically. Find the angular acceleration when the motion is opposed by a constant friction couple G.

If the string is released from the shaft after the wheel has turned through an angle θ from rest, and the wheel then turns through a further angle ϕ before it is brought to rest by the frictional couple, shew that

$$G=\frac{KMgr\theta}{K\theta+(K+Mr^2)\phi}.$$

15. A uniform trap-door, swinging about a horizontal hinge, is closed by a spring coiled about the hinge. The spring is coiled so that it is just able to hold the door closed in the horizontal position. The horizontal opening which the door closes is in a body which is mounting with uniform acceleration f. Shew that if

$f=\left(0{\cdot}57+\frac{1{\cdot}23}{\alpha}\right)g$, the door starting from the vertical position will just reach the horizontal position, α being the angle through which the spring is coiled when the door is in the horizontal position.

16. A circular disc, of radius R and mass M, can turn freely about its centre which is fixed on a smooth plane and another disc, of radius r and mass m, moves in the plane with velocity v without rotation and impinges on the former. If both discs have milled edges, shew that the kinetic energy lost by the impact of the discs is

$$\tfrac{1}{2}mv^2\cos^2\alpha+\tfrac{1}{2}v^2\sin^2\alpha\Big/\left[\frac{1}{m}\left(1+\frac{r^2}{k^2}\right)+\frac{R^2}{MK^2}\right],$$

where α is the angle of incidence, and K, k are the radii of gyration of the two discs about axes through their centres perpendicular to their planes.

17. A solid cylinder and a solid sphere, both uniform and having the same mass and the same radius, are at rest on a plane horizontal board which is rough enough to prevent sliding. Prove that if the board be suddenly moved in its own plane and in a direction perpendicular to the axis of the cylinder, the resulting velocity of the centre of the sphere will be $\frac{6}{7}$ times that of the axis of the cylinder.

18. To the end of a fine string wound on a solid reel of circular cross section with its axis fixed is attached a particle which is projected at right angles to the straight piece of string initially leading from the particle to the reel. Shew that the square of the length of the string subsequently unwrapped is a quadratic function of the time, and that the unwrapped string can at most turn through an angle $\frac{\pi}{2}\sqrt{1+\frac{m}{2p}}$, where m and p are the masses of the reel and particle.

19. A cubical box, with a uniform square lid of side $2a$ smoothly hinged to it along the edge, is placed on a smooth table. The lid is raised to a vertical position, and allowed to fall backward from rest. The mass of the lid is m and the mass of the box without the lid is M. Assuming that the box does not tilt, shew that the velocity of the box when the lid has turned through an angle θ is

$$u=\pm\frac{2m\sqrt{ga}\sin\frac{\theta}{2}}{\sqrt{[(M+m)\{\frac{1}{3}(4M+m)\sec^2\theta+m\tan^2\theta\}]}}.$$

Shew also that the horizontal and vertical components of the reaction between the box and lid are

$$u\sec\theta\frac{M(M+m)}{ma}\frac{du}{d\theta}\quad\text{and}\quad mg-u\sec\theta\frac{(M+m)^2}{ma}\frac{d}{d\theta}(u\tan\theta).$$

20. A door, of uniform thickness and of width $2a$, has its line of hinges inclined at an angle α to the vertical. It is opened through a right angle and shuts itself under gravity in time t; the hinges being assumed to be smooth, shew that

$$t=\sqrt{\frac{a}{6g\sin\alpha}}\cdot\frac{\Gamma(\frac{1}{4})\Gamma(\frac{1}{2})}{\Gamma(\frac{3}{4})}=\sqrt{\frac{a}{6g\sin\alpha}}\times 5{\cdot}244.$$

21. Three uniform rods, AB, BC, and CD, each of length a, are freely jointed at B and C and suspended from points A and D which are in the same horizontal line and at a distance a apart. Shew that, when the rods move in a vertical plane, the length of the simple equivalent pendulum is $\frac{5a}{6}$.

22. A mass m hangs from a fixed point by a light string of length l, and a mass m' hangs from m by a second string of length l'. For oscillations in a vertical plane, shew that the periods of the principal oscillations are the values of $\frac{2\pi}{n}$ given by the equation

$$n^4 - n^2 \frac{m+m'}{m} g \left(\frac{1}{l}+\frac{1}{l'}\right) + g^2 \frac{m+m'}{mll'} = 0.$$

23. A mass M hangs from a fixed point at the end of a very long string whose length is a; to M is suspended a mass m by a string whose length l is small compared with a; prove that the time of oscillation of m is

$$2\pi\sqrt{\frac{M}{M+m}\cdot\frac{l}{g}}.$$

24. A particle m is contained in a uniform smooth circular tube of small cross-section a which is free to rotate about a vertical diameter. The tube is rotating with constant angular velocity, and the particle is in relative equilibrium at an angular distance α from the lowest point. If equilibrium is slightly disturbed, shew that the time of a small oscillation is

$$\frac{2\pi}{\omega \sin \alpha}\left\{\frac{Mk^2 + ma^2 \sin^2 \alpha}{Mk^2 + ma^2(1+3\cos^2 \alpha)}\right\}^{\frac{1}{2}},$$

where Mk^2 is the moment of inertia of the tube about a vertical diameter.

Consider the cases when $M=0$ and ∞ respectively.

25. A fine string has masses M, M' ($M>M'$) attached to its ends and passes over a rough pulley with its centre fixed; shew that if m, mk^2 and a are respectively the mass, moment of inertia about the axis, and radius of the pulley, then, to prevent slip, the coefficient of friction must be greater than

$$\frac{1}{\pi}\log_e \frac{M(2M'a^2+mk^2)}{M'(2Ma^2+mk^2)}.$$

26. The point of suspension of a compound pendulum is made to move backwards and forwards in a horizontal line, the displacement at time t being ξ. Prove that the equation of angular motion of the pendulum has the form

$$l\frac{d^2\theta}{dt^2} + g\sin\theta = -\frac{d^2\xi}{dt^2}\cos\theta.$$

If the motion of the point of suspension be a very rapid simple harmonic motion of small amplitude, the centre of oscillation of the pendulum will be almost stationary so far as the forced oscillation is concerned.

27. A solid hemisphere, of mass M, has its curved surface rough and its plane surface smooth, and rests with the latter in contact with a smooth horizontal table. A rough sphere is dropped without rotation so as to strike the hemisphere. Shew that the kinetic energy before impact is to the kinetic energy after impact in the ratio

$$1+\frac{k^2}{a^2}-\frac{m}{M+m}\cos^2\alpha : \sin^2\alpha,$$

where α is the angle the common normal at the point of impact makes with the vertical.

28. A rough perfectly elastic sphere is dropped without rotation on to a horizontal cylinder which is free to turn about its axis. There is no slipping at the point of contact during the impact, and the sphere starts moving horizontally after the impact. If θ is the angle which the radius of the cylinder through the point of contact makes with the vertical, prove that

$$\tan^2\theta = 1 + \frac{1}{\dfrac{ma^2}{I} + \dfrac{ma'^2}{I'}},$$

where a and a' are the radii of the cylinder and sphere, I and I' are their moments of inertia about their centres and m is the mass of the sphere.

Shew also that the coefficient of friction between the cylinder and sphere must be greater than $\frac{1}{2}(\tan\theta - \cot\theta)$ in order that there may be no slipping during the impact.

29. A thin hemispherical bowl, of radius a and mass m, rests on a rough table, and is struck a vertical blow I at a point of its rim. Shew that it will roll over on to its base if $I^2 > \dfrac{2(\sqrt{5}-1)m^2ga}{3}$. If it rolls over, shew that the angular velocity, when the rim strikes the ground, is

$$\frac{3}{\sqrt{10}} \cdot \frac{I}{ma}.$$

30. A uniform flat rod, of length $2a$, rests on a rough horizontal plane with its weight uniformly distributed. A horizontal force P large enough to produce motion is applied suddenly at one end perpendicularly to the length of the rod. Shew that initially the rod begins to turn about a point distant x from the middle point, where x is given by the positive root of the equation

$$x^3 - \left(\frac{1}{3} - \frac{2P}{\mu W}\right)a^2x - \frac{2}{3}\frac{P}{\mu W}a^3 = 0,$$

W being the weight of the rod and μ the coefficient of friction.

31. On a smooth table there lies a straight bar with one end pivoted to the table. A uniform disc of radius a rests on the table in contact with the bar, the point of contact being at distance b from the pivot. The disc can slide freely on the table, but friction prevents any slipping between the disc and bar. The bar starts revolving with uniform angular velocity ω about the pivot. If $b^2 = \frac{3}{2}a^2$, shew that the point of contact will describe an equiangular spiral.

32. A solid spherical ball rests in equilibrium at the bottom of a fixed spherical globe whose inner surface is perfectly rough. The ball is struck a horizontal blow of such magnitude that the initial speed of its centre is v. Shew that, if v lies between $\sqrt{\dfrac{10dg}{7}}$ and $\sqrt{\dfrac{27dg}{7}}$, the ball will leave the globe, d being the difference between the radii of the ball and globe.

33. A vertical hoop travelling with velocity V parallel to the ground, and at the same time spinning with angular velocity ω, comes into contact with the ground; find the condition on which depends whether it will roll forward or backward.

34. A sphere is projected with underhand spin Ω up a slope of angle α; shew that, if the velocity v of projection be large, the sphere will turn back after a time

$$\frac{v - \frac{2}{5}a\Omega}{g\sin\alpha}.$$

35. A sphere, of mass m, rolls down the rough face of an inclined plane, of mass M and angle a, which is free to slide on a smooth horizontal plane in a direction perpendicular to its edge. Investigate the motion, and shew that the pressure between the sphere and plane is

$$\frac{m(2m+7M)g\cos\alpha}{(2+5\sin^2\alpha)m+7M}.$$

36. A billiard ball sliding on a smooth table with velocity u and carrying "side," *i.e.* rotating about a vertical axis with angular velocity ω, impinges directly on an equal ball at rest. Calculate the deviation produced in terms of u, ω and the coefficients of friction and restitution. Shew that as the amount of side varies, u remaining constant, the deviation increases with the side up to a certain value, and then remains constant.

37. A uniform sphere, of radius a, is projected on a rough horizontal plane with velocity V and spin Ω about a horizontal axis making an angle θ with the direction of projection. Shew that, while slipping occurs, the centre of the sphere will describe a parabola of latus-rectum

$$\frac{2a^2V^2\Omega^2\cos^2\theta}{\mu g\{V^2+2aV\Omega\sin\theta+a^2\Omega^2\}},$$

where μ is the coefficient of friction between the sphere and the plane.

38. A heavy homogeneous sphere moves on a uniformly rough horizontal plane; prove that, whilst the sphere slips, the direction of the velocity of the particle of the sphere in contact with the plane always remains the same.

Shew also that if the initial value of the velocity is V, the sphere will lose kinetic energy of amount $\frac{1}{7}MV^2$ before it begins to roll, where M is its mass.

39. A solid homogeneous sphere, of radius b, makes small oscillations at the bottom of a thin spherical shell, of radius a, the surfaces being sufficiently rough to prevent sliding and the motion being in a vertical plane. Shew that, when the shell is fixed, the length of the simple equivalent pendulum is $(a-b)\left(1+\frac{k^2}{b^2}\right)$, and that, when the shell is free to roll on a horizontal plane, the corresponding length is

$$(a-b)\frac{M(K^2+a^2)(k^2+b^2)}{Mb^2(K^2+a^2)+ma^2(k^2+b^2)},$$

where m, M are the masses of the sphere and shell, and mk^2 and MK^2 their moments of inertia about a diameter.

40. A straight weightless rod, of length $2a$, can turn freely in a horizontal plane about its centre, which is fixed. At one end it carries a particle, of mass m, while the other end is tangential to a circular wire of mass μm and radius λa, which is suspended with freedom to swing about this tangent. The circle is drawn on one side so that its plane makes an angle α with the vertical, and is then released, the system being at rest. Shew that the rod will oscillate through an angle

$$4\sqrt{\frac{\mu}{3(2+\lambda^2\mu+2\mu)}}\tan^{-1}\left[\sqrt{\frac{3\lambda^2\mu}{2+\lambda^2\mu+2\mu}}\sin\alpha\right].$$

41. A uniform plank, of length $2a$ and thickness $2h$, rests in equilibrium on the top of a fixed rough cylinder, of radius r, whose axis is horizontal. Prove that, if r is greater than h, the equilibrium is stable; and that, if the plank is slightly disturbed, the period of an oscillation is that of a simple pendulum of length

$$\frac{a^2+4h^2}{3(r-h)}.$$

42. Two equal uniform rods, AB and BC, are freely jointed at B, which is attached by means of an elastic string, of length l and unstretched length l_0, to a fixed point.

The ends A and C are in contact with a smooth horizontal plane, and the system is in equilibrium in a vertical plane. Shew that the time of oscillation in this plane in which B moves vertically is $2\pi\sqrt{\dfrac{2(l-l_0)}{3g\cos^2\alpha}}$, where α is the angle BAC or BCA.

43. A uniform bar, of length $2a$, is hung from a fixed point by a string, of length b, fastened to one end of the bar; show that, when the system makes small normal oscillations in a vertical plane, the length l of the simple equivalent pendulum is a root of the quadratic

$$l^2-\left(\frac{4a}{3}+b\right)l+\frac{1}{3}ab=0.$$

If the system is let go from rest in a displaced position with the bar and string in a straight line, shew that it cannot continue to swing in this mode.

44. A particle rests at the lowest point of a uniform smooth circular tube which is free to turn about a horizontal axis through its highest point perpendicular to its plane. The system being slightly disturbed, find the periods of small oscillations, and shew that for one principal mode of oscillation the particle remains at rest relative to the tube, and that, for the other, the centre of gravity of the particle and the tube remains at rest.

45. A uniform rod AB, of length $8a$, is suspended from a fixed point C by means of a light inextensible string, of length $13a$, attached to B. If the system is slightly disturbed in a vertical plane, shew that $\theta+3\phi$ and $12\theta-13\phi$ are principal coordinates, where θ and ϕ are the angles which the rod and string respectively make with the vertical.

46. A rod, of length $2a$, is suspended from a fixed point by a string, of length b, fastened to the rod at a distance c from the centre; find equations to give the modes and periods of principal oscillation.

Solve the problem completely if $b=\dfrac{4a}{3}$ and $c=\dfrac{a}{3}$, proving that in one mode of oscillation the uppermost end A of the rod will be approximately at rest and in the other a point C at a distance c below the centre.

47. A motor car, of total mass M, is actuated by a couple G on the hind axle and there are two pairs of wheels, of radius a, such that the moment of inertia of either pair with its axle about its axis of rotation is mk^2. Shew that, neglecting the friction of the axles on their bearings, the acceleration of the car is $G\div[a(M+2mk^2/a^2)]$, and that, for a coefficient of adhesion μ, the greatest value of G that will not cause skidding is given by $G[M(\mu a-\mu h+d)+dmk^2/a^2]=Mg\mu ac[M+2mk^2/a^2]$, where d is the distance between the axles, h the height of the centre of gravity, and c its horizontal distance behind the front axle.

48. The speed of a railway truck, weighing 5 tons, is reduced uniformly from 25 to 20 miles per hour on the level in a distance of $695\frac{3}{4}$ ft. by the brakes. Shew that, if no slipping takes place between the wheels and the rails, the normal pressure between the rails and each of the front wheels is 50 lbs. wt. greater than the corresponding pressures on the back wheels, given that the distance between the axles is 12 ft. and that the centre of gravity of the truck is $4\frac{1}{2}$ ft. above the ground and equidistant from the axles while the diameter of each wheel is 3 ft. and the moment of inertia of each pair of wheels and axle about its axis is 3600 lb.-ft.^{2} units.

49. A uniform straight tube, of length $2a$, contains a particle of mass equal to its own. The particle is placed at the middle point of the tube which is started rotating about this point on a horizontal plane with angular velocity ω. Shew that the velocity of the particle relative to the tube, when it leaves it, is $a\omega\sqrt{\frac{2}{3}}$.

50. A particle, of mass m, is attached by a string AB of length a to a point on the rim of a circular disc, of mass $2m$ and radius a. The disc can turn freely about its centre O in its own plane, and the system is at rest on a smooth table with O, A, B in a straight line. An impulse is communicated to the particle at right angles to the string. Shew that the system will be instantaneously moving as a rigid body when AB has turned through an angle of 60° relative to the disc.

51. A rough inelastic sphere rolls down over the rungs of a sloping ladder, without slipping or jumping, leaving each rung in turn as it impinges on the next. Shew that the descent may be made, without gathering or losing speed, only if the slope θ of the ladder is less than the acute angle θ_0 given by the equation $\tan(\theta_0+\alpha)\cot\alpha = 2-\frac{r^2\sin^2\alpha}{r^2+k^2}$, and greater than the acute angle θ_1 given by the equation $\tan\frac{\theta_1}{2} = \frac{r^2\sin\alpha(1-\cos\alpha)}{r^2\cos^2\alpha+k^2}$, r being the radius of the sphere, k its radius of gyration about a diameter, and $2r\sin\alpha$ the distance between consecutive rungs of the ladder.

52. A circular disc, with n spikes projecting from it in its plane at equal angular intervals, is projected with its plane vertical, and impinges on a rough horizontal inelastic plane, so that the line joining the centre of the disc with the point of contact makes an angle π/n with the vertical. Shew that, if at that instant the angular velocity of the disc is ω and the velocity of its centre perpendicular to the spike is V, the number of its spikes which strike the ground is $p+2$, where p is the greatest integer in the value of m given by

$$\left(1-\frac{2a^2}{k^2}\sin^2\frac{\pi}{n}\right)^m[(k^2-a^2)\omega+aV]=2k\sqrt{ga}\sin\frac{\pi}{2n},$$

where a is the radius of the circle on which the tips of the spikes lie, k is the radius of gyration about the tip of one of these spikes and the radius of the disc is less than $a\cos\pi/n$.

53. A homogeneous circular cylinder is divided by an axial plane and kept in shape by a band round it. If the cylinder is placed on a smooth horizontal plane with the plane of separation vertical and the band is then cut, prove that the pressure on the plane is instantaneously reduced by the fraction $\frac{32}{27\pi^2}$ of itself.

54. Two equal particles are attached by light inextensible strings, each of length a, to points at the same level which are at a distance $2\sqrt{3}a$ apart. The particles are connected by a light elastic string, of length l, and of such elasticity that they are in equilibrium when the two former strings make an angle of 60° with the vertical. Find the free periods of small vibrations, and describe the normal modes.

If $l=\frac{2a}{\sqrt{3}}$, shew that it is possible for one particle to vibrate without disturbing the other from its equilibrium position, and that its vibration has then the same frequency as a simple pendulum of length $\frac{2a}{7}$.

55. A light wire, of length l, is tightly stretched with tension T between two fixed points; to its middle point is attached a particle of mass m which makes small lateral oscillations; shew that the period of these oscillations is $\pi\sqrt{\frac{ml}{T}}$.

56. Two equal particles m are fastened at points distant a from the ends of a light string, of length $4a$, which is stretched between two fixed points with tension

T; the centre of the string is attached to a spring of strength μ, whose inertia is negligible. Prove that, if $p_0^2=\frac{2T}{ma}$ and $n=2+\frac{a\mu}{T}$, the periods of transverse principal oscillation are $\frac{2\pi}{p_0}$ and $\frac{2\pi}{p_1}$, where

$$p_1^2=p_0^2\left(1-\frac{1}{n}\right).$$

If n be large, and the first particle is displaced and let go from rest, prove that after n complete oscillations the vibration will be transferred to the second particle.

57. To an elastic string, of negligible mass, stretched between two fixed points massive particles are attached at the point of trisection and they oscillate under no forces except the tension of the string. Determine the character of the principal modes (1) when the masses are equal, (2) when one is much greater than the other, and illustrate the transition by discussing the case when the masses are in the ratio 5:8.

58. A light string, of length $6l$, is stretched between two fixed points with tension T; two particles, each of mass m, are attached to the points of trisection and a particle, of mass M, to the middle point of the string. Shew that in small transverse oscillations one period is $2\pi\sqrt{\frac{2ml}{3T}}$; and that the other two periods cannot lie between this value and $2\pi\sqrt{\frac{Ml}{2T}}$.

59. A light string, of length $4l$, is stretched between two fixed points with tension T, $(=mla^2)$, and particles of masses $3m$, $4m$, $3m$ are attached at equal intervals l; shew that the periods of small transverse oscillations are $\frac{2\pi}{a}$, $\pi\frac{\sqrt{6}}{a}$, and $2\pi\frac{\sqrt{6}}{a}$.

The particles are originally at rest with the string straight, and one of the smaller particles is set in motion with velocity V. Shew that at time t the displacement of the middle particle is $\frac{3}{10}\frac{V}{a}\left\{\sqrt{6}\sin\frac{at}{\sqrt{6}}-\sin at,\right\}$ and find the displacements of the other particles at this time.

60. Find the equation for the longitudinal vibrations of a stretched elastic string in the form $\frac{d^2\xi}{dt^2}=a^2\frac{d^2\xi}{dx^2}$, where $a^2=\frac{\lambda l}{\rho l_0}$, λ being the modulus of elasticity, l and ρ the length and line density of the string when in equilibrium, and l_0 the unstretched length.

One end of the string is fastened to a point on a smooth horizontal table, and the other end is fastened by a light inextensible string to a mass M, which hangs over the edge of the table. Prove that the system can execute longitudinal vibrations of period $\frac{2\pi}{p}$, where $\frac{pl}{a}\tan\frac{pl}{a}=\frac{m}{M}$, m being the mass of the string.

APPENDIX

ON THE SOLUTION OF SOME OF THE MORE COMMON FORMS OF DIFFERENTIAL EQUATIONS

I. $\dfrac{dy}{dx}+Py=Q$, where P and Q are functions of x.

[Linear equation of the first order.]

Multiply the equation by $e^{\int P dx}$, and it becomes

$$\frac{d}{dx}[ye^{\int P dx}]=Qe^{\int P dx}.$$

Hence $ye^{\int P dx}=\int Qe^{\int P dx}dx+$ a constant.

Ex. $\dfrac{dy}{dx}+y\tan x=\sec x$.

Here $$e^{\int P dx}=e^{\int \tan x\, dx}=e^{-\log \cos x}=\frac{1}{\cos x}.$$

Hence the equation becomes

$$\frac{1}{\cos x}\frac{dy}{dx}+y\frac{\sin x}{\cos^2 x}=\sec^2 x.$$

$$\therefore \frac{y}{\cos x}=\tan x+C.$$

II. $\dfrac{d^2y}{dx^2}+P\left(\dfrac{dy}{dx}\right)^2=Q$, where P and Q are functions of y.

On putting $\left(\dfrac{dy}{dx}\right)^2=T$, we have $2\dfrac{dy}{dx}\dfrac{d^2y}{dx^2}=\dfrac{dT}{dx}$, so that $\dfrac{d^2y}{dx^2}=\dfrac{1}{2}\dfrac{dT}{dy}$.

The equation then becomes

$$\frac{dT}{dy}+2P.T=2Q,$$

a linear equation between T and y, and is thus reduced to the form I.

III. $\dfrac{d^2y}{dx^2}=-n^2y$.

Multiplying by $2\dfrac{dy}{dx}$ and integrating, we have

$$\left(\frac{dy}{dx}\right)^2=-n^2y^2+\text{const.}=n^2(C^2-y^2).$$

$$\therefore nx=\int\frac{dy}{\sqrt{C^2-y^2}}=\sin^{-1}\frac{y}{C}+\text{const.}$$

$$\therefore y=C\sin(nx+D)=L\sin nx+M\cos nx,$$

where C, D, L, and M are arbitrary constants.

IV. $\frac{d^2y}{dx^2}=n^2y.$

We obtain, as in III,

$$\left(\frac{dy}{dx}\right)^2=n^2y^2+\text{a constant}=n^2(y^2-C^2).$$

$$\therefore\ nx=\int\frac{dy}{\sqrt{y^2-C^2}}=\cosh^{-1}\frac{y}{C}+\text{const.}$$

$$\therefore\ y=C\cosh(nx+D)=Le^{nx}+Me^{-nx},$$

where C, D, L, and M are arbitrary constants.

V. $\frac{d^2y}{dx^2}=f(y).$

Similarly, we have in this case

$$\left(\frac{dy}{dx}\right)^2=2\int f(y)\frac{dy}{dx}dx=2\int f(y)dy.$$

VI. Linear equation with constant coefficients, such as

$$\frac{d^3y}{dx^3}+a\frac{d^2y}{dx^2}+b\frac{dy}{dx}+cy=f(x).$$

[The methods which follow are the same, whatever be the order of the equation.] Let η be *any* solution of this equation, so that

$$(D^3+aD^2+bD+c)\eta=f(x)\ \ldots\ldots\ldots\ldots\ldots\ldots(1).$$

On putting $y=Y+\eta$, we then have

$$(D^3+aD^2+bD+c)Y=0\ \ldots\ldots\ldots\ldots\ldots\ldots(2).$$

To solve (2), put $Y=e^{px}$, and we have

$$p^3+ap^2+bp+c=0\ \ldots\ldots\ldots\ldots\ldots\ldots(3),$$

an equation whose roots are p_1, p_2, and p_3.

Hence Ae^{p_1x}, Be^{p_2x}, Ce^{p_3x} (where A, B, and C are arbitrary constants) are solutions of (2), and hence $Ae^{p_1x}+Be^{p_2x}+Ce^{p_3x}$ is a solution also.

This solution, since it contains three arbitrary and independent constants, is the most general solution that an equation of the third order, such as (2), can have.

Hence $$Y=Ae^{p_1x}+Be^{p_2x}+Ce^{p_3x}\ \ldots\ldots\ldots\ldots\ldots\ldots(4).$$

This part of the solution is called the COMPLEMENTARY FUNCTION.

If some of the roots of equation (3) are imaginary, the equation (4) takes another form.

For let $\alpha+\beta\sqrt{-1}$, $\alpha-\beta\sqrt{-1}$, and p_3 be the roots.

Then
$$y=Ae^{(\alpha+\beta\sqrt{-1})x}+Be^{(\alpha-\beta\sqrt{-1})x}+Ce^{p_3x}$$
$$Ae^{\alpha x}[\cos\beta x+i\sin\beta x]+Be^{\alpha x}[\cos\beta x-i\sin\beta x]+Ce^{p_3x}$$
$$=e^{\alpha x}[A_1\cos\beta x+B_1\sin\beta x]+Ce^{p_3x},$$

where A_1 and B_1 are new arbitrary constants.

In some cases two of the quantities p_1, p_2, p_3 are equal, and then the form (4) for the Complementary Function must be modified.

Let $p_2 = p_1 + \gamma$, where γ is ultimately to be zero.

Then the form (4)

$$= Ae^{p_1 x} + Be^{(p_1+\gamma)x} + Ce^{p_3 x}$$

$$= Ae^{p_1 x} + Be^{p_1 x}\left[1 + \gamma x + \frac{\gamma^2 x^2}{2!} + \dots\right] + Ce^{p_3 x}$$

$$= A_1 e^{p_1 x} + B_1 e^{p_1 x}\left[x + \frac{\gamma x^2}{2!} + \dots\right] + Ce^{p_3 x},$$

where A_1, B_1 are fresh arbitrary constants.

If γ be now made equal to zero, this becomes

$$(A_1 + B_1 x)e^{p_1 x} + Ce^{p_3 x}.$$

If three roots p_1, p_2, p_3 are all equal, we have, similarly,

$$(A_1 + B_1 x + C_1 x^2)e^{p_1 x}$$

as the form of the Complementary Function.

The value of η given by (1) is called the PARTICULAR INTEGRAL.

The method of obtaining η depends on the form of $f(x)$. The only forms we need consider are x^n, $e^{\lambda x}$, $\begin{smallmatrix}\sin\\ \cos\end{smallmatrix}\lambda x$, and $e^{\mu x}\begin{smallmatrix}\sin\\ \cos\end{smallmatrix}\lambda x$.

(i) $f(x) = x^n$.

Here, by the principles of operators,

$$\eta = \frac{1}{D^3 + aD^2 + bD + c}\cdot x^n$$

$$= [A_0 + A_1 D + A_2 D^2 + \dots + A_n D^n + \dots]x^n,$$

on expanding the operator in powers of D.

Every term is now known, and hence

$$\eta = A_0 x^n + A_1 . nx^{n-1} + A_2 . n(n-1)x^{n-2} + \dots + A_n 1.2\dots n.$$

(ii) $f(x) = e^{\lambda x}$.

We easily see that $D^r e^{\lambda x} = \lambda^r e^{\lambda x}$.

$$\therefore \eta = \frac{1}{D^3 + aD^2 + bD + c}\cdot e^{\lambda x}$$

$$= (A_0 + A_1 D + A_2 D^2 + \dots)e^{\lambda x}$$

$$= (A_0 + A_1 \lambda + A_2 \lambda^2 + \dots)e^{\lambda x}$$

$$= \frac{1}{\lambda^3 + a\lambda^2 + b\lambda + c}e^{\lambda x},$$

so that in this case η is obtained by substituting λ for D.

(iii) $f(x) = \sin \lambda x$.

We know that $D^2 \sin \lambda x = -\lambda^2 \sin \lambda x$, and that

$$D^{2r} \sin \lambda x = (-\lambda^2)^r \sin \lambda x,$$

and in general that

$$F(D^2) \sin \lambda x = F(-\lambda^2) \sin \lambda x.$$

Hence
$$\eta=\frac{1}{D^3+aD^2+bD+c}\sin\lambda x$$
$$=(D^3-aD^2+bD-c)\cdot\frac{1}{D^2(D^2+b)^2-(aD^2+c)^2}\sin\lambda x$$
$$=(D^3-aD^2+bD-c)\cdot\frac{1}{-\lambda^2(b-\lambda^2)^2-(-a\lambda^2+c)^2}\sin\lambda x$$
$$=-\frac{1}{\lambda^2(\lambda^2-b)^2+(a\lambda^2-c)^2}\cdot(-\lambda^3\cos\lambda x+a\lambda^2\sin\lambda x+b\lambda\cos\lambda x-c\sin\lambda x)$$
$$=\frac{(\lambda^3-b\lambda)\cos\lambda x-(a\lambda^2-c)\sin\lambda x}{\lambda^2(\lambda^2-b)^2+(a\lambda^2-c)^2}.$$

(iv) $f(x)=e^{\mu x}\sin\lambda x$.

We easily obtain
$$D(e^{\mu x}\sin\lambda x)=e^{\mu x}(D+\mu)\sin\lambda x,$$
$$D^2(e^{\mu x}\sin\lambda x)=e^{\mu x}(D+\mu)^2\sin\lambda x,$$
$$\cdots\cdots\cdots\cdots\cdots\cdots$$
$$D^r(e^{\mu x}\sin\lambda x)=e^{\mu x}(D+\mu)^r\sin\lambda x,$$
and, generally,
$$F(D)(e^{\mu x}\sin\lambda x)=e^{\mu x}F(D+\mu)\sin\lambda x.$$

Hence
$$\eta=\frac{1}{D^3+aD^2+bD+c}e^{\mu x}\sin\lambda x$$
$$=e^{\mu x}\frac{1}{(D+\mu)^3+a(D+\mu)^2+b(D+\mu)+c}\sin\lambda x,$$
the value of which is obtained as in (iii).

In some cases we have to adjust the form of the Particular Integral.

Thus, in the equation
$$(D-1)(D-2)(D-3)y=e^{2x},$$
the particular integral obtained as above becomes infinite; to get the corrected form we may proceed as follows:
$$\eta=\frac{1}{(D-1)(D-2)(D-3)}e^{2x}$$
$$=\frac{1}{D-2}\cdot\frac{1}{(D-1)(D-3)}\cdot e^{2x}$$
$$=\frac{1}{D-2}\cdot\frac{1}{1.(-1)}e^{2x},\text{ by the result of (ii),}$$
$$=-\underset{\gamma=0}{\mathrm{Lt}}\frac{1}{D-2}{}^{(2+\gamma)x}$$
$$=-\underset{\gamma=0}{\mathrm{Lt}}\frac{1}{\gamma}e^{2x}.e^{\gamma x}$$
$$=-e^{2x}\underset{\gamma=0}{\mathrm{Lt}}\frac{1}{\gamma}\left[1+\gamma x+\frac{\gamma^2x^2}{1.2}+\ldots\right]$$
=something infinite which may be included in the Complementary Function $-xe^{2x}$.

Hence the complete solution is

$$y=Ae^{x}+Be^{2x}+Ce^{3x}-xe^{2x}.$$

As another example take the equation

$$(D^2+4)\,(D-3)\,y=\cos 2x.$$

The Complementary Function $=A\cos 2x+B\sin 2x+Ce^{3x}$.

The Particular Integral as found by the rule of (iii) becomes infinite.

But we may write

$$\eta=\frac{1}{D^2+4}\cdot\frac{D+3}{D^2-9}\cos 2x$$

$$=-\frac{1}{13}\frac{1}{D^2+4}[3\cos 2x-2\sin 2x]$$

$$=-\frac{1}{13}\underset{\gamma=0}{\text{Lt}}\frac{1}{D^2+4}\,[3\cos(2+\gamma)\,x-2\sin(2+\gamma)\,x]$$

$$=-\frac{1}{13}\underset{\gamma=0}{\text{Lt}}\frac{1}{4-(2+\gamma)^2}\,[(3\cos 2x-2\sin 2x)\cos\gamma x-(3\sin 2x+2\cos 2x)\sin\gamma x]$$

$$=-\frac{1}{13}\underset{\gamma=0}{\text{Lt}}\frac{1}{-4\gamma-\gamma^2}\Big[(3\cos 2x-2\sin 2x)\left(1-\frac{\gamma^2x^2}{2!}+\dots\right)$$

$$-(3\sin 2x+2\cos 2x)\left(\gamma x-\frac{\gamma^3x^3}{3!}+\dots\right)\Big]$$

$=$ something infinite included in the Complementary Function

$$-\frac{1}{52}(3\sin 2x+2\cos 2x).x.$$

VII. Linear equations with two independent variables, *e.g.*

$$f_1(D)y+f_2(D)z=0\ \dots\dots\dots\dots(1),$$

$$F_1(D)y+F_2(D)z=0\ \dots\dots\dots\dots(2),$$

where

$$D\equiv\frac{d}{dx}.$$

Perform the operation $F_2(D)$ on (1) and $f_2(D)$ on (2) and subtract; we thus have

$$\{f_1(D).F_2(D)-f_2(D)F_1(D)\}y=0,$$

a linear equation which is soluble as in VI.

Substitute the solution for y thus obtained in (1), and we have a linear equation for z.

Ex.

$$\left.\begin{aligned}&\frac{d^2y}{dx^2}+y+6\frac{dz}{dx}=0\\&\frac{dy}{dx}+\frac{d^2z}{dx^2}+2z=0\end{aligned}\right\}\quad\begin{aligned}&\dots\dots\dots\dots(1),\\&\dots\dots\dots\dots(2),\end{aligned}$$

i.e. and

$$\left.\begin{aligned}(D^2+1)y+6Dz=0,\\Dy+(D^2+2)z=0.\end{aligned}\right\}$$

$$\therefore\ [(D^2+2)(D^2+1)-D.6D]y=0,$$

i.e.

$$(D^2-1)\,(D^2-2)\,y=0.$$

$$\therefore\ y=Ae^{x}+Be^{-x}+Ce^{\sqrt{2}x}+De^{-\sqrt{2}x}.$$

Hence (1) gives

$$6\frac{dz}{dx}+2Ae^{x}+2Be^{-x}+3Ce^{\sqrt{2}x}+3De^{-\sqrt{2}x}=0,$$

and hence we have the value of z, viz.

$$z=-\frac{A}{3}e^{x}+\frac{B}{3}e^{-x}-\frac{C}{2\sqrt{2}}e^{\sqrt{2}x}+\frac{D}{2\sqrt{2}}e^{-\sqrt{2}x}+E.$$

On substituting in (2), we find that $E=0$.

Printed in the United States
By Bookmasters

Printed in the United States
By Bookmasters